Chernobyl Children

In the wake of the Chernobyl nuclear disaster, more than a million Belarusian, Ukrainian, and Russian children were sent abroad. Aided by the unprecedented efforts of transnational NGOs and private individuals, these children were meant to escape and recover from radiation exposure, but also from the increasing hardships of everyday life in post-Soviet society. Through this opening of the Soviet Union, hundreds of thousands of people in more than forty countries witnessed the ecological, medical, social, and political consequences of the disaster for the human beings involved. This awareness transformed the accident into a global catastrophe which could happen anywhere and have widespread impact. In this brilliantly insightful work, Melanie Arndt demonstrates that the Chernobyl children were both witness to and representative of a vanishing bipolar world order and the future of life in the Anthropocene, an age in which the human impact on the planet is increasingly borderless.

Melanie Arndt is Professor of Economic, Social, and Environmental History and Vice Rector at the University of Freiburg, Germany.

Studies in Environment and History

Editors

J. R. McNeill, *Georgetown University*
Ling Zhang, *University of Cambridge*

Editors Emeriti

Alfred W. Crosby, *University of Texas at Austin*
Edmund P. Russell, *Carnegie Mellon University*
Donald Worster, *University of Kansas*

Other Books in the Series

Sverker Sörlin and Eric Paglia *Stockholm and the Rise of Global Environmental Governance: The Human Environment*

Zozan Pehlivan *The Political Ecology of Violence: Peasants and Pastoralists in the Last Ottoman Century*

Matthew P. Johnson *Hydropower in Authoritarian Brazil: An Environmental History of Low-Carbon Energy, 1960s–1990s*

Ellen Arnold *Medieval Riverscapes: Environment and Memory in Northwest Europe, c. 300–1100*

Richard C. Hoffmann *The Catch: An Environmental History of Medieval European Fisheries*

Samuel Dolbee *Locusts of Power: Borders, Empire, and Environment in the Modern Middle East*

Andy Bruno *Tunguska: A Siberian Mystery and Its Environmental Legacy*

Lionel Frost et al. *Cities in a Sunburnt Country: Water and the Making of Urban Australia*

Adam Sundberg *Natural Disaster at the Closing of the Dutch Golden Age: Floods, Worms, and Cattle Plague*

Germán Vergara *Fueling Mexico: Energy and Environment, 1850–1950*

Peder Anker *The Power of the Periphery: How Norway Became an Environmental Pioneer for the World*

David Moon *The American Steppes: The Unexpected Russian Roots of Great Plains Agriculture, 1870s–1930s*

James L. A. Webb, Jr. *The Guts of the Matter: A Global Environmental History of Human Waste and Infectious Intestinal Disease*

Maya K. Peterson *Pipe Dreams: Water and Empire in Central Asia's Aral Sea Basin*

Thomas M. Wickman *Snowshoe Country: An Environmental and Cultural History of Winter in the Early American Northeast*

Debjani Bhattacharyya *Empire and Ecology in the Bengal Delta: The Making of Calcutta*

Chris Courtney *The Nature of Disaster in China: The 1931 Yangzi River Flood*

Dagomar Degroot *The Frigid Golden Age: Climate Change, the Little Ice Age, and the Dutch Republic, 1560–1720*

Continued after the index

Chernobyl Children

A Transnational History of a Nuclear Disaster

MELANIE ARNDT

University of Freiburg

Translated by

ALASTAIR MATTHEWS

CAMBRIDGE
UNIVERSITY PRESS

Shaftesbury Road, Cambridge, CB2 8EA, United Kingdom

One Liberty Plaza, 20th Floor, New York, NY 10006, USA

477 Williamstown Road, Port Melbourne, VIC 3207, Australia

314–321, 3rd Floor, Plot 3, Splendor Forum, Jasola District Centre,
New Delhi – 110025, India

103 Penang Road, #05–06/07, Visioncrest Commercial, Singapore 238467

Cambridge University Press is part of Cambridge University Press & Assessment,
a department of the University of Cambridge.

We share the University's mission to contribute to society through the pursuit of
education, learning and research at the highest international levels of excellence.

www.cambridge.org
Information on this title: www.cambridge.org/9781009457767

DOI: 10.1017/9781009457750

Originally published in German as © Brill Deutschland GmbH, Vandenhoeck & Ruprecht,
Tschernobylkinder. Die transnationale Geschichte einer nuklearen Katastrophe, Melanie Arndt
(Göttingen, 2020)

The translation of this work was funded by Geisteswissenschaften International – Translation
Funding for Humanities and Social Sciences from Germany, a joint initiative of the Fritz Thyssen
Foundation, the German Federal Foreign Office, the collecting society VG WORT and the
Börsenverein des Deutschen Buchhandels (German Publishers & Booksellers Association).

© Melanie Arndt 2025

First published 2025

Printed in the United Kingdom by CPI Group Ltd, Croydon CR0 4YY

A catalogue record for this publication is available from the British Library

Library of Congress Cataloging-in-Publication Data
NAMES: Arndt, Melanie, author.
TITLE: Chernobyl children : a transnational history of a nuclear disaster / Melanie Arndt.
DESCRIPTION: Cambridge ; New York, NY : Cambridge University Press, 2025. | Series: Studies in
environment and history | Includes bibliographical references and index.
IDENTIFIERS: LCCN 2024052526 | ISBN 9781009457767 (hardback) | ISBN 9781009457774
(paperback) | ISBN 9781009457750 (ebook)
SUBJECTS: LCSH: Chernobyl Nuclear Accident, Chornobyl', Ukraine, 1986 – Social aspects. |
Child disaster victims – Ukraine – Chornobyl'. | Child welfare – Soviet Union. | Soviet Union –
Social conditions – 1970–1991. | Humanitarian assistance – International cooperation.
CLASSIFICATION: LCC HV623 1986.U38 A76 2025 | DDC 363.17/99094777–dc23/eng/20241226
LC record available at https://lccn.loc.gov/2024052526

ISBN 978-1-009-45776-7 Hardback

For Filipp and all the other Chernobyl children

Contents

Figures

The authors and publishers acknowledge the aforementioned sources of copyright material and are grateful for the permissions granted. While every effort has been made, it has not always been possible to identify the sources of all the material used, or to trace all copyright holders. If any omissions are brought to our notice, we will be happy to include the appropriate acknowledgements on reprinting and in the next update to the digital edition, as applicable.

Archives

CDAVO: Central State Archive of the Supreme Governmental and Administrative Bodies of Ukraine (Tsentral'nyi derzhavnyi arkhiv vyshchykh orhaniv vlady ta upravlinnia Ukrainy)
    Council of Ministers of Ukraine (fond R2)

GARF: State Archives of the Russian Federation (Gosudarstvennyi arkhiv Rossiiskoi Federatsii)
    Council of Ministers of the Soviet Union (fond R5446)
    Ministry of Health of the Soviet Union (fond R8009)

HA: Hoover Archives
    Center for Civil Society International
    Enid Schreibman Papers
    Francis U. Macy Papers
    James K. Asselstine Papers
    Russian Subject Collection

HA/ACPSS: Hoover Archives/Archives of the Communist Party and Soviet State
    Fond 89
    Volkogonov Files

LYA: Lithuanian Special Archive (KGB archive, Lietuvos Ypatingas Archyvas)
    Central Departments of the KGB of the LSSR (fond K-1)

NARB: National Archives of the Republic of Belarus (Natsional'nyi arkhiv Respubliki Belarus'/Natsyianal'ny arkhiu Respubliki Belarus')
  Party Archive (fond 4p)
  State Committee on the Consequences of the Chernobyl Nuclear Power Plant Disaster (Goskomchernobyl', fond 507)

NRC: Nuclear Regulatory Commission
  Commission Documents

Private Archive: Iourie Pankrats

Private Archive: Irina Grushevaia

Private Archive: Joanna Macy

RRL: Ronald Reagan Library
  European and Soviet Directorate Files
  Jack F. Matlock Jr. Files
  White House Office of Records Management (WHORM) Subject File
    Atomic Energy (AT)
  WHORM Subject File Countries (CO165: USSR)

SBU: KGB Archive of Ukraine (Arkhiv Sluzhby bezpeky Ukrainy)
  4th Directorate, Ukrainian SSR KGB (fond 29)
  6th Directorate, Ukrainian SSR KGB (fond 31)
  Operational Matters (fond 65)
  Secretariat, GPU/KGB Ukrainian SSR (fond 16)

SCARC OSUL: Special Collections & Archives Research Center, Oregon State University Libraries
  Eugene Starr Papers
  Theodore Rockwell Papers

UWLSC: University of Washington Libraries Special Collections
  Hanford Litigation Office Records (HLOR)

Acknowledgements

This book was written over many years in many locations on (at least) two continents. During this long period, numerous friends, colleagues, contemporary witnesses, and staff at research institutions, libraries, and archives have been extraordinarily generous in their support. Without them, it would have been impossible to write it. I would like to thank them from the very bottom of my heart. There were so many stages on the journey and so many people that what follows is inevitably incomplete; I apologize if some names do not appear. The full list of those who have given their backing includes many more than the names and institutions mentioned herein. Three of them will never get to hold the English translation in their hands. Rudi Saule, Julia Obertreis, and Maya K. Peterson – you are missed.

The book is a revised version of a German book that was published by Vandenhoeck & Ruprecht (now Brill) in 2020 and for which I received the 2021 Geisteswissenschaften International: Translation Funding for Humanities and Social Sciences from Germany award from the German Publishers and Booksellers Association, which made the translation into English possible. Alastair Matthews took on this mammoth task; his feeling for language, accuracy, and patience remain incomprehensible to me to this day. One could not wish for a better translator; thank you, Alastair, the book is so much better because of you! It goes without saying that any remaining errors are my own. I would like to thank the Studies in Environment and History series editors, John McNeill and Ling Zhang, for accepting the volume. Many thanks also to Lucy Rhymer and Rosa Martin at Cambridge University Press, and Jessica Dietz at Brill, for all their efforts and support.

The Volkswagen Foundation, the German Research Foundation, and the Agence nationale de la recherche funded the research for the book. I am very grateful that they considered the project worthy of funding at different stages and in different contexts. I am also deeply indebted to the Leibniz Institute of East and Southeast European Studies (IOS), the Graduate School of East and Southeast European Studies, and the University of Regensburg, which gave me six generous years of freedom to work on the book and provided an extraordinarily stimulating intellectual environment in which to do so. My special thanks go to Ulf Brunnbauer, Volker Depkat, Heidrun Hamersky, and Martin Schulze Wessel, and to everyone else at the IOS.

The first steps towards the book were made from 2008 to 2012 with 'Politics and Society after Chernobyl', the international research project that I led at the Potsdam Centre for Contemporary History (ZZF). Exchanges with Thomas Bohn and the project members – Aliaksandr Dalhouski, Evgeniia Ivanova, Tatiana Kasperski, Anastasiia Leukhina, and Andrei Stepanov – as well as everyone else who contributed to our work enriched the book immensely. I would also like to thank all my colleagues at the ZZF for their support, especially Thomas Lindenberger, Jürgen Danyel, Martin Sabrow, Stefan-Ludwig Hoffmann, Annette Vowinckel, and Frank Bösch.

My fellowship at the Rachel Carson Center for Environment and Society (RCC) at Ludwig Maximilian University of Munich in 2012 opened my eyes in many ways: without the exchanges and dialogues at the RCC, North America would never have come to play such a crucial role in the book. I am very grateful to Christof Mauch and Helmuth Trischler for making both this change of perspective and international networking with the brightest of the brightest in environmental history possible. My fellow Fellows and the staff at the RCC deserve the greatest thanks for their inspiration, their questions, their encouragement, and all the little and big adventures beyond research: Sarah Cameron, Chiara Certomà, Wilko Graf von Hardenberg, Carmel Finley, Amy Hay, Arielle Helmick, Matthew Kelly, Andrea Kiss, Siddharta Krishnan, Claudia Leal, Agnes Limmer-Kneitz, Andrew Isenberg, Uwe Lübken, Michelle Mart, Giacomo Parrinello, Christopher Pastore, Maya K. Peterson, James Rice, Katie Ritson, Edmund P. Russell, John Sandlos, Fei Sheng, Bron Taylor, Franziska Torma, Louis S. Warren, Frank Zelko, and all the others – thank you very much!

Another fellowship, this time at the Stanford Humanities Center (SHC), provided a veritable paradise in which to deepen the international and interdisciplinary discussion of my ideas and utilize the unique resources of the Hoover Archives. Here, too, I was fortunate to meet colleagues who I can only look up to both academically and personally, and from whom I learnt a tremendous amount. Very special thanks to the then-director of the SHC, Caroline Winterer, the staff of the centre, and my fellow Fellows and colleagues there, especially Uğur Zekeriya Pece (Valentine will never be forgotten!), Lilla Balint, Elizabeth Anker, Keith Baker, Amanda R. Greene, David Holloway, S. Lochlann Jain, Tanya Luhrmann, Matthew Kaiser, Dorinne Kondo, Regina Kunzel, Norman M. Naimark, Yi-Ping Ong, Daniel Rosenberg, Dara Strolovich, Bronwen Tate, Amir Weiner, and Richard White.

Returning from California was made much easier by the inspiring cooperation in the Franco-German research project 'Contemporary Environmental History of the Soviet Union and its Successor States', which I led together with Klaus Gestwa and Marc Elie. I would like to thank all the project members – Laurent Coumel, Alexander Ananyev, Raphael Schulte-Kellinghaus, and Katja Doose – for the stimulating discussions in offices, trains, and snowy mountains between Paris and Moscow!

I am deeply indebted to all the contemporary witnesses who responded to my – sometimes last-minute, often tenacious – enquiries, and who entrusted me with stories, memories, and source material, even though their everyday lives demanded quite different things of them, even though remembering was sometimes painful. In my head, I could not switch off the warning voice of reproach that I had so often encountered in sources and conversations: those affected by the disaster had all too often found that their fate was being exploited – at no benefit whatsoever to them – to advance careers far away from radioactively contaminated landscapes. I hope that I have succeeded in expressing my humility and my profound respect for all those who have suffered and are still suffering from the consequences of the disaster. I would also like to pay tribute to all those who have endeavoured to help the Chernobyl children. I hope to give something back with this book. My special thanks go to Irina Grushevaia, Joanna Macy, and Iouri Pankrats, who also provided me with material from their personal archives.

As a historian, one is always part of the history that one writes. I was so in a special way: from April 1996 to September 1997 I worked as a volunteer for Children of Chernobyl and at an orphanage in Minsk. This direct connection to the subject of my book was a huge challenge. Above all, however, I regarded it as an advantage, as it allowed me to see things to which I might otherwise have remained blind. I would like to thank all my friends in Minsk who have given me insights and shelter for more than twenty-five years, above all Elena Kasko and Filipp Chmyr, along with Kim, Aleksei Khatskevich, and Lana Krel. Without my pen-pal relationship with Tatiana Kovalevskaia, which began when I was a fifth-grader in 1988, I might never have become interested in Belarus – *spasibo*, Tania! Today, unfortunately, many things that I took for granted when I was finalizing the manuscript seem impossible. My book is about suffering, but also about resistance and perseverance. May both contribute to us all being able to sit round a Minsk kitchen table again without fear.

Many thanks to all those who supported my research in archives and libraries, especially the staff of the National Archives of the Republic of Belarus, the Hoover Archives, the State Archives of the Russian Federation, the KGB Archives of Ukraine, the Archives of the Nuclear Regulatory Commission (special thanks to Thomas Wellock for his help here!), and the Ronald Reagan Library.

I have been very fortunate to have had the best student assistants one could wish for over the years. Thank you to Florian Krug, Gulnora Usmanova, Aksana Yankovich, Alexander Legler, Melanie Hussinger, Anselm Schmidt, and Julia Garbe for all your support! I would also like to thank the patient physicists who taught me about nuclear power plants, their construction and the processes that take place in them, and the nature of radioactivity: Henning von Philipsborn, Ingo Wiesler, and Tobias Cronert.

Many thanks are also due to all the colleagues and friends who contributed ideas and criticism, invited me to give presentations, and/or read parts of the text – even though their own desks were overloaded. Their wise comments and advice have enriched the book, and I have benefited greatly from their work: Kathleen Beger, Nebi Bardhoshi, Kate Brown, Petra Goedde, Guido Hausmann, Julia Herzberg, Friederike Kind-Kovács, Jó Klanovicz, Caitlin Murdock, Paul R. Josephson, Julia Obertreis, H. Glenn Penny, Tanja Penter, Andreas Renner, Astrid Sahm, Nikolai Snegurski, and Elena Vishlenkova.

I owe an immense debt of gratitude to my parents, my family, and my closest friends. Thank you for bearing with me, for being there for me even though 'real' life was so often elsewhere – with an open ear and heart, with apples, wine, and *Pudding* at midnight, with fresh air and sunshine for the road. What a gift to have you in my life! Luise Alt, Racha Kirakosian, Michaela Maria Müller, Rudi Saule, Ruth Wunnicke, and Irmgard

Zündorf. And, of course, especially to you, Walter Bersorger, thank you. Thank you for everything – not just the desk with the best view.

The translation of this work was funded by Geisteswissenschaften International – Translation Funding for Humanities and Social Sciences from Germany, a joint initiative of the Fritz Thyssen Foundation, the German Federal Foreign Office, the collecting society VG WORT and the Börsenverein des Deutschen Buchhandels (German Publishers & Booksellers Association).

Note on Languages

The area covered in this book has undergone substantial changes during the last four decades; the Russian war on Ukraine is only the most recent of these. The book attempts to account for the complex linguistic situation that has emerged as a result.

To aid readability while facilitating consultation of sources, soft (') and hard (") signs are not included in transliterations from Belarusian, Ukrainian, and Russian in the text but are retained in the references. Original-language forms are retained where sources are reproduced; otherwise, the names of places and individuals are generally given in the form that is current today. An exception is made if a particular form was routinely used in the period under consideration and/or preferred by a particular individual: for example, 'Gennadii Grushevoi' (Ru.) instead of 'Henadz Hrushavy' (Bel.). Variants in other languages are given in parentheses when names are first mentioned.

The book uses 'Belarus' for the period after independence in 1991; the Belorussian Soviet Socialist Republic (BSSR) is referred to as such, following contemporary usage. Translations from Russian, Belarusian, and Ukrainian are original to the book unless otherwise indicated.

Abbreviations

AES	*atomnaia elektrostantsia* (nuclear power plant)
AHP	Association for Humanistic Psychology
AMN	Akademiia meditsinskikh nauk (Academy of Medical Sciences)
AN	Akademiia nauk (Academy of Sciences)
ANI	American Nuclear Insurers
ark.	*arkush* (page)
BAG	Bundesarbeitsgemeinschaft 'Den Kindern von Tschernobyl in Deutschland' e.V.
BBF	Belorusskii Blagotovritel'nyi Fond 'Detiam Chernobylia' (Bel. Belaruski Dabrachynny Fond 'Dzetsiam Charnobylia'; Belarusian Charitable Fund 'For the Children of Chernobyl')
Bel.	Belarusian
BelaPAN	Belaruskae pryvatnae agentstva navin (Belarusian Private News Agency)
BNF	Belaruski Narodny Front 'Adradzhen'ne' (Belarusian People's Front 'Rebirth')
BSSR	Belorusskaia Sovetskaia Sotsialisticheskaia Respublika (Belorussian Soviet Socialist Republic)
ChAES	*Chernobylskaia atomnaia elektrostantsiia* (Chernobyl nuclear power plant)
Ci	curie
CIA	Central Intelligence Agency
CIS	Community of Independent States
CK	Central'nyi Komitet (Central Committee)

CNN	Cable News Network
CofCUSA	Children of Chernobyl United States Alliance
CORE	Cooperation for Rehabilitation
d.	*delo* (file)
DOD	Department of Defense
EPRI	Electric Power Research Institute
EPZ	Emergency Planning Zone
f.	*fond* (collection)
F.	folder
FEMA	Federal Emergency Management Agency
FRG	Federal Republic of Germany
g./gg.	*god/y* (year/s)
GMT	Greenwich Mean Time
IAEA	International Atomic Energy Agency
INES	International Nuclear Event Scale
INS	Immigration and Naturalization Service
l./ll.	*list/y* (sheets)
LTO	*lager truda i otdykha* (labour and recreation camp)
mSv	millisievert
NCIV	National Council for International Visitors
NEA	Nuclear Energy Agency
NGO	non-governmental organization
NIRS	Nuclear Information Resource and Service
NPP	nuclear power plant
NSC	National Security Council
op.	*opis'* (inventory)
rem	roentgen equivalent in man
Ru.	Russian
SDI	Strategic Defense Initiative
UCC	United Church of Christ
Ukr.	Ukrainian
USSR	Union of Soviet Socialist Republics
UN	United Nations
USCEA	US Council on Energy Awareness
USDA	US Department of Agriculture
WANO	World Association of Nuclear Operators
WHO	World Health Organization
YMCA	Young Men's Christian Association

Prologue

Prypiat – Artek – Boston

PRYPIAT

Blue skies. Sunshine. A fresh spring breeze rustles through the vibrant green leaves of young birches and tugs at women's skirts. Children in shorts and dresses romp around a playground, scamper back and forth, and kick a ball about (Fig. 0.1). Mothers and fathers push strollers through the bright streets. It is hard not to notice how many children there are; the town seems so young, the spring day so beautiful. Even the two civil defence officials pacing through the idyll could not appear more relaxed if they tried. And yet – with their drab green overalls and grey-green respirators, they look as if they have come from another world. When someone enquires as to the reason for these strange outfits, they pause briefly to explain that it's an exercise. It is just an exercise.

These fragmentary impressions are part of the ten-minute documentary *Nezabyvaemoe* (*The Unforgettable*) by Mikhail Nazarenko, the official photographer of Prypiat and director of the amateur film studio Pripiat-Film.[1] The film contains the few surviving recordings of everyday life in

[1] Mikhail Nazarenko, *Nezabyvaemoe* (Pripiat'-Fil'm, 1988), www.youtube.com/watch?v= fg6q7uCsv3E (1 May 2024 [dates of last access given in parentheses throughout]). It is no longer possible to determine with certainty what footage is from 26 April 1986 and what might have been filmed earlier; in any event, it matters little in the present context. Edward Geist reports, based on the experiences of a contemporary witness, that 25 April marked the beginning of a civil defence exercise in the Kyiv oblast. It is not known whether the exercise involved Prypiat. Darina Pustovaia, *Chelovek, kotoryi ostalsia za kadrom*, https:// bit.ly/4g3iGtw (1 May 2024); Edward Geist, 'Political Fallout: The Failure of Emergency Management at Chernobyl', *Slavic Review* 74, no. 1 (2015), 104–26, here 113.

FIGURE 0.1   Film still: Children playing.

'one of Ukraine's youngest towns'.[2] According to Nazarenko, he recorded the material on 26 April 1986 – the afternoon after the fateful night when reactor no. 4 at the Vladimir Ilich Lenin Nuclear Power Plant exploded. Prypiat, which had only been founded in 1970, with high hopes for its future as a model socialist town, had not even come of age.

Nazarenko's documentary has also preserved for posterity the announcements made over public loudspeakers on 27 April, almost thirty-six hours after the explosion. Residents (there were nearly 45,000 of them) were instructed to leave in order to ensure the safety of the children. Visual records of the evacuation are few and far between. Nazarenko later said that he had shot a great deal of footage but that members of the KGB had rendered most of it worthless by exposing the film to light. What little did survive largely shows convoys of Ikarus buses, pictures of children and drawings by children on the walls of deserted homes, and abandoned toys.

[2] Thus, the title of a collection of photographs published not long before the disaster: *Prypiat'. Fotorozpovid pro odne z naymolodshykh mist Ukrainy* (Kyiv: n. p., 1986). See Anna V. Wendland, 'Tschernobyl: (K)eine visuelle Geschichte: Nukleare Bilderwelten in der Sowjetunion und ihren Nachfolgestaaten' in Melanie Arndt (ed.), *Politik und Gesellschaft nach Tschernobyl:(Ost-)Europäische Perspektiven* (Berlin: Ch. Links, 2016), 182–210, here 193.

By recording these motifs in 1986, Nazarenko created an imagery that continues to define the visual history of the Chernobyl disaster today. For the purposes of this book, 'Chernobyl' means not simply the physical accident site or the disaster as an *event* that took place on 26 April 1986, but also the disaster as a *process* leading up to the explosion and continuing after it.[3] The visual language of that process is centred on legacies associated with childhood – a childhood that was suddenly ruptured, a lost childhood. Its most enduring symbol is the Prypiat Ferris wheel. In all likelihood, the attraction never went into operation; the amusement park was due to be opened as part of the May Day holiday, but by that point the children's presence in the town was over. This has not stopped thousands of photographers, researchers, tourists, and thrill-seekers from capturing the stationary ride with their cameras. It has become deeply ingrained in the visual memory of the disaster.

ARTEK

This second scene is also part of a documentary by Nazarenko.[4] Several weeks after the accident, he was filming at the flagship Soviet Pioneer camps of Molodaia Gvardiia [Young Guard] and Artek on the Black Sea (Fig. 0.2). Some of the roughly 7,000 schoolchildren evacuated from Prypiat had been accommodated there in May 1986; their stay was to last up to several months. Nazarenko had been sent to visit the children – who had been separated from their families abruptly – by the trade union committee at the power plant. His brief was twofold. First, he was to bring them photographs from home, from Prypiat, to help them cope with their homesickness. Second, he was to return with footage that would reassure parents that their children were safe and well.[5]

Accordingly, Nazarenko filmed numerous scenes with orderly children dressed in smart Pioneer uniforms or matching summer clothes; they seem healthy, happy, and lively, and give every appearance of enjoying their time in the Crimean sun. This framing is completely in line with convention, which required pictures of socialist holiday camps to show

---

[3] By convention, this book uses the Russian form 'Chernobyl' in the context of the disaster; the Ukrainian form 'Chornobyl' is employed when referring specifically to the town of that name.

[4] Mikhail Nazarenko, *Mne dorogi eti mesta* (Pripiat'-Fil'm, 1986), www.youtube.com/watch?v=-M_tYuGfXdY (1 May 2024).

[5] Pustovaia, *Chelovek, kotoryi ostalsia za kadrom.*

FIGURE O.2 Film still: A group of children at Artek (including Alena Oginets).

untroubled, universal happiness.[6] Not the slightest hint of distress, hardly any sign of sadness – even though these children had often not seen their parents for weeks and in many cases did not even know where their parents were now.[7] Alena Oginets, who was twelve at the time and spent more than a month at Artek, recalls that Nazarenko had urged them to look relaxed.[8]

Yet, however tentatively, Nazarenko began finding cracks in the official idyll. In passing, the camera captures melancholic faces; if one listens carefully, children's voices that suggest all is not well can be made out. One boy remarks that it is hard 'to act normally so as not to give Prypiat

---

[6] See 'Children on Display: Children's History, Socialism, and Photography', a special issue of the *Jahrbücher für Geschichte Osteuropas*, edited by Martina Winkler, esp. Monica Rüthers, 'Picturing Soviet Childhood: Photo Albums of Pioneer Camps', *Jahrbücher für Geschichte Osteuropas* 67, no. 1 (2019), 65–95.

[7] Aleksandr Demidov, the owner of a Prypiat disco who accompanied Nazarenko to Artek, said in a later interview that the children cried a lot and were beset with insecurity, fear, and hopelessness. Cf. Pustovaia, *Chelovek, kotoryi ostalsia za kadrom*.

[8] Elena Otriaskina was born as Alena Oginets. She prefers the latter name and uses it in her creative work; I follow that preference here. Alena Oginets, interviewed by Melanie Arndt, 27 March 2017. Online.

a bad name'. Toward the end of the film, the children strike up a song they have written about their home town: 'We'll overcome the distance!' ('My razluku pobedim!').

It was not just Nazarenko who got to speak to the Chernobyl children at Artek. The Soviet leadership gave selected press representatives access to the camp, even extending this to foreign journalists such as Felicity Barringer from the *New York Times*. It was hoped that carefully planned group interviews would show them how well the Soviet Union was coping with the aftermath of the disaster.[9] Using the children seemed to provide the perfect way to do this.

BOSTON

This third scene is preserved in a black-and-white photograph from April 1991 (Fig. 0.3).[10] An endearing, visibly sick boy smiles proudly, if weakly, at the camera. The oversized baseball cap on his little head cannot hide the fact that he has lost his hair. Someone has lifted him onto a lectern at an event in Boston to mark the fifth anniversary of the Chernobyl disaster. On his right, Stojan 'Stojko' Vranković – a Croatian basketball player whose athletic physique looks even larger next to the child – is gently propping him up. On his left, Edward 'Ted' Kennedy, Democratic senator for Massachusetts, is holding the boy's arm aloft triumphantly. Vladimir Malofienko – or Vova, as he was endearingly known by the US media and everyone else involved – was a seven-year-old Ukrainian who had started a new form of cancer treatment in the United States a year earlier. It was clearly working. Together with guests of honour and other cancer patients, he helped plant a garden of hope at the event.[11]

Vova had been part of the first group of Chernobyl children to fly to the United States. They had arrived in 1990 and gone to the Hole in the Wall Camp, a Connecticut summer camp for children with cancer, for a period of respite. Vova had developed leukaemia three years after the Chernobyl

[9] Felicity Barringer, 'From Children of Chernobyl, Stories of Flight and of Fears', *New York Times*, 5 June 1986, https://bit.ly/3NSrGoR (1 May 2024). The Soviet plan was not particularly effective.

[10] The photograph, which was printed with a report on the commemoration in the *Boston Herald*, was taken by Arthur Pollock. Zachary R. Dowdy, 'City Rally Remembers Chernobyl', *Boston Herald*, 27 April 1991. I would like to thank the *Boston Herald* for providing me with a copy of the photograph, and for permission to use it here.

[11] Vladimir Malofienko to Melanie Arndt, Facebook chat, 12 and 13 December 2017.

FIGURE 0.3 Vova Malofienko.

explosion. He and his family had been living just under 60 km from the power plant when the accident occurred.

While Vova was at the camp, one of the doctors there arranged for him to access an experimental therapy. He decided to stay in the United States for it. The seven other boys who had been part of his group died after returning to Ukraine, but Vova's leukaemia was soon in remission.[12] As his recovery progressed and he grew up into a teenager with top marks at

[12] 'Chernobyl Victim's Immigration Status Left Hanging by Congress', *CNN*, 22 October 1998, http://edition.cnn.com/WORLD/europe/9810/22/chernobyl.boy/index.html (1 May 2024).

FIGURE 0.4 Alexander Kuzma (Children of Chernobyl Relief Fund), First Lady Hillary Clinton and Vova Malofienko at a commemorative event in the White House on the tenth anniversary of the disaster. Washington, DC, 1 May 1996.

school, CNN described him as 'a spokesman for the children of Chernobyl'.[13] He became involved in efforts to keep the memory of the disaster alive in the United States, and publicized the plight of those it had affected in the former Soviet Union. The Democratic senator Frank Lautenberg, who praised him for displaying 'the qualities all Americans value', personally intervened in an effort to obtain permanent residency for him and his parents.[14] Vova has since become a US citizen (Fig. 0.4).

[13] Phil Hirschkorn, 'Chernobyl Boy's Family Fears Deportation Could Mean Death', *CNN*, 20 October 1998, http://edition.cnn.com/WORLD/europe/9810/20/chernobyl.boy/ (1 May 2024).
[14] 'Chernobyl Victim's Immigration Status'; 'A Bill for the Relief of Alexandre Malofienko, Olga Matsko, And Their Son, Vladimir Malofienko', *Congressional Bills 106th Congress, 2nd session*, 8 September 1999, www.congress.gov/bill/106th-congress/senate-bill/199/text (1 May 2024).

I

A Transnational Disaster: Currents and Connections

The three scenes in the Prologue are representative of the life 'on a damaged planet' with which this book is concerned.[1] The Chernobyl children lie at its heart – but who are they? There is no universally accepted definition. The term 'children of Chernobyl' first appeared shortly after the disaster, as a description applied to them in the Western press in the title of Barringer's report on the Artek children in the *New York Times*.[2] In the Soviet Union, the equivalent term *deti Chernobylia* only came to prominence when the censorship of Chernobyl-related themes was lifted in spring 1989. Concerned parents and activists were the first to employ it, but soon the term was appearing everywhere. It was even the name of the programme that the centralist Soviet state drew up in its first, and only, attempt to mitigate the effects of the disaster on affected minors. Today, the term has long since been appropriated by those it was coined to categorize: children are not simply classed as such but also describe themselves as children of Chernobyl, as do those who have by now been adults for many years – once a Chernobyl child, always a Chernobyl child.[3] Use of the term proliferated in all areas of politics and society at home and abroad, in no small part because, following the disaster, more than a million Belarusian, Ukrainian, and Russian children were sent,

---

[1] Anna Tsing et al. (eds.), *Arts of Living on a Damaged Planet: Ghosts of the Anthropocene* (Minneapolis: University of Minnesota Press, 2017).

[2] Barringer, 'From Children of Chernobyl'.

[3] The same is true of other such designations: e.g. 'liquidator' (*likvidator*), which refers to those involved in the clean-up and decontamination work and measures to make the site secure.

together with thousands of accompanying adults, to other parts of the Soviet Union and other countries for periods of several weeks or months. The idea was that they would not just be able to recover from exposure to radiation but also, increasingly, find respite from the challenges of everyday life in the turmoil of late and post-Soviet society. These children are, for me, the Chernobyl children; I am aware that using the term risks stigmatizing individuals by reducing them to the disaster, and seek throughout the book to avoid this.

Initially, the Soviet state arranged for the Chernobyl children to go to holiday camps, sanatoriums, and spartan 'tourist bases' inside the Soviet Union. From the end of 1989, as the socialist empire opened up in the late stages of perestroika and the economic crisis deepened, more and more children began to go abroad – in no small part because a number of go-to destinations could no longer be used: they were located in what were now former Soviet republics, in some cases ones that had become embroiled in civil war. The trips abroad were arranged by the state at first, but it was hardly in a position to pay for them. Instead, the costs were covered by the children's hosts – in most cases, civil society initiatives of all sizes that sprang up in practically every country in Western Europe and North America. In parallel to this, newly founded non-state initiatives in the East increasingly stepped in to provide assistance that had previously been the purview of the state – most of all, trips for the Chernobyl children. In large part, these initiatives passed the burden on again – to foreign partner organizations – because they did not have the financial resources to bear it alone either. The result was an increasingly transnational network of individuals and non-state organizations that is inseparably linked to the Chernobyl children and the role they played in the story of this transnational disaster.

The present study is concerned with the social and political consequences of the fact that one of the showcase disasters of the Anthropocene coincided with the passing of a world order: the end of the Cold War. By approaching the Chernobyl children from the perspective of a 'global microhistory',[4] I aim to show that transnational non-state networks had a crucial influence on strategies for confronting the legacy of the disaster in the late Soviet Union and its successor republics. Equally, what happened in those republics was still frequently determined by perceptions and practices that

---

[4] Angelika Epple, 'Globale Mikrogeschichte: Auf dem Weg zu einer Geschichte der Relationen', in Ewald Hiebl, Ernst Langthaler (eds.), *Im Kleinen das Große suchen: Mikrogeschichte in Theorie und Praxis* (Innsbruck: Studienverlag, 2012), 37–47.

originated in the Cold War and before the Anthropocene; it proved impossible to shake them off completely.

Three interrelated questions are central to my analysis of the currents and connections that made Chernobyl a transnational disaster. First, I am interested in how the disaster was perceived and what the Soviet state did to safeguard children – the most vulnerable among those affected by the accident. By the end of the 1980s, the Soviet Union was – despite what it claimed – no longer able to ensure that these children were protected and properly looked after. For many affected by this, as well as for many who witnessed it, the failings indicated the extent of the state's collapse. I am interested in the courses of action open to state actors, and the knowledge to which they were privy, in this unprecedented situation. What interests were they guided by? How did they try to shield the children, and thereby themselves, from radiation and insecurity at a time when familiar sources of meaning and certainty were no longer as stable they had been, at a time when several crises – environmental, political, social, and economic – coincided?

Second, I am interested in the humanitarian movement that emerged against the background of these multiple crises, a movement whose activities extended far across national borders and depended in large part on donations from the West.[5] What made this possible? Why did individuals and groups start to take an interest in the youngest victims of a disaster in a land on the other side of the world that had once been their enemy? What ideas and expectations lay behind the acts of compassion that ensued? Foreign initiatives made the Chernobyl children part of their everyday lives in a very literal sense, thereby giving them a central place in how the Chernobyl disaster and the collapse of the Soviet Union were perceived. This book is about the role that the children's 'power of innocence' played in this process.[6] It was not uncommon for differing ideas of childhood, the state, society, and nature to clash; the various parties involved found themselves perforce using different compasses to navigate

---

[5] For definitions of movements, see Melanie Arndt, *Tschernobylkinder: Die transnationale Geschichte einer nuklearen Katastrophe* (Göttingen: Vandenhoeck & Ruprecht, 2020), 23; Dieter Rucht, *Modernisierung und neue soziale Bewegungen: Deutschland, Frankreich und USA im Vergleich* (Frankfurt a. M.: Campus, 1994), 13.

[6] Doris Bühler-Niederberger, 'Einleitung: Der Blick auf das Kind – gilt der Gesellschaft', in Doris Bühler-Niederberger (ed.), *Macht der Unschuld: Das Kind als Chiffre* (Wiesbaden: Verlag für Sozialwissenschaften, 2015), 9–22, here 14–15. On the naturalization of children, see Lee Edelman, *No Future: Queer Theory and the Death Drive* (Durham: Duke University Press, 2004), 57.

the uncertain territory between politics and humanitarianism. I understand humanitarian aid as political – as a form of foreign and internal politics that is primarily pursued by non-state actors. This became increasingly common around the world from the early 1990s onwards and is correlated with the increasing significance of 'human security': an extended concept of security that foregrounds individual people instead of the state.[7]

Third, I am interested – as I have been since first starting to work on the Chernobyl children – in what it means to live (or indeed, go on living) with the disaster, not least because precisely this aspect has tended to be overlooked in existing research, which concentrates primarily on the accident sequence, its immediate consequences, and the scientific and medical specialists involved. Disasters, though, are not events; instead, they play out over extended periods of time.[8] Those who have to live with Chernobyl on a daily basis for the rest of their lives, with the vague sense of uncertainty and the existential fear of illness and pain that this brings, have often been rendered invisible, left behind after the initial headlines faded. Only by recognizing the importance of the long view – the seemingly unspectacular personal trajectories and the intimate relationships between the body, health, and the environment[9] – can we start to understand what living on a damaged planet entails. It may have begun, or seem to have begun, with an iconic event, but the story of the Chernobyl children is nonetheless one of 'slow violence'.[10]

I do not, however, intend to reduce this story to a narrative of victimhood. Classifying people as victims has rightly been criticized because of the passive quality it implies.[11] The Chernobyl children are a good

---

[7] See, e.g., Cornel Zwierlein, Rüdiger Graf, Magnus Ressel (eds.), 'The Production of Human Security in Premodern and Contemporary History', special issue of *Historical Social Research* 35, no. 4 (2010).

[8] See Sara B. Pritchard, Carl A. Zimring (eds.), *Technology and the Environment in History* (Baltimore: Johns Hopkins University Press, 2020), 102 ff.

[9] See Joachim Radkau, *Nature and Power: A Global History of the Environment* (New York: Cambridge University Press, 2008), 6.

[10] For 'slow violence', see Rob Nixon, *Slow Violence and the Environmentalism of the Poor* (Cambridge, MA: Harvard University Press, 2011); Pritchard and Zimring, *Technology and the Environment*, 99–114.

[11] For the 'victimhood' discourse, see Didier Fassin, Richard Rechtman, *The Empire of Trauma: An Inquiry into the Condition of Victimhood* (Princeton: Princeton University Press, 2009); Jean-Michel Chaumont, *Die Konkurrenz der Opfer: Genozid, Identität und Anerkennung* (Lüneburg: zu Klampen, 2001); Alyson M. Cole, *The Cult of True Victimhood: From the War on Welfare to the War on Terror* (Stanford: Stanford University Press, 2007); Svenja Goltermann, *Opfer: Die Wahrnehmung von Krieg und Gewalt in der Moderne*

example of how individuals, even if they have been directly affected by a disaster and/or constantly considered victims of it, need not – or at least not all the time – see themselves as victims, and can even choose, more or less freely, to turn 'victimhood' to their own advantage.[12] I am concerned with such self-perceptions and self-representations, which I reconstruct primarily on the basis of interviews, less formal conversations, my own observations, and questionnaires. Many Chernobyl children went abroad for several summers, usually to the same host family. By the early 1990s, if not before, they had become a firmly established and widely used symbol of the (ongoing) disaster in both East and West. The lives of several million people unfolded against this background,[13] yet the Chernobyl children and their significance have largely been ignored by research in history and the social sciences.

It is not the aim of this book (and it would not be practical anyway) to 'check' whether a given Chernobyl child really is one, or whether the suffering of someone who considers themselves a victim was indeed caused by the disaster. The boundaries between the nuclear and the non-nuclear are always fluid and contested, as Gabrielle Hecht has convincingly shown with her category of 'nuclearity'.[14] What counts as nuclear – and, one might add here, who counts as a Chernobyl child – is always affected by social relationships as well. Nuclearity can be 'a tool of empowerment or disempowerment'.[15] I am interested in the nexus of relationships in which 'Chernobyl child' originated as a way of referring to oneself and being referred to, as well as in who appropriated it (and to what ends) and gave it an ethical dimension. I further consider the real-life consequences of identifying or being identified as a Chernobyl child; this

(Frankfurt a. M.: S. Fischer, 2017); Tanja Penter : 'Vernichtungskrieg, Besatzung und juristische Aufarbeitung: Opferperspektiven', *Jahrbücher für Geschichte Osteuropas* 68, no. 3–4 (2020).

[12] See Adriana Petryna, *Life Exposed: Biological Citizens after Chernobyl* (Princeton, NJ: Princeton University Press, 2002).

[13] See Melanie Arndt, 'From Nuclear to "Human Security"? Prerequisites and Motives for the German Chernobyl Commitment in Belarus', *Historical Social Research* 35, no. 4 (2010), 289–308; Arndt, 'Tschernobyl in Deutschland', in Bernd Greiner, Tim B. Müller, Klaas Voss (eds.), *Erbe des Kalten Krieges* (Hamburg: Hamburger Edition, 2013), 364–82; Arndt (ed.), 'Memories, Commemorations, and Representations of Chernobyl', special issue *Anthropology of East Europe Review* 30 (2012); Arndt, *Tschernobyl: Auswirkungen des Reaktorunfalls auf die Bundesrepublik Deutschland und die DDR*, 3rd rev. ed. (Erfurt: Landeszentrale für politische Bildung, 2012).

[14] Gabrielle Hecht, *Being Nuclear: Africans and the Global Uranium Trade* (Cambridge, MA: MIT Press, 2012).

[15] Hecht, *Being Nuclear*, 16.

includes efforts by the Chernobyl children themselves to influence how they are perceived.

Like all the other *Chernobyltsy* ('Chernobylers') – as Chernobyl 'victims' often refer to themselves or are referred to by others – Chernobyl children were the subject of myriad scientific studies. Medical scientists and other researchers used their bodies to produce findings from which those affected did not, as a rule, benefit; instead, the knowledge obtained was swallowed up, so to speak, by scientific discourses.[16] Individuals became statistical groups, 'people who are counted, but who do not count'[17] – a perception that I encountered time after time in conversations and interviews. These experiences on the part of the Chernobyl children and their families demonstrate that their vulnerability has more than one dimension – an idea that will be central to the rest of this book. Building on the anthropological work of Virginia García-Acosta,[18] I associate the Chernobyl children with a specific position of vulnerability that stems from a number of interrelated factors, such as the fact that they were born in a socialist state (and on its periphery, at that), their age, and their gender. None of these vulnerabilities have previously been considered in detail with reference to envirotechnical disasters – that is, disaster processes in which environmental and technical systems overlap dynamically with one another.[19] Greater vulnerability does not necessarily mean passivity and victimhood: women have often played a significant role in their communities during the aftermath of natural disasters, despite remaining particularly exposed to dangers such as the risk of violence and having to face the entrenchment of traditional gender expectations.[20]

---

[16] The *Chernobyltsy* are not alone in this respect; a similar treatment of people harmed by a nuclear tragedy is described in work on the *hibakusha*, the survivors of the bombings of Hiroshima and Nagasaki. For the *hibakusha*, see Robert A. Jacobs, Mick Broderick, *The Global Hibakusha Project*, https://bit.ly/4f9jdd6 (1 May 2024).

[17] Volha Piotukh, *Biopolitics, Governmentality and Humanitarianism: 'Caring' for the Population in Afghanistan and Belarus* (Hoboken: Routledge, 2015), 152.

[18] Virginia García-Acosta, 'Historical Disaster Research', in Susanna Hoffman, Anthony Oliver-Smith (eds.), *Catastrophe and Culture: The Anthropology of Disaster* (Santa Fe: School of American Research Press, 2002), 49–66, here 60.

[19] See Sara B. Pritchard, 'An Envirotechnical Disaster: Nature, Technology, and Politics at Fukushima', *Environmental History* 17, no. 2 (2012), 219–43, here 223.

[20] See Fordham, 'Gendering Vulnerability Analysis', 175–6; Susanna M. Hoffman, 'The Regenesis of Traditional Gender Patterns in the Wake of Disaster', in Anthony Oliver-Smith, Susanna Hoffman (eds.), *The Angry Earth: Disaster in Anthropological Perspective* (New York: Routledge, 2008), 173–91.

1.1 THE CHERNOBYL CHILDREN: CONCEPTS AND CONTEXTS

Writing about the Chernobyl children means having to operate within a number of fluid constructs around which the various discourses are constructed and which have come to be understood in different ways in different contexts. The uncertainties surrounding what is meant by the terms 'Chernobyl child' and 'affected' follow in large part from the nature of radioactivity itself: there is no clear (or generally accepted) definition of the point at which exposure becomes harmful. At the time of the disaster, around seven million people, including roughly three million children, were living in the area that was to be contaminated with radiation. Between 200,000 and 350,000 of those seven million were evacuated or resettled or decided to leave the area on their own initiative. Even today, around five million people are still living on ground that is contaminated; more than one million of them are under the age of eighteen.[21]

It was networks and their social practices that determined who counted as a Chernobyl child and who did not. Whether Tatsiana from Minsk – which received less contamination than some parts of Bavaria (Germany) – was going to get a place on the list of those going abroad or not would depend on the ideas and expectations of those inviting her and their partner organization(s) in her home country, as well as a whole range of other factors that are also considered in the present study. With only a few exceptions, the people involved in supporting the children were not experts in nuclear physics or radiation biology; they had no choice but to radically simplify the formidable complexities involved in weighing up the impact of radiation exposure. This was the only way they could establish a basis on which to operate and ensure that the work in which they had become so invested could go on. Even specialists were often unable to provide unambiguous answers. The medical and environmental effects of the disaster remain controversial today. I am interested in how people embedded the relevant knowledge (or lack thereof) into their social practices and strategies for mobilizing a receptive public that was ready to contribute to a good cause.[22] What knowledge did they consider relevant, and why?

---

[21] These figures, like almost all those relating to the disaster, vary. But even the first and only Soviet state programme for protecting the Chernobyl children assumed in 1990 that there were around one million 'affected children' who needed help. Cf. GARF, f. R8009, op. 51, d. 5138, l. 129–96. The programme is discussed in detail in Section 3.3.2.

[22] On 'hybrid' forms of knowledge, see John Krige, 'Hybrid Knowledge: The Transnational Co-production of the Gas Centrifuge for Uranium Enrichment in the 1960s', *The British Journal for the History of Science* 45, no. 3 (2012), 337–57; on hegemonies of knowledge

How did they themselves generate knowledge (or ignorance)? The factors impinging on the production of knowledge (or ignorance) included not only the material legacies of the Cold War but also its conceptual and ideological heritage. This was a crucial factor in explaining why certain decision-making frameworks (e.g. knowledge that did not fit an individual's or a group's perception of reality) and courses of action (e.g. collaboration with state institutions) were not adopted.[23]

Being described as a 'Chernobyl child', therefore, in no way always implies that someone was *directly* affected by the disaster – that, for example, they came from Prypiat or the 'zone' and were evacuated from there. It does not help matters in this respect that the concept of 'zones' is itself not unambiguous.[24] Similar issues are associated with describing someone as 'affected': it is entirely possible for Chernobyl children and their families to have been directly affected (i.e. exposed to a high level of radiation) even if they lived outside the evacuated area. They might, in fact, have received a higher dose than they would have in some places inside the evacuation zone; radioactive particles were dispersed far beyond the immediate surroundings of the accident reactor. I will show in more detail later that the term 'Chernobyl children' also included children who cannot be shown to have been directly affected: Chernobyl initiatives and interested parties increasingly preferred to employ a broad and inclusive concept in which all children from the three most affected states were treated as 'at risk'.[25] There were good reasons for this, too, given that contaminated food was widely distributed, usually intermingled with 'clean' products, throughout the Soviet Union in the early years

and currents of knowledge spanning different cultures, see Krige, *American Hegemony and the Postwar Reconstruction of Science in Europe* (Cambridge, MA: MIT Press, 2006).

[23] On ignorance, see Robert N. Proctor, 'Agnotology: A Missing Term to Describe the Cultural Production of Ignorance (and Its Study)', in Robert Proctor, Londa L. Schiebinger (eds.), *Agnotology: The Making and Unmaking of Ignorance* (Stanford: Stanford University Press, 2008) 1–33, here 2–3.

[24] The term 'zone' (*zona*) had various layers of meaning in Soviet history. It was initially used in the context of the forced-labour camp system. *Zona* acquired a new cultural status in the 1970s with the science fiction novel *Piknik na obochine* (*Roadside Picnic*) by the brothers Arkadii and Boris Strugatskii (1972) and Andrei Tarkovskii's film version, *Stalker* (1979). For the individuals and institutions covered in the present book, 'zone' always referred to Chernobyl: mostly to the evacuated area, but also to contaminated areas and places outside the exclusion zone. There were actually several Chernobyl 'zones'; on this, and the differences between them, see Section 2.3.

[25] For the concept of 'children at risk', see Sharon Stephens, 'Children and the Politics of Culture in "Late Capitalism"', in Stephens (ed.), *Children and the Politics of Culture* (Princeton, NJ: Princeton University Press, 1995), 3–48, here 9–14.

after the disaster. Everyday items and parts of buildings now used as construction materials were also smuggled out of evacuated areas and are still circulating throughout the former Soviet Union today. If first-hand accounts are anything to go by, efforts to put a stop to this commenced too late and were only partially effective.

The story of the Chernobyl children is a perfect example of how the various parties involved in the process of confronting this disaster produced 'new maps and categories of entitlement'.[26] People can find themselves classified by terms such as 'Chernobyl child' or can actively choose to refer to themselves as such; rather than being clearly distinct from each other, the two usages are akin to two sides of the same coin and circulate together in a 'community of shared suffering'.[27] The mental map on which children of Chernobyl are to be found has been expanding constantly since 1990. The Ukrainian boxer Wladimir Klitschko captured this in an interview with the *New York Times* on the eve of the 2012 European Football Championship, which was jointly hosted by Ukraine: 'Right after the Soviet Union, people didn't know if Ukraine was a city or a country[;] the easiest way to explain was to say, "We are the children of Chernobyl."'[28]

Time has proved just as elastic as space when it comes to understandings of the categories at stake. A Chernobyl child is not simply someone born before or immediately after the accident: until recently, many children from Belarus, Ukraine, and the Russian Federation were still being invited to spend time abroad as Chernobyl children by various initiatives. The invasion of Ukraine may have put a stop to that, but the disaster continues nonetheless – irrespective of what 'Chernobyl tours' and similar offerings, as well as campaigns to 'revive' contaminated areas, might suggest. Chernobyl children are still being born; in fact, medical radiation experts such as Iury Bandazheuski believe that children born in the 2000s will actually suffer greater harm because of their parents' exposure to radiation than their parents did themselves.[29]

---

[26] Adriana Petryna, 'Biological Citizenship: The Science and Politics of Chernobyl-exposed Populations', *Osiris* 19 (2004), 250–265, here 250.

[27] Nancy Ries, *Russian Talk: Culture and Conversation during Perestroika* (Ithaca, NY: Cornell University Press, 1997), 87.

[28] Jere Longman, 'Ukraine's Poisoned Past', *New York Times*, 12 June 2012, www.nytimes .com/2012/06/13/sports/soccer/a-visit-to-chernobyl-which-some-athletes-can-recall-first hand.html (1 May 2024).

[29] Adar'ia Hushtyn, 'Professor Bandazhevskii: Segodniashnie deti podverzheny radiatsii bol'she, chem ikh roditeli', *Tut.by*, 26 April 2019, https://bit.ly/4iyR4OI (1 May 2024).

The question of who counts as a Chernobyl child is closely related to the ongoing debates about victims and victimhood. The number of people negatively impacted by the disaster was controversial from the outset and remains so today.[30] Estimates of the death toll range from 'around fifty' to a million. Assessing the medical and biological effects of the disaster, in particular, is not straightforward; when combined with efforts to lobby and influence for or against nuclear power, this leaves ample room for agendas to be pushed, as Kate Brown, for example, has recently demonstrated with her *Manual for Survival*.[31]

The most common Russian term – in both the source material and in everyday usage – for sending the Chernobyl children away is *ozdorovlenie*, roughly meaning 'betterment of health'. This was often something of a misnomer, particularly where the early years after Chernobyl are concerned.[32] *Ozdorovlenie*, which in most cases was not linked to specific medical measures, could take various forms in practice – for example, temporary evacuation, recuperation and rehabilitation, holidays for leisure or therapeutic trips, cultural exchange, or even trips as a reward or mark of distinction. In general terms, the practice of sending children away to the country can be traced back to the nineteenth century. It has figured prominently in the European imagination because of measures such as the evacuation of children to rural areas in Nazi Germany during the Second World War.[33] German children continued to be sent away in the post-war period; the Scandinavian countries were a particularly common destination.[34] This form of humanitarian action was marked from the outset by two essential aspects: passivity and paternalism. Children

---

[30] See also Arndt, 'Tschernobyl in Deutschland'.

[31] Kate Brown, *Manual for Survival: A Chernobyl Guide to the Future* (New York: Norton, 2019).

[32] Soviet health and leisure travel as a whole was marked by terminological variation and a lack of universal definitions. See Monika Henningsen, *Der Freizeit- und Fremdenverkehr in der (ehemaligen) Sowjetunion unter besonderer Berücksichtigung des Baltischen Raums* (Frankfurt a. M.: Lang, 1994), 22.

[33] The increasing bombing raids towards the end of the war led the Nazis to evacuate a total of around five million children (i.e. almost every second child), predominantly to rural areas. See, e.g., 'Dokumente und Berichte zur Erweiterten Kinderlandverschickung', a series of studies and accounts published by Projektverlag.

[34] See, e.g., Jörg Lindner, *Den svenska Tysklands-hjälpen 1945–1954* (Umeå: Universitet Umeå, Almqvist & Wiksell International, 1988); Tor Nyseter, Axel Valen-Sendstad, *Forsoningens bru: Vest-Berlins flyktningebarn kommer til Norge* (Bonn: n. p., 1986). For Austrian children, see Gudrun Springer, 'Das Erbe von 1945: Kinder, quer durch Europa verschickt', *Der Standard*, 25 April 2015, www.derstandard.at/story/20000146 93464/kinder-quer-durch-europa-verschickt (1 May 2024).

were sent away because state or non-state actors considered it necessary. In the case of Chernobyl, the children's wishes – and, in the early years, even those of their parents – often had little, if any, role to play.

By considering the spectrum of support for the Chernobyl children at home and abroad, and bringing the children's own experiences into focus, the present study offers a panoramic account that spans the final period of the Cold War and life in the midst of the Anthropocene. When considered in relation to the Chernobyl children in their inter- and transnational 'networks of concern',[35] the societies in question no longer present themselves as self-contained, static entities; instead, it becomes possible to understand them dynamically in terms of the unceasing processes of transfer and interplay that bind them to their environmental and social contexts. The Chernobyl children also open up new perspectives on history as subjectively experienced and recalled, yielding profound insights into how the effects of this 'slow disaster' have been approached,[36] as well as into transnational supporter networks and the role they have played in this multidimensional process.

Countries that faced one another as enemies in the Cold War stand at the centre of attention in what follows; the focus lies on the United States and the (former) Soviet Union with their rival ideologies. Covering the three countries worst affected by the accident – Belarus, Ukraine, and the Russian Federation (and their antecedent Soviet republics) – in equal depth would be beyond the scope of the book. Instead, particularly where the post-Soviet period is concerned, I concentrate on Belarus. It is there that contamination levels were highest, the Chernobyl children movement most extensive, and the legacy of the ideas and institutions of the Soviet Union most obvious – not least after President Aliaksandr Lukashenka (Ru. Aleksandr Lukashenko) took office in 1994.

1.2 CHERNOBYL (HI)STORIES

A new wave of interest in Chernobyl appeared suddenly in 2019. The highlight in the public imagination was the US mini-series *Chernobyl*, produced by HBO.[37] The show achieved better ratings than any other

---

[35] Rebecca Gill, 'Networks of Concern, Boundaries of Compassion: British Relief in the South African War', *The Journal of Imperial and Commonwealth History* 40, no. 5 (2012), 827–44.

[36] Pritchard and Zimring, *Technology and the Environment*, 102.

[37] Johan Renck, *Chernobyl*, USA, 2019: Sister Pictures, The Mighty Mint. HBO.

HBO series and sparked a short-lived but intense fascination with Chernobyl; the Wikipedia article on the accident had never had so many visits.[38] In April 2022, the Russian state broadcaster NTV sought to emulate the success of the HBO mini-series with a production of its own that offered an 'alternative history'.[39]

Shortly before the HBO series was broadcast, three significant new studies were published almost simultaneously by authors based in the United States. Serhii Plokhy and Adam Higginbotham[40] add to earlier research on the accident sequence and its immediate consequences, particularly for Ukraine.[41] Plokhy also considers the beginnings of environmental mobilization among writers of the Ukrainian *intelligentsiia* in

[38] See International Movie Database, *Chernobyl: TV Mini-Series 2019*, www.imdb.com/tit le/tt7366338/ (1 May 2024); '"Chernobyl" Is the Highest-rated TV Series Ever: HBO's Show Rates Higher on IMDb than Other Historical Dramas and Even "Breaking Bad"', *The Economist*, 4 June 2019, https://bit.ly/3AzudRK (1 May 2024).

[39] The director, Aleksei Muradov, picks up an older conspiracy theory that revolves around a CIA agent being present at the plant before the accident. Anzhela Zhitniuk, *Serial "Chernobyl"': Novaia versiia ot NTV*, 7 August 2019, https://kinofilmpro.ru/cherno byil-ot-ntv.html (1 May 2024); 'Chernobyl (serial 2022)', https://www.kinopoisk.ru/ser ies/1169262/ (1 May 2024). The alleged CIA sabotage was 'exposed' by Anatoly Tkachuk, who claims to have been a KGB agent. Anatoly N. Tkachuk, *Ich war im Sarkophag von Tschernobyl. Der Bericht des Überlebenden* (Vienna: Styria Premium, 2011). See also Melanie Arndt, 'Merle Hilbk: Tschernobyl Baby: Wie wir lernten, das Atom zu lieben. – Anatoly N. Tkachuk : Ich war im Sarkophag von Tschernobyl. Der Bericht des Überlebenden. – Alla A. Yaroshinskaya: Chernobyl. Crime without Punishment', *Osteuropa* 62, no. 6–8 (2012), 542–44.

[40] Serhii Plokhy, *Chernobyl: History of a Tragedy* (London: Basic Books, 2019); Adam Higginbotham, *Midnight in Chernobyl: The Untold Story of the World's Greatest Nuclear Disaster* (New York: Simon & Schuster, 2019).

[41] See the works of David R. Marples, for example, *Chernobyl and Nuclear Power in the USSR* (New York: St. Martin's Press, 1986); Jane I. Dawson, *Eco-nationalism: Antinuclear Activism and National Identity in Russia, Lithuania, and Ukraine* (Durham: Duke University Press, 1996); Zhores A. Medvedev, *The Legacy of Chernobyl* (New York: W. W. Norton, 1992); Paul R. Josephson, *Red Atom: Russia's Nuclear Power Program from Stalin to Today* (Pittsburgh: University of Pittsburgh Press, 2005); Sonja D. Schmid, *Producing Power: The Pre-Chernobyl History of the Soviet Nuclear Industry* (Cambridge, MA: MIT Press, 2015). For a detailed research review, see Arndt, *Tschernobylkinder*, 33–8, and note also various other studies, most of which, however, have a regional focus: e.g., Dolores L. Augustine, *Taking on Technocracy: Nuclear Power in Germany, 1945 to the Present* (New York: Berghahn Books, 2018); Katrin Jordan, *Ausgestrahlt: Die mediale Debatte um 'Tschernobyl' in der Bundesrepublik und in Frankreich 1986/1987* (Göttingen: Wallstein-Verlag, 2018); Karena Kalmbach, *The Meanings of a Disaster: Chernobyl and Its Afterlives in Britain and France* (New York: Berghahn Books, 2021); Tatiana Kasperski, *Les politiques de la radioactivité: Tchernobyl et la mémoire nationale en Biélorussie contemporaine* (Paris: Éditions Petra, 2020); Susanne Bauer, Tanja Penter (eds.), *Tracing the Atom: Nuclear Legacies in Russia and Central Asia* (New York: Routledge, 2022).

the late 1980s.[42] Kate Brown examines the mechanisms by means of which large international organizations, above all the United Nations, regulated efforts to understand and communicate the medical and environmental consequences of the accident.[43] She argues that they were guided by two principles in particular: the idea that low-level radiation does not pose a health risk, and an interest in continuing to support the use of nuclear power. The individual plights and experiences (including illness) of those affected were of less, if any, importance; the objective was clearly to keep the number of victims as low as possible.[44]

For Brown, Chernobyl was not so much an accident as 'an acceleration on a time line of exposures'.[45] In other words, the world's population had long since been exposed to increasing amounts of radiation anyway – by accident, of course, but also by decades of nuclear weapons testing (which is now widely considered to mark the beginning of the Anthropocene) and as a result of inadequate precautions in the production of nuclear weapons and the construction of nuclear power stations. Chernobyl merely meant that this came to be appreciated more rapidly. The Anthropocene is characterized by a general build-up of harmful substances in the environment that makes it harder to establish cause–effect relationships in the case of individual events. The ensuing uncertainties and ambiguities underpin all the stories told in this book. The anthropologist Adriana Petryna introduced the concept of 'biological citizenship' to describe how radiation exposure was exploited as a survival strategy (e.g. as a way of asserting a right to medical treatment) in Ukraine.[46] My work picks up her ideas, but goes far beyond the context of a single (former) Soviet republic by setting efforts to meet the challenges posed by the disaster in a transnational context. The world grew smaller: in environmental terms, with

---

[42] See Plokhy, *Chernobyl*, 285–316. Writing in German, Astrid Sahm had already published a detailed study of mobilization after Chernobyl in 1999; it focuses on environmental and energy politics in Ukraine and Belarus, and remains a standard point of reference today. Astrid Sahm, *Transformation im Schatten von Tschernobyl: Umwelt- und Energiepolitik im gesellschaftlichen Wandel von Belarus und der Ukraine* (Münster: Lit, 1999).

[43] Brown, *Manual for Survival*.

[44] Olga Kuchinskaya has likewise challenged the way in which the disaster has been kept out of sight, showing how its effects, and the memory of it, have been displaced from public consciousness in Belarus. Olga Kuchinskaya, *The Politics of Invisibility: Public Knowledge about Radiation Health Effects after Chernobyl* (Cambridge, MA: MIT Press, 2014). See further Andrej Stepanov, 'Tschernobyl ist niemals passiert? Praktiken der Legitimierung des technopolitischen Regimes in Belarus' in Arndt (ed.), *Politik und Gesellschaft nach Tschernobyl*, 256–82.

[45] Brown, *Manual for Survival*, 302–3.

[46] Petryna, *Life Exposed*.

the unwished-for global dispersal of radionuclides, and in social terms, with the new interconnections that developed in response to this.[47] My aim is to bring these two aspects together, thereby making it possible to move beyond the spatial and temporal boundaries that have previously characterized the study of disasters.

Existing work on Chernobyl considers the Chernobyl children in no more than a cursory fashion, if at all. Discussion is limited to the predicament of the (mainly Ukrainian) children in the vicinity of the power plant, and mention is made of the evacuations or the accommodation of children at holiday camps inside the Soviet Union in the immediate aftermath of the explosion.[48] Psychological and sociological studies are few in number, but the findings of those that do exist have informed the picture drawn in this book. They cover the impact that international trips had on the Chernobyl children, but they almost entirely elide the people at home and abroad who made these visits possible.[49] Such individuals have written numerous, often autobiographical, narratives about their efforts to help the Chernobyl children.[50] The Chernobyl children themselves, by contrast, had hardly published any accounts on their own initiative at the time of writing;[51] a number of shorter recollections are to be found scattered across volumes produced to commemorate the visits by the initiatives that organized them.[52]

---

[47] I am alluding here to the understanding of globalization as a shrinking of the world (for critical discussion, see Roland Wenzlhuemer, 'Globalization, Communication and the Concept of Space in Global History', *Historical Social Research* 35, no. 1 (2010), 19–47).

[48] Cf. Marples, *Chernobyl*, 150–1; Plokhy, *Chernobyl*, 212–14; Higginbotham, *Midnight in Chernobyl*, 252–3; Brown, *Manual for Survival*, 61–2.

[49] Cf. Galina V. Gatal'skaia, Nina M. Zaitseva, 'Sotsial'no-psikhologicheskii analiz opyta ozdorovleniia belorusskikh detei za rubezhom v postchernobyl'skii period' *Psikhologicheskii zhurnal*, no. 1 (2008), 44–54; V. Vorona, E. Golovacha, Iu. Saenko, *Sotsial'ni naslidky Chornobyl's'koi katastrofy: Rezul'taty sotsiolohichnykh doslidzhen' 1986–1995 rr.* (Charkiv: Folio, 1996).

[50] See: Michelle Carter, Michael J. Christensen, *Children of Chernobyl: Raising Hope from the Ashes* (Minneapolis: Augsburg Fortress, 1993); Aliaksandr Tamkovich (ed.), *Filasofiia dabryni: Ad katastrofy – da sada nadzei* (Minsk: self-pub., 2016); Erika Schuchardt, Lev Z. Kopelev, *Die Stimmen der Kinder von Tschernobyl: Geschichte einer stillen Revolution*, 2nd ed. (Freiburg: Herder, 1996); Giuseppina Torricelli, Carla Baroncelli, *Piccoli ospiti e parenti del cuore: Non chiamiamoli i bambini di Chernobyl* (Modena: Edizioni del Loggione, 2016).

[51] Exceptions include Svetlana Bodrunova, 'Chernobyl in the Eyes: Mythology as a Basis of Individual Memories and Social Imaginaries of a "Chernobyl Child"' in Arndt (ed.), *Memories, Commemorations, and Representations of Chernobyl*, 13–24; Valentina Petrochenkova, *Ostanovis', mgnovenie!* (Kyiv: self-pub., 2010).

[52] See: Tschernobyl-Initiative der Propstei Schöppenstedt e. V., *Die Spur der schwarzen Wolke: Die Katastrophe von Tschernobyl mit den Augen der betroffenen Kinder und*

1.3 STRUCTURE AND SOURCES: FROM PERESTROIKA
TO POST-SOCIALISM

My study focuses on the first twenty-two years after the Chernobyl accident; its scope ends in 2008, when President Lukashenka put an end to trips abroad for Belarusian children, at least in their established form. The Chernobyl children movement, which began in the heyday of glasnost and perestroika,[53] consolidated its position between 1989 and 1998. Around the turn of the new millennium, many humanitarian organizations restructured their activity; in several cases, winding up operations seemed on the cards. The nail in the coffin for efforts to support the Belarusian Chernobyl children in the United States came in the form of the controversy surrounding the teenager Tatstsiana Kazyra (Ru. Tatiana Kozyro), who, at the end of her ninth consecutive summer in California, decided that she no longer wanted to go back to Belarus. Lukashenka ordained in response that Chernobyl children would be allowed to travel only to countries that had concluded an intergovernmental agreement with Belarus. A number of countries, above all the United States, saw Lukashenka's directive as state interference in the concerns of civil society, and consequently declined to enter into such an arrangement.

It is impossible to tell the story of the Chernobyl children without considering the event with which it all began. On the night of 25/26 April 1986, there came to pass something that advocates of nuclear power – as well as some who were undecided – had communicated as a residual risk, if not an outright impossibility. Against this background, Chapter 2 surveys the context in which efforts to help the Chernobyl children subsequently took shape. It does so by considering responses to the disaster in the Soviet Union and in the West, as well as early Soviet interactions with Western governments and civil societies. As well as reactions to the Soviet handling of the disaster at home and abroad, the chapter considers what it means to live in 'irradiated landscapes'.[54]

Eindrücke einer deutsch-weissrussischen Reisegruppe (Klitzschen: Elbe-Dnjepr-Verlag, 2000).

[53] English-language accounts of this final phase in the history of the Soviet Union include Archie Brown, *Seven Years that Changed the World: Perestroika in Perspective* (Oxford: Oxford University Press, 2008); Brown, *The Gorbachev Factor* (Oxford: Oxford University Press, 1997); Ries, *Russian Talk*; Stephen Kotkin, *Armageddon Averted: The Soviet Collapse, 1970–2000* (Oxford: Oxford University Press, 2001).

[54] The term 'irradiated landscape' is drawn from Nicholas Lezard, 'Chernobyl Prayer by Svetlana Alexievich Review: Witnesses Speak', *The Guardian*, 27 April 2016, https://bit .ly/3ZO3ZVo (1 May 2024).

Directly or indirectly, many Chernobyl children have had to face this experience – a 'living in prognosis' that is filled with contradictions and the challenges of coping with illness, or the fear of it.[55]

Chapter 3 turns to the initial measures (or the lack thereof) taken to safeguard the Soviet Chernobyl children – 'Soviet', because it was initially the Soviet state that took action and evacuated them from the 'zone' and/or sent them to holiday camps and similar facilities without their parents on a temporary basis. The first mass protests took place in 1988, when it became known that the towns and villages to which many children had to return for the start of the new school year in September 1986 (unless they were inside the 30 km exclusion zone) had been contaminated. The second part of the chapter illuminates the processes through which such 'rebellious parents'[56] and their supporters were mobilized and came to realize that they had reason to be concerned. The state responded to the protests with a Union-wide strategy for addressing the legacy of the disaster – one that envisaged not just further evacuations and forms of redress, but also a Chernobyl children programme. The third part of the chapter analyses that programme and the ensuing, but ultimately inadequate efforts on the part of the state to help those harmed by the disaster, specifically the children.

Chapter 4 shows how the Soviet Chernobyl children became 'children of the whole planet'. It examines the ways of involving other countries in the humanitarian effort that were explored, and considers the first children's trips abroad, to the socialist island of Cuba, as well as the visits to the United States – on paper, still enemy territory – that soon followed. Non-state initiatives that sent Chernobyl children to other countries are also discussed, in particular, the Belarusian Charitable Fund 'For the Children of Chernobyl' in Minsk and its overseas partners.[57] As well as initiatives that operated in the context of this fund's network, the chapter covers a number of other non-state institutions, including diaspora associations and religious bodies. The national and international networks that developed as a side effect, as it were, of this work also come into focus. As the number of Chernobyl children wanting to go abroad – and initiatives arranging visits for them – kept growing, so too did friction

---

[55] S. Lochlann Jain, *Malignant: How Cancer Becomes Us* (Berkeley: University of California Press, 2013), 27.

[56] Katalin Fábián, Elzbieta Korolczuk (eds.), *Rebellious Parents: Parental Movements in Central-Eastern Europe and Russia* (Bloomington: Indiana University Press, 2017).

[57] Despite its name, the organization's legal status was actually that of a registered association, not a charitable fund.

with Lukashenka and his authoritarian rule. The chapter considers the implications of this for Chernobyl children and humanitarian organizations, and shows how initiatives at home and abroad responded to the challenges it posed.

Chapter 5 gives centre stage to the voices of the Chernobyl children, their own families, their host families abroad, and the chaperones who accompanied them. It becomes clear how those involved created and moulded the category 'Chernobyl child', adapting it to changing political and social contexts. The often acrimonious arguments about the medical and cultural benefits – or, conversely, the 'culture shock' – of sending the children abroad are traced and analysed as well. The chapter also focuses on the trips themselves. How did these visits to other countries unfold? What were the interrelationships between the children, their host families, and the organizations involved at home and abroad? The humanitarian effort was dependent on public support; accordingly, the chapter illuminates the strategies that were employed in order to get noticed. Most of all, though, the perspectives of the Chernobyl children themselves are foregrounded: how do they assess the impact of their time abroad on the rest of their lives, and on the way they see the world? Finally, the chapter covers the end of efforts to support Belarusian Chernobyl children in the United States, which became inevitable after the controversy surrounding Tatstsiana Kazyro.

My account draws on secondary literature, contemporary accounts, and grey literature, as well as on published records and my own research in a number of post-Soviet and US archives. These traditional sources are supplemented using methods from the social sciences and oral history: I carried out qualitative interviews (face to face, and using Skype and social media) with individuals from all the relevant groups: Belarusian and Ukrainian Chernobyl children, representatives of state and non-state institutions, North American host families, and a variety of experts from the former Soviet Union and the United States. A number of interviewees asked to remain anonymous due to the deeply personal nature of the information they shared.

The book is constructed around three interrelated contexts that are crucial to any understanding of the Chernobyl children movement. First, the obsession with technology during the struggle for supremacy in the Cold War and the 'great acceleration'[58] is of central importance: it led to

---

[58] John Robert McNeill, Peter Engelke, *The Great Acceleration: An Environmental History of the Anthropocene since 1945* (Cambridge, MA: Harvard University Press, 2016).

risks being disregarded and safety being relegated to second place. Second, in the mid-1980s, the widening gulf between the trope of 'happy Soviet childhood' and reality became apparent. Third, a growing awareness of living on a single planet lent itself to efforts to approach the other side with 'organized compassion'[59] in both directions across the Iron Curtain; it is this interconnected world that made it possible for an accident in the Soviet republic of Ukraine to be treated as a disaster of global proportions that demanded an international humanitarian response.

1.4 CONTEXTS

1.4.1 'Atomic Kids' and the Nuclear Adventure Playground of the Cold War

The development of nuclear power in the Cold War was embedded in a 'cult of science and technology that was promoted with considerable energy in both East and West';[60] this cult was also driven by other cutting-edge technologies, in particular, space exploration.[61] The nuclear age saw the two world powers engaged in an unprecedented arms race,[62] in the context of which they attached considerable importance to demonstrating their military capabilities.[63] Putting scientific know-how and techno-logical superiority on display was a hardly less important source of prestige. The use of nuclear power in civilian contexts was interconnected with potential military applications to such an extent that they are, to all intents and purposes, one and the same thing; the distinction is, however, retained here for the sake of convenience.

---

[59] Michael N. Barnett, *Empire of Humanity: A History of Humanitarianism* (Ithaca, NY: Cornell University Press, 2011), 50.

[60] Klaus Gestwa, Stefan Rohdewald, 'Verflechtungsstudien. Naturwissenschaft und Technik im Kalten Krieg', *Osteuropa* 59, no. 10 (2009), 5–14, here 9. 'Atomic kids' is a quotation from Robert A. Jacobs, *The Dragon's Tail: Americans Face the Atomic Age* (Amherst: University of Massachusetts Press, 2010), 99.

[61] See Julia Richers, 'Die erste Kosmonautin: Valentina Tereshkova und der transkontinentale Geschlechterkampf im Kalten Krieg' in Martina Ineichen et al. (eds.), *Gender in Trans-it: Transkulturelle und transnationale Perspektiven* (Zürich: Chronos, 2009), 235–45; Karsten Werth, *Ersatzkrieg im Weltraum: Das US-Raumfahrtprogramm in der Öffentlichkeit der 1960er Jahre* (Frankfurt a. M.: Campus, 2006).

[62] For a relatively recent overview of work on the nuclear age, see Karena Kalmbach, 'Revisiting the Nuclear Age: State of the Art Research in Nuclear History', *Neue Politische Literatur* 62, no. 1 (2017), 49–69.

[63] See Andreas Renner, 'Globale Ikone des Kalten Krieges? Der Atompilz und die sowjetische Nuklearkultur', *Osteuropa* 66, no. 6–7 (2016), 215–36.

The United States and the Soviet Union both led the way in the use of nuclear power. The first reactor in the world to generate energy – just enough to illuminate four lightbulbs – went operational in Idaho in December 1951. Two years later, Dwight D. Eisenhower's 'Atoms for Peace' speech to the UN General Assembly paved the way for the commercial generation of nuclear power. The Cold War was now at its height, and the president used the speech to present a vision of a future without fear that held out the promise of peace and well-being.[64] The Soviet Union promptly inscribed the prospect of a bright future into the name of the first reactor to produce electricity on a commercial scale: the Mirnii atom ('Peaceful atom') reactor in Obninsk, south-west of Moscow, which was connected to the grid in June 1954. The Soviets thereby went one step ahead, and it took the United States four years to catch up with them.[65] Despite appearances, the development of nuclear power during the Cold War was not grounded solely in isolationism and secrecy: from the outset, exchange took place within the scientific communities involved.[66]

The 1950s witnessed an unprecedented acceleration in a number of environmentally significant fields all at once: the use of fossil fuels, population growth, and urbanization.[67] Riding the crest of this wave, the industrialized world lost no time in expanding the use of nuclear power. The political, economic, and scientific decision-makers managed – for a time, at least, in the West, and much more successfully so in the East – to steer the discourse in such a way that the ominous military atom of Hiroshima and Nagasaki was kept separate from a civilian atom that was 'too cheap to meter'.[68] In the eyes of most advocates of nuclear technology, electricity generation was only the beginning: scientists, politicians,

---

[64] The speech can be read in, e.g., Philip L. Cantelon, Richard G. Hewlett, Robert C. Williams (eds.), *The American Atom: A Documentary History of Nuclear Policies from the Discovery of Fission to the Present*, 2nd ed. (Philadelphia: University of Pennsylvania Press, 1991), 96–104.

[65] On Obninsk, see Josephson, *Red Atom*, 20–32.

[66] Thomas R. Wellock, 'The Children of Chernobyl: Engineers and the Campaign for Safety in Soviet-Designed Reactors in Central and Eastern Europe', *History and Technology* 29, no. 1 (2013), 3–23.

[67] Melanie Arndt, 'Environmental History', *Docupedia-Zeitgeschichte*, 23 August 2016, h ttps://docupedia.de/zg/zgArndt_environmental_history_v3_en_2016 (1 May 2024).

[68] Lewis Strauss, chairman of the Atomic Energy Commission, used this much-quoted catchphrase at a gathering of science writers in 1954. On the suggestion that he actually had a more distant future – nuclear fusion reactors – in mind, rather than the generation of energy by nuclear fission that had just been achieved, see Thomas Wellock, *'Too Cheap to Meter': A History of the Phrase*, 3 June 2016, bit.ly/4fhyw2n (14 December 2024). On the 'display value' of fusion power in the Soviet Union, see Josephson, *Red Atom*, 167–202.

writers, artists, and musicians were all enthralled by ionizing radiation and its seemingly endless potential applications in biology, medicine, agriculture, and even tourism. There was more to the myth of the nuclear age than just the prospect of cheaper electricity; nuclear power became the harbinger of a new era because of the wider cultural significance attached to it. A veritable flood of literary and other output celebrated the possibilities of the new technology.[69] Soviet nuclear euphoria, which became positively ecstatic after the death of Stalin, fitted in perfectly with the human subjugation of nature according to the Marxist worldview. The Communist restructuring of the Russian Empire had been defined by an 'energy imperative' characterized by technological 'gigantomania'.[70] With science and technology as panaceas for economic and social problems,[71] the Party leadership embraced a 'boundless optimism' that trivialized any possible dangers.[72] Safety issues were not prioritized in the United States either, notwithstanding the effort put into addressing private-sector liability for the use of nuclear power and associated insurance issues.[73]

The threat of nuclear war had not been forgotten, but it did nothing to detract from the promises attached to the peaceful atom. This is apparent from contemporary messaging to children. On the one hand, schoolchildren in the United States in the 1950s were able to write 'atom' and 'bomb' before they learnt to spell 'mother'.[74] These 'atomic kids' had to crouch

[69] See Jacobs, *Dragon's Tail*; cf. Bill Geerhart, Ken Sitz, *Atomic Platters: Cold War Music from the Golden Age of Homeland Security*, Bear Family Records 2005, 4 Sound Discs.

[70] Klaus Gestwa, *Die Stalinschen Großbauten des Kommunismus. Sowjetische Technik- und Umweltgeschichte, 1948–1967* (Munich: Oldenbourg, 2010), 11; Paul. R. Josephson, 'Atomic-Powered Communism: Nuclear Culture in the Postwar USSR', *Slavic Review* 55, no. 2 (1996), 297–324, here 300.

[71] For Eastern Europe and the Soviet Union, see esp. Susanne Schattenberg, *Stalins Ingenieure: Lebenswelten zwischen Technik und Terror in den 1930er Jahren*, re-print 2014 (Berlin, Boston: Oldenbourg, 2002); Gestwa, *Die Stalinschen Großbauten*; Paul R. Josephson, '"Projects of the Century" in Soviet History: Large-scale Technologies from Lenin to Gorbachev', *Technology and Culture* 36, no. 2 (1995), 519–59; Josephson, *Industrialized Nature: Brute Force Technology and the Transformation of the Natural World* (Washington, DC: Island Press 2002); Josephson, *Totalitarian Science and Technology*, 2nd ed. (Amherst, NY: Humanities Press, 2005); Josephson, *Resources under Regimes: Technology, Environment, and the State* (Cambridge, MA: Harvard University Press, 2006); Josephson, *Lenin's Laureate: Zhores Alferov's Life in Communist Science* (Cambridge, MA: MIT Press, 2010); Josephson, *Would Trotsky Wear a Bluetooth: Technological Utopianism under Socialism 1917–1989* (Baltimore: Johns Hopkins University Press, 2010).

[72] Klaus Gestwa, 'Ökologischer Notstand und sozialer Protest: Ein umwelthistorischer Blick auf die Reformunfähigkeit und den Zerfall der Sowjetunion', *Archiv für Sozialgeschichte* 43 (2003), 349–83, here 358.

[73] Martin V. Melosi, *Atomic Age America* (Boston: Pearson, 2013), 103.

[74] From a teacher quoted in Jacobs, *Dragon's Tail*, 99.

down on the ground in repeated duck-and-cover drills that were supposed
to guarantee safety in the event of a real nuclear attack.[75] On the other
hand, Walt Disney depicted nuclear power in the vibrant and atmospheric
colours of a 1957 animation programmatically entitled 'Our Friend the
Atom'; the happy fable about the new technology was brought into people's
living rooms by the colour television sets that were entering widespread use
at this time. The animation depicts nuclear power as a frightful genie that
comes to seem more and more benign; tamed by science, he turns into an
amenable, generous servant of humanity.[76]

In a similar manner, the Soviet Union used the documentary series
Mirnii atom ('The peaceful atom') to try to popularize 'good' nuclear
energy and dissociate it from its 'evil' military counterpart. The covers
of countless magazines, in both East and West, were filled with models of
futuristic nuclear-powered cars and household appliances.[77] The unset-
tling, anxiety-inducing side of nuclear energy – 'the nuclear uncanny'[78] –
seemed to have slipped into the background where most of the general
public were concerned.

In contrast to the West, where the euphoria was quickly joined by
scepticism, people in the East seemed largely to accept the assurances of
safety and a bright future. The credo of the scientific-technological revolution, which was grounded in a glorification of science and technology and
became a central element in the Soviet Union's confident self-image, became
firmly ingrained in the minds of the population – all the more so because
there was no public discourse where opinions and expert views that
diverged from the official doctrine might have given pause for thought.
Individual specialists did raise the alarm and call for changes, but these
discussions never filtered through into wider society or made any difference
to those in power.[79] From the outset, the decisions behind 'atomic-powered
communism' were,[80] like its dangers, carefully kept secret.[81]

---

[75] Ibid., 99–117; Melosi, *Atomic Age America*, 127–9.

[76] Spencer R. Weart, *The Rise of Nuclear Fear* (Cambridge, MA: Harvard University Press,
2012), 87.

[77] See ibid., 86–95.

[78] Joseph Masco, *The Nuclear Borderlands: The Manhattan Project in Post-Cold War New
Mexico* (Princeton, NJ: Princeton University Press, 2006), 28.

[79] See Gestwa, 'Ökologischer Notstand'; Douglas R. Weiner, *A Little Corner of Freedom:
Russian Nature Protection from Stalin to Gorbachëv* (Berkeley: University of California
Press, 1999).

[80] Josephson, *Red Atom*, 1.

[81] See Gestwa, 'Ökologischer Notstand'; Josephson, *Red Atom*.

The case of a Japanese fishing boat whose crew experienced acute radiation sickness following a US nuclear bomb test at Bikini Atoll in 1954 was the catalyst for a change in how 'peaceful' nuclear energy was seen in the West.[82] This was the first time that US society realized that innocent people could be affected; the fear of being next in line began to figure in everyday conversations, and now encompassed the potential risks associated with nuclear power plants as well. Critical sentiment in the United States first culminated in the controversy surrounding the nuclear power plant planned for Bodega Bay, north of San Francisco, in the early 1960s. In the years that followed, the anti-nuclear movement grew into one of the largest and most influential social movements in post-war Western history. Nuclear power figured prominently throughout the environmental debates of the 1970s, but the first real setback for the nuclear industry did not occur until the Three Mile Island accident in March 1979.[83] The construction and planning of new nuclear power plants in the United States was subsequently suspended for thirty years,[84] even though the accident did not officially have any victims or represent a health hazard to the local population.[85]

1.4.1.1 *Spoiling the Fun: Incidents and Technological Mishaps*

The idea of having subjugated nature and gone through a scientific-technological revolution was a key element in how actually existing socialism defined itself; accordingly, disasters were 'an event-type alien to the system'.[86] Thus, the 'collapse of ... cultural protections', which Lowell Carr, one of the founding fathers of sociological disaster research, identified as constitutive of disasters,[87] could not happen in the Soviet

---

[82] See Yuka Tsuchiya, '"I Was Not Afraid of the Atom Bomb": Young Japanese Tuna Fishermen and Thermonuclear Tests in the Pacific, 1953–1962'. Paper presented at the *8th Biennial Society for the History of Children and Youth Conference*, Vancouver, 2015.

[83] On Three Mile Island, see Natasha Zaretsky, *Radiation Nation: Three Mile Island and the Political Transformation of the 1970s* (New York: Columbia University Press, 2018).

[84] Constructing and commissioning new power plants in the United States only became possible again with the Nuclear Power Program 2010, launched by George W. Bush in April 2002. Cf. Shane Johnson, *Nuclear Power 2010: Program Overview* (Office of Nuclear Energy, Science and Technology, 2002), https://bit.ly/4gctJ3y (1 May 2024).

[85] Various local initiatives that remain active today see things very differently. Mary Stamos even puts the accident on a par with Chernobyl. Mary Stamos (previously Osborn), interviewed by Melanie Arndt, 25 April 2014. Harrisburg, PA.

[86] Andreas Guski, 'Die Stimme der Opfer: Vom Umgang mit Katastrophen in Russland', *Osteuropa* 58, no. 4–5 (2008), 69–80, here 74.

[87] Lowell J. Carr, 'Disaster and the Sequence-Pattern Concept of Social Change', *The American Journal of Sociology* 38, no. 2 (1932), 207–18, here 211.

Union. This self-image meant that the only real course of action available was to play down disasters or sweep them under the carpet entirely. The only acceptable disaster narrative was that of disaster averted, a 'tale of heroism made to fit a happy ending'.[88] If incidents could not be communicated in this form, they had to be kept quiet. The continued use of this approach, which was by now anachronistic, demonstrates the 'persistence of well-practised mechanisms for controlling information', despite the changed political context of glasnost.[89]

The truth is that there were problems everywhere, including – even before the accident – Chernobyl. As early as winter 1979, the KGB identified serious shortcomings in the construction of the power plant, including (among other things) where the monitoring systems were concerned.[90] Its first four years of electricity generation witnessed twenty-nine incidents (*avariinye obstanovki*); the KGB attributed eight of them to operator error, the rest to technical issues.[91] Almost exactly five years before the explosion, in spring 1981, a significant incident occurred, resulting in an area of 180 m² being contaminated with radiation, according to another report from the Ukrainian KGB. This was not publicly disclosed, so it is hardly surprising that the KGB did not detect any 'panic-spreading rumours' or 'negative sentiments'.[92] Even during normal operations, the environs of the power plant were exposed to troubling levels of radiation, as can be seen from the contaminated fish that gave the KGB cause to reprimand the nearby state fish farm: it had been using the plant's cooling-water pond and selling the catch in the local area.[93]

Problems plagued the Soviet nuclear industry throughout its history.[94] The list begins with the quality of construction material and discipline at construction sites.[95] Building the power plants was the responsibility of

---

[88] Guski, 'Die Stimme der Opfer', 74.

[89] Ibid., 76.

[90] SBU, f. 16, sp. 1072, ark. 98–99; HA/ACPSS, f. 89, op. 53, d. 30; N. Vakulenko, 'O nedostatochnoi nadezhnosti', 16 October 1981, in 'Chornobyl'ska trahediia v dokumentakh ta materialakh' (Kyiv: Sfera, 2001), special issue *Z arkhiviv VUChK, GPU, NKVD, KGB* 16, no. 1 (2001), 41–3.

[91] Vakulenko, 'O nedostatochnoi nadezhnosti', 41.

[92] N. Vakulenko, 'Spetsial'noe soobshchenie', 20 April 1981, *Z arkhiviv VUChK, GPU, NKVD, KGB* 16, no. 1 (2001), 40.

[93] N. Vakulenko, 'Dokladnaia zapiska', 12 March 1981, *Z arkhiviv VUChK, GPU, NKVD, KGB* 16, no. 1 (2001), 37–9.

[94] See Kate Brown, *Plutopia: Nuclear Families, Atomic Cities, and the Great Soviet and American Plutonium Disasters* (Oxford: Oxford University Press, 2013), 75–123; Josephson, *Red Atom*; Schmid, *Producing Power*.

[95] On the Soviets' use of concrete, see Josephson, *Red Atom*, 81–108.

the Ministry for Medium Machine-Building Industry (Minsredmash), which used soldiers to do the work. Criminal backgrounds were widespread, and violations of military discipline, as well as alcohol and drug abuse, were a regular occurrence. Even the KGB had to conclude that all this had an impact on the safety of the power plants.[96] Dealing with radioactive waste was a significant challenge in the Soviet Union, as in all countries with nuclear power plants. Some power stations did not have storage facilities that complied with the safety standards of the time; in some cases, it did not even prove possible to build enough new spent-fuel pools to keep up with the flow of used fuel rods.[97]

Six years before the Chernobyl reactor exploded, a team of authors – including Valerii Legasov, first deputy director of the Kurchatov Institute of Nuclear Power and later a well-known critical voice within the system[98] – acknowledged in the popular-science magazine *Priroda* ('Nature') that the safety measures in the Soviet nuclear power system could not rule out accidents and the 'release of a small amount of radioactive elements into the atmosphere'.[99] The catastrophic meltdown at Chernobyl, however, came as a surprise to everyone.

1.4.2 'Happy Soviet Childhood'

'Happy are those born under the Soviet star!', proclaims a poster on which a thriving infant wrapped in thick cloth smiles beneath the red star.[100] The poster is from 1936, when promoting the idea of 'happy Soviet childhood' became a Soviet propaganda priority.[101] The message was directed at the

---

[96] LYA, f. 46, d. 2449, l. 48.

[97] SBU, f. 16, spr. 1114, ark. 251.

[98] Legasov later became a key figure in efforts to address the causes and consequences of the disaster. As a member of the government commission set up following the accident he presented a report to the IAEA in 1986. Shortly after the second anniversary of the explosion, he hanged himself in his Moscow apartment. He had recorded his views on the causes of the accident on tape; among other things, he held a wide range of people responsible for the explosion and criticized an overriding need for secrecy that had meant that even staff at nuclear power plants were not informed about accidents at other facilities. See Josephson, *Red Atom*, 261.

[99] N. S. Babaev et al., 'Problemy bezopasnosti na atomnykh elektrostantsiiakh', *Priroda*, no. 6 (1980), 30–43, here 43.

[100] The poster is reproduced in the picture section of Alla A. Sal'nikova, *Rossiiskoe detstvo v XX veke: Istoriia, teoriia i praktika issledovaniia* (Kazan': Kazanskii Gosudarstvennyi Universitet, 2007), no page numbers.

[101] A number of groundbreaking works on Soviet childhood have been published in recent years, but they focus primarily on the early Soviet Union. See: Catriona Kelly, *Children's*

Soviet public but also further afield, at the ideological enemy in the West. This idealized childhood was overseen by Stalin as a 'wise yet stern father'.[102] It became common for propaganda pictures to depict him together with children, above all endearing girls.[103] Even after the end of Stalinism, however, children and young people continued to be presented as 'loving, docile beneficiaries of the leader' everywhere. This image underpinned the understanding of 'adults as subjects and beneficiaries of the state, rather than as empowered citizens', as Catriona Kelly has shown in her study of Soviet childhood. In Stalin's place, Lenin returned to serve as a kindly role model for children.[104]

Like everywhere else in the world, experiences of childhood in the Soviet Union varied by generation, social status, gender, and age. Even so, some distinctive features can be identified. 'The Soviet cult of childhood' combined 'two apparently contradictory elements … authoritarianism and the sanctity of childhood happiness'.[105] Of crucial importance for the study of the Chernobyl children is the fact that the state saw itself as responsible for this uniquely 'happy Soviet childhood'. There was political indoctrination and a propensity to disempowerment that manifested itself in children being brought up with a firm focus on being part of the collective consciousness, conformity, and submission to authority – but there were also cases in which self-empowerment was encouraged, particularly in the context of education and fostering talent. This was later put to good use by the Chernobyl aid movement – for instance, when dance groups were dispatched to help solicit donations abroad. Alongside dance and sport, chess, literature, and music, imparting practical and technological knowledge was highly valued – for all children, even those from the most modest backgrounds. Elite holiday camps appeared at an

---

World: *Growing Up in Russia, 1890–1991* (New Haven, CN: Yale University Press, 2007); Anne E. Gorsuch, *Youth in Revolutionary Russia: Enthusiasts, Bohemians, Delinquents* (Bloomington: Indiana University Press, 2000); Lisa A. Kirschenbaum, *Small Comrades: Revolutionizing Childhood in Soviet Russia, 1917–1932* (New York: RoutledgeFalmer, 2001); Donald J. Raleigh, *Soviet Baby Boomers: An Oral History of Russia's Cold War Generation* (Oxford: Oxford University Press, 2012); Margaret Peacock, *Innocent Weapons: The Soviet and American Politics of Childhood in the Cold War* (Chapel Hill: The University of North Carolina Press, 2014); Sal'nikova, *Rossiiskoe detstvo v XX veke.*

[102] Kelly, *Children's World*, 129.

[103] Cf. the illustrations in Sal'nikova, *Rossiiskoe detstvo v XX veke.*

[104] Kelly, *Children's World*, 2, 130. See further Serguei A. Oushakine, 'Realism with Gaze-Appeal: Lenin, Children, and Photomontage', *Jahrbücher für Geschichte Osteuropas* 67, no. 1 (2019), 11–64.

[105] Kelly, *Children's World*, 104.

early date; the legendary Artek in Crimea was one such case.[106] Being chosen to go there was a mark of special distinction for Pioneers from all the socialist countries, not least because children from the West were invited to the camp as well. The hope was that the Western visitors would see with their own eyes the superiority of the Soviet system and the architectural and natural splendour of socialist landscapes. That is not always what transpired in reality – by the early 1980s, at the latest, the accommodation and leisure activities on offer at Artek had fallen far behind the standard of Western establishments.[107]

The importance attached to science and technology is apparent from the figures chosen to serve as role models for children: after war heroes, it was primarily representatives of the world of science and technology who served as beacons showing the way for 'happy Soviet childhood'.[108] It was consequently all the more difficult, even after the Chernobyl disaster, to bring such figures down from the pedestals on which they had been placed.

Some provisions for children's well-being in the early and middle periods of the Soviet Union can be considered progressive, even when measured against contemporary standards in the rest of the world.[109] Fundamental changes to the healthcare system resulted in a fall in child mortality – albeit primarily in comparison to the war years, and more slowly than in the West.[110] Comprehensive preventative care and vaccination programmes provided protection against the serious childhood illnesses of earlier times. Only during perestroika – and not least when it came to the effects of the Chernobyl accident – did the weaknesses in the state healthcare system become apparent. Staffing levels and resources had been neglected during the preceding years; investment had been inadequate.

Instead of facing up to the Soviet Union's shortcomings, functionaries and decision-makers repeatedly turned to othering in order to demonstrate

[106] On Artek, see Kathleen Beger, *Erziehung und 'Unerziehung' in der Sowjetunion: Das Pionierlager Artek und die Archangelsker Arbeitskolonie im Vergleich* (Göttingen: Vandenhoeck & Ruprecht, 2019).

[107] Holger Nehring kindly shared his experience of Artek as a Western child with me.

[108] See Kelly, *Children's World*, 132–4; John McCannon, 'Technological and Scientific Utopias in Soviet Children's Literature, 1921–1932', *The Journal of Popular Culture* 34, no. 4 (2001) 4, 153–69.

[109] See Urie Bronfenbrenner, John C. Condry, *Two Worlds of Childhood* (London: Allen & Unwin, 1971).

[110] See S. N. Zatravkin, E. A. Vishlenkova, *'Kluby' i 'getto' sovetskogo zdravookhraneniia* (Moscow: ShIKO, 2022).

the blessings of being a Soviet child.[111] Throughout the history of actually existing socialism, the 'children of the West' were portrayed as individuals who were being morally corrupted and exploited.[112] The usual strategy was to depict the miserable fate of Western children who had been failed by family and state; from the 1980s, this often took the form of grim images of drug addiction, homelessness, and unemployment among young people. The contrasting pictures of Soviet Pioneers in their crisp school uniforms were meant to radiate confidence and optimism – especially the girls wearing dapper white aprons over dark brown dresses, complete with the unmissable white ribbons in their hair.

There was, however, deafening silence when it came to those children who did not live up to the image of the model Soviet Pioneer or Komsomol member. The 'happy Soviet childhood' was largely inaccessible to the children of politically disloyal parents and children with physical or cognitive disabilities. Children with disabilities were just as incompatible with the system as disasters where Soviet ideology was concerned. It is true that there had been progressive efforts to support and integrate people with disabilities following the Revolution, but they never ended up being fully implemented and increasingly ceased to be a priority.[113] Only veterans who had been disabled while serving their country heroically during the Second World War had any social standing;[114] there was no place for children with disabilities in a view of life where being physically and cognitively 'sound' was the mark of a fulfilled socialist existence and the state saw itself as responsible for children's happiness. Accordingly, the preferred course of action was for Soviet children with disabilities to be shut away in homes. Allowing them to be visible in public would have exposed the

---

[111] The concept of 'othering' originates in post-colonial studies, where the term was coined by Edward Said in *Orientalismus*, 5th ed. (Frankfurt a. M.: S. Fischer, 2010). On the early application of the concept to post-Communism, see Dorota Kołodziejczyk, Cristina Şandru, *Postcolonial Perspectives on Postcommunism in Central and Eastern Europe* (London: Routledge, 2016).

[112] *Deti zapada. Sbornik dlia detei i iuniushchestva* (Moscow: n. p., 1926).

[113] The Soviet approach to deafness was particularly progressive, allowing the Soviet deaf community to participate actively in social life, at least for a time. See Claire L. Shaw, *Deaf in the USSR: Marginality, Community, and Soviet Identity, 1917–1991* (Ithaca, NY: Cornell University Press, 2017).

[114] See Beate Fieseler, *Arme Sieger: Die Invaliden des 'Großen Vaterländischen Krieges' der Sowjetunion 1941–1991* (Cologne: Böhlau, 2008); Mark Edele, *Soviet Veterans of the Second World War: A Popular Movement in an Authoritarian Society 1941–1991* (Oxford: Oxford University Press 2008).

fact that the state had, by its own standards, increasingly failed to look after them.[115]

Like other countries, the Soviet Union witnessed a leisure boom during the 1970s and 1980s. Children and young people with disabilities did not see anything of this. Everyone else experienced it at the Pioneer houses and palaces, particularly in new towns and cities such as the *atomgrady* ('nuclear towns'). Not only were there more of these venues; they were now, more than ever, a place for hobbies, pastimes, and other leisure activities, rather than political education.[116] The young people who frequented them increasingly modelled themselves on what the anthropologist Alexei Yurchak has called 'symbols of the imaginary West' – the specific image that the new Soviet generation had of the West, even though it was not always congruent with the reality there.[117] A penchant for the trappings of 'Western' culture did not, however, necessarily entail rebellion against the prevailing order – quite the opposite. Western jeans (or trousers that were meant to look like them), Western music, and a colloquial idiom that was peppered with Western (or Western-sounding) terms and nicknames had an appeal that co-existed with conformism in other aspects of the life under the Soviet system. By the late 1970s, displays of political loyalty had long since become empty rituals. It was entirely possible to compose speeches heaping praise on the socialist system and

---

[115] See Kelly, *Children's World*, 8. Life with disabilities, particularly for children, under socialism – especially in its late phase – has still not received much attention in the literature. For an overview, see Sarah D. Phillips, '"There Are No Invalids in the USSR!": A Missing Soviet Chapter in the New Disability History', *Disability Studies Quarterly* 29, no. 3 (2009). Studies that point the way for further research include William O. McCagg, Lewis Siegelbaum (eds.), *The Disabled in the Soviet Union: Past and Present, Theory and Practice* (Pittsburgh: University of Pittsburgh Press, 2009); Michael Rasell, Elena Iarskaia-Smirnova, *Disability in Eastern Europe and the Former Soviet Union. History, Policy, and Everyday Life* (London: Routledge, 2014); Alexander Friedman, Rainer Hudemann, *Diskriminiert, vernichtet, vergessen: Behinderte in der Sowjetunion, unter nationalsozialistischer Besatzung und im Ostblock 1917–1991* (Stuttgart: Franz Steiner Verlag, 2016); Shaw, *Deaf in the USSR*; Shaw, '"We Have No Need to Lock Ourselves Away": Space, Marginality, and the Negotiation of Deaf Identity in Late Soviet Moscow', *Slavic Review* 74, no. 1 (2015), 57–78.

[116] See Kelly, *Children's World*, 557.

[117] Yurchak describes the Russian concept of *zagranitsa* ('across the border', 'abroad') as 'a Soviet imaginary "elsewhere" that was not necessarily about any real place. The "West" (*zapad*) was its archetypal manifestation. It was produced locally and existed only at the time when the real West could not be encountered.' He introduced the term 'Imaginary West' to refer to this. Yurchak, *Everything Was Forever*, 158–206, here 193, 159.

attend Pioneer and Komsomol meetings on a regular basis while wearing (pseudo-)Western outfits.[118]

After Leonid Brezhnev's death in 1982, the Party leadership once again started to intervene more directly in how young people chose to lead their lives. The Central Committee plenum in June 1984 took issue with 'ideological slackness' in the Soviet youth and the Komsomol; the bone of contention, as it had been in the 1940s, was music and the youth culture associated with it.[119] The Party thought that an enemy 'psychological warfare' campaign was under way here, aimed at undermining the Communist values of the Soviet youth.[120] The discourse of the state-supporting *intelligentsiia* at the end of the 1980s began to recall that of the *Zhdanovshchina* in the 1940s:[121] Soviet children and young people were seen as 'victims of Western influence'.[122] It was now not just rock fans who were undermining the traditional image of socialism's future champions in the eyes of officials; young people expressing themselves creatively, individuals with an interest in history and politics, and people concerned about the environment had also come together in groups that were either acting completely independently or, at best, were only loosely connected to mass organizations and other state institutions.

The twentieth Komsomol congress in April 1987 marked a turning point. The organization was attempting not least to save itself : hundreds of thousands of young people had turned their back on it since the beginning of perestroika.[123] Members openly discussed the new informal groups, which had also been appearing inside the organization for some time. The congress officially recognized some of them, and even decided to provide them with material support (e.g. environmental groups).[124] The twenty-first congress, in April 1990, consequently adopted a new programme that reflected how the organization saw its new political function: its role now was to stand up for the interests of young people, rather than

---

[118] Ibid., 204.

[119] Pilkington, *Russia's Youth*, 79 (with quotation). Cf. also the list of bands and artists with 'ideologically harmful' repertoires drawn up by the Komsomol's Nikolaev Oblast Committee; it included, for example, the Scorpions, Pink Floyd, Tina Turner, Depeche Mode, and Julio Iglesias; Yurchak, *Everything Was Forever*, 214.

[120] Pilkington, *Russia's Youth*, 80 (with quotation).

[121] The *Zhdanovshchina* was a repressive cultural policy named after Andrei Zhdanov, Politburo member and confidant of Stalin.

[122] See Pilkington, *Russia's Youth*, 64–6.

[123] By the end of 1989, ten million members had left the organization. Pilkington, *Russia's Youth*, 162.

[124] On this, see Weiner, *Little Corner of Freedom*, 429–39.

to indoctrinate them ideologically.[125] Perestroika had come to the official Soviet youth organization. A contemporary commentator underlined the significance of the change: 'Like any political project, perestroika is seeking a space for itself in the future, and the future, according to the iron logic of the change of generations, belongs to the young.'[126]

In actual fact, though, the twentieth and twenty-first Komsomol congresses were centred not on the future but, above all, on the present – a present in which young people were increasingly being seen as a 'lost generation'.[127] For the Komsomol, the Soviet youth was no longer the mighty avant-garde of socialism but a 'sickly child in need of its protection'.[128] The socialist state, however, had very limited means at its disposal to help its youth. This is readily apparent from the efforts to safeguard the Chernobyl children that will be discussed in this book: what were originally planned as wide-ranging measures ended up being implemented on a much more modest basis.

Young people were not simply considered lost or 'uncoupled':[129] during perestroika, public perceptions of childhood and youth were marked most of all by associations with pain, vulnerability, and suffering. This perception was accompanied by a 'traumatic reassessment of the actual achievements of the Soviet regime with regard to ensuring children's happiness'.[130] The way children and young people were treated by the state after the Chernobyl explosion was one reason for this. Children were also considered the main victims of a wider situation of economic and social upheaval that was increasingly perceived as *katastroika*.[131] The

---

[125] See Pilkington, *Russia's Youth*, 126–50, 166–75.

[126] Fadeev, quoted in ibid., 192–3.

[127] Pilkington, *Russia's Youth*, 193 (with quotation).

[128] Ibid., 176.

[129] Wolfgang Schlott, 'Abgekoppelt, auf anderen Gleisen: Wohin strebt Rußlands Jugend?' in Wolfgang Schlott (ed.), *Die enterbte Generation: Russische Jugend nach der Perestroika* (Leipzig: Reclam Leipzig, 1994), 10–19.

[130] Kelly, *Children's World*, 153.

[131] The concept of *katastroika*, originally coined in 1988 by the dissident Aleksandr Zinovev in the context of Gorbachev's economic reforms, was extended by Klaus Gestwa to reflect a more general sense of catastrophe and awareness of disasters; the media of glasnost were filled with revelations about catastrophic conditions in all aspects of Soviet life, most of all the environment – one of the main reasons why this became the 'self-diagnosed hallmark' of late perestroika. Klaus Gestwa, 'Katastrojka und Super-GAU. Die Nuklearmoderne in Zeiten von Tschernobyl und Fukushima' in Katharina Kucher, Gregor Thum, Sören Urbansky (eds.), *Stille Revolutionen: Die Neuformierung der Welt seit 1989* (Frankfurt a. M.: Campus, 2013), 57–68, here 58; Alexander Sinowjew, *Katastroika: Gorbatschows Potemkinsche Dörfer* (Frankfurt a. M.: Ullstein, 1988).

general feeling was that children were faced with risks from all sides, even to the point, it was reported in all seriousness, of being abducted by extraterrestrials.[132] The historian Alla Sal'nikova called the generation born in the 1980s and 1990s the 'grandchildren of the Soviet empire'.[133] They lie at the heart of this book.

1.4.3 Spaceship Earth: Competing Claims to One World

Many people around the world experienced a sense of awe when the first photographs of Earth, the Blue Planet, in its entirety appeared in the late 1960s. In the industrialized countries, these pictures contributed to a change in how the world and the shared human presence in it were perceived: they brought home not only the Earth's beauty and uniqueness, but also its vulnerability. The photographs were one of the factors that made it possible for a One World vision that transcended different political systems to emerge from the 1970s onwards.[134] The images were obtained with the help of cutting-edge engineering.[135] The economist and environmental activist Barbara Ward was the first to draw a connection between modern technology and the new vision of the vulnerable Earth; the titles of her books *Spaceship Earth* (1966) and *Only One Earth* (1971) became political catchphrases.[136]

In the West, the New Social Movements, above all the environmental and peace movements, made plentiful use of the slogans and images.[137] The communication strategies and social practices that were refined and handed down in these movements shaped the processes of social change that began in the 'long 1970s',[138] including a new awareness of the risks of living in modern society that went hand in hand with the articulation of

---

[132] See Kelly, *Children's World*, 153. Mysticism and superstition were widespread in the Soviet Union of late perestroika.

[133] Sal'nikova, Rossiiskoe detstvo v XX veke, 212.

[134] See Joachim Radkau, *The Age of Ecology: A Global History*, tr. Patrick Camiller (Cambridge: Polity Press, 2014), 92–4.

[135] It is worth noting in the present context that *Voyager 1*, which has now been under way in space for over forty years, draws its electrical power from three nuclear batteries.

[136] Barbara E. Ward, *Spaceship Earth* (Boston: Columbia University Press, 1966); Ward, René Jules Dubos, *Only One Earth: The Care and Maintenance of a Small Planet* (New York: Norton, 1972) (co-authored with a microbiologist).

[137] On the New Social Movements, see Rucht, *Modernisierung*; Ansgar Klein, Hans-Josef Legrand, Thomas Leif, *Neue soziale Bewegungen: Impulse, Bilanzen und Perspektiven* (Opladen: Westdeutscher Verlag, 1999).

[138] See Anselm Doering-Manteuffel, Lutz Raphael, *Nach dem Boom: Perspektiven auf die Zeitgeschichte seit 1970* (Göttingen: Vandenhoeck & Ruprecht, 2008); for the two

individual emotions and experiences resulting from them.[139] This 'new subjectivity' in the West stood in marked contrast to developments in Eastern Europe. One World slogans were also very popular there, but the accompanying prominence of the subject was largely absent: the primacy of the collective remained unchallenged.

In the New Social Movements of the West, on the other hand, personal matters, particularly fears, could now be articulated openly.[140] This went not only for adults but also for children and young people – and their views should be taken seriously. As one 1985 psychological study concluded, they were growing up in an 'environment completely hostile to children'.[141] This study of 'our children's fear in the nuclear age' involved asking 4,214 young people aged 14–17 in 17 countries, including socialist ones, about their views of the nuclear age. It is not only the children's responses that are of interest; the way in which the questions were posed also reveals something about adults and the general mood in society at the time.[142] The authors of the study made no secret of the fact that their account 'was based on the idea that we're "all in the same boat" today on an Earth under threat'.[143]

Fears of nuclear war were joined by new fears of illness, particularly cancer and AIDS. Talking emotionally about such conditions in public, and taking a stance on perceived or actual risks, became commonplace.

Germanies, Frank Bösch (ed.), *Geteilte Geschichte: Ost- und Westdeutschland 1970–2000* (Göttingen: Vandenhoeck & Ruprecht, 2015).

[139] Frank Biess, 'Die Sensibilisierung des Subjekts: Angst und "neue Subjektivität" in den 1970er Jahren', *Werkstatt Geschichte* 49 (2008), 51–71. See further Virgina Berridge, Alex Mold, 'Professionalisation, New Social Movements and Voluntary Action in the 1960s and 1970s' in McKay Hilton, James McKay (eds.), *The Ages of Voluntarism: How We Got to the Big Society* (Oxford: Oxford University Press, 2011), 114–34; Frank Biess, '"Everybody Has a Chance": Nuclear Angst, Civil Defence, and the History of Emotions in Postwar West Germany', *German History* 27, no. 2 (2009), 215–43.

[140] See Susanne Schregel, 'Konjunktur der Angst: "Politik der Subjektivität" und "neue Friedensbewegung"' in Bernd Greiner, Christian T. Müller, Dierk Walter (eds.), *Angst im Kalten Krieg* (Hamburg: Hamburger Edition, 2009), 495–520; Bernd Greiner, 'Angst im Kalten Krieg: Bilanz und Ausblick' in Greiner et al. (eds.), *Angst im Kalten Krieg*, 7–31; Eckart Conze, Martin Klimke, Jeremy Varon, *Nuclear Threats, Nuclear Fear and the Cold War of the 1980s* (New York: Cambridge University Press, 2016).

[141] Renate Biermann, Gerd Biermann, Heiner Biermann, *Die Angst unserer Kinder im Atomzeitalter* (Frankfurt a. M.: Fischer, 1988), 9.

[142] The questionnaire began by asking whether subjects had seen the film *The Day After*, and also contained a number of leading questions: e.g. '6. Did you know that, if there is a nuclear war, your children might die later on because of the effects of radiation exposure?' and '11. Is the destruction of the environment by people the bigger problem?'; ibid., 107.

[143] Ibid., 7.

Connections were increasingly being drawn between environmental pollution and cancer. Communication about the relationship between humans and nature changed, as did perceptions of it. In the Soviet Union, however, such matters had no place in public. Articulating personal suffering and personal fears was socially frowned upon and politically undesirable.[144]

It was practically impossible to identify with the One World vision espoused in the West without also taking an interest in that part of the world that was hidden behind the Iron Curtain. The aforementioned study of children's fears is just one among many examples of this. In the United States, there was a growing interest in reaching out on the level of citizens' diplomacy (citizens, not just state representatives, engaging with one another).[145] It was against this background that more and more Americans ventured behind the curtain, not least to make a point of demonstrating solidarity after Ronald Reagan's 'evil empire' speech in March 1983.[146] Work to support the Chernobyl children often began in the milieus of these citizens' diplomacy groups and drew on contacts that had been made during such visits to the Soviet Union.

1.4.3.1 'A Humanitarian's World'

The vision of One World was increasingly accompanied by a humanitarian interest in making a difference.[147] Living in *one* world meant not just

---

[144] See Ries, *Russian Talk*.

[145] On this concept, see John L. Davies, Edward Kaufman, 'Second Track / Citizens' Diplomacy: An Overview' in John L. Davies, Edward Kaufman (eds.), *Second Track Diplomacy for Ethnic and Nationalist Conflicts: Applied Techniques for Conflict Transformation* (Lanham, MD: Rowman & Littlefield, 2002), 1–12.

[146] See Melanie Arndt, *Nostalgic Bonfires and Nuclear Burnups: West Meets East in the Post-Soviet Garden, 1985–1995*, unpub. manuscript, 2018; Martin Bradford, 'Musical Cultural Exchanges in the Age of Détente: Cultural Fixation, Trust, and the Permeability of Culture', *Journal of Contemporary History* 51, no. 2 (2016), 364–84; Yale Richmond, *Cultural Exchange and the Cold War: Raising the Iron Curtain* (University Park: Pennsylvania State University Press, 2003); Simo Mikkonen, Pekka Suutari (eds.), *Music, Art and Diplomacy: East–West Cultural Interactions and the Cold War* (London: Routledge, 2016). For early encounters, see Patryk Babiracki, Kenyon Zimmer (eds.), *Cold War Crossings: International Travel and Exchange across the Soviet Bloc, 1940s–1960s* (College Station: Texas A & M University Press, 2014). For Reagan's speech, cf. 'Ronald Reagan, Evil Empire Speech', 8 March 1983, *Voices of Democracy: The US Oratory Project*, https://voicesofdemocracy.umd.edu/reagan-evil-empire-speech-text/ (1 May 2024).

[147] On the United States, see Peter D. Hall, 'Philanthropie, Wohlfahrtsstaat und die Transformation der öffentlichen Institutionen in den USA, 1945–2000' in Thomas Adam, Simone Lässig, Gabriele Lingelbach (eds.), *Stifter, Spender und Mäzene:*

contemplating one's own concerns and fears, but also remembering the plight of others, feeling with them, and taking action to reduce suffering that was not one's own. The history of humanitarianism and charitable activity, closely intertwined with the history of human rights, is a booming research field at present.[148] Michael Barnett has identified three different 'ages of humanitarianism', starting in the early nineteenth century.[149] His periodization is underpinned by the idea that an 'infantilizing civilizing ideology', dominant up to the end of the Second World War, was increasingly displaced by the sense of an obligation 'to bring progress and modernity to the backward populations'.[150] For Barnett, the end of the Cold War marked the transition to a 'liberal humanitarianism'[151] whose characteristics – for example, a marked shift away from governmental involvement, the evolution of informal groups into more formal organizations, and competition for resources – remain significant in the present day.

Analysis of the Chernobyl children can draw profitably on five 'essential dilemmas' identified by Johannes Paulmann as 'inherent in … international humanitarian aid … since the beginning of the twentieth century, if not earlier'. First, Paulmann explains, spatial distance led to a 'fundamental gap between spectatorship and agency'. Second, media

---

USA und Deutschland im historischen Vergleich (Stuttgart: Steiner, 2009), 69–98. The quotation in the section title is from Barnett, *Empire*, 161.

[148] See, e.g., Johannes Paulmann (ed.), *Dilemmas of Humanitarian Aid in the Twentieth Century* (Oxford: Oxford University Press, 2016); Daniel Laqua, Charlotte Alston (eds.), 'Ideas, Practices and Histories of Humanitarianism', special issue *Journal of Modern European History* 2 (2014); Warren F. Ilchman, Stanley N. Katz, Edward L. Queen II (eds.), *Philanthropy in the World's Traditions* (Bloomington: Indiana University Press, 1998); Hilton, McKay, *The Ages of Voluntarism*; Kevin O'Sullivan, Matthew Hilton, Juliano Fiori, 'Humanitarianisms in Context', *European Review of History / Revue européenne d'histoire* 23, no. 1–2 (2016), 1–15; Matthew Hilton et al. (eds.), *A Historical Guide to NGOs in Britain: Charities, Civil Society and the Voluntary Sector since 1945* (Houndmills: Palgrave Macmillan, 2012); Thomas Adam, Simone Lässig, Gabriele Lingelbach (eds.), *Stifter, Spender und Mäzene: USA und Deutschland im historischen Vergleich* (Stuttgart: Steiner, 2009); Gabriele Lingelbach, *Spenden und Sammeln: Der westdeutsche Spendenmarkt bis in die 1980er Jahre* (Göttingen: Wallstein, 2009); Piotukh, *Biopolitics*. On the history of human rights, see Stefan-Ludwig Hoffmann (ed.), *Human Rights in the Twentieth Century* (Cambridge, MA: Cambridge University Press, 2010); Samuel Moyn, *The Last Utopia: Human Rights in History* (Cambridge, MA: Belknap Press of Harvard University Press, 2010); Akira Iriye, Petra Goedde, William I. Hitchcock (eds.), *The Human Rights Revolution: An International History* (Oxford: Oxford University Press, 2012).

[149] Barnett, *Empire*, 29.
[150] Ibid., 31.
[151] Ibid., 29.

involvement played an ambivalent role: it was, he argues, necessary to bring in donations, but the media's fixation on disasters could also run counter to the actual, often long-term goals of aid organizations. Third, Paulmann identifies the 'politics of empathy' as problematic: 'narratives of suffering and relief often focused on events and actions' rather than the everyday. This was often accompanied by alarmist, dramatizing, and gendered representations that focused on women – particularly mothers – and children; 'structural causes of suffering were often left out'. Fourth, Paulmann points to the political instrumentalization of humanitarian aid, specifically its manipulation by state and military actors to further the pursuit of foreign- and economic-policy goals. Fifth, Paulmann notes, 'aid organizations pursued their own politics' because of the need to stand out in the 'competition between NGOs'.[152] This competition intensified in the 1990s,[153] as is reflected not least in efforts to support the Chernobyl children.

The growing technologization of the world from the mid-twentieth century onwards was a double-edged sword. Humanitarianism was no exception to this. Technologization, after all, was associated with environmental destruction and could cause suffering – but, at the same time, it was possible to put new technologies to use in the context of humanitarian aid: aircraft, fax machines, medical devices, and the first computers all made providing humanitarian assistance easier.[154]

For ideological reasons, there was no place for humanitarianism in the Western sense in the Soviet Union before perestroika.[155] Charity was considered bourgeois, something that could only be found in a class-based society,[156] as the extensive articles in the first two editions of the *Great Soviet Encyclopedia* (1927, 1950) make clear. In a classless society, the encyclopedia explained, nobody could be without the means to get by, even if they were unable to participate in the production process.[157] The

[152] Johannes Paulmann, 'The Dilemmas of Humanitarian Aid: Historical Perspectives' in Paulmann (ed.), *Dilemmas of Humanitarian Aid in the Twentieth Century* (Oxford: Oxford University Press, 2016), 1–31, here 28–9.

[153] Piotukh, *Biopolitics*, 74.

[154] Barnett, *Empire*, 29.

[155] See Anne White, 'Charity, Self-help and Politics in Russia, 1985–91', *Europe-Asia Studies* 45, no. 5 (1993), 787–811; Adele Lindenmeyr, 'From Repression to Revival: Philanthropy in Twentieth-century Russia' in Ilchman, Katz, Queen II (eds.), *Philanthropy in the World's Traditions*, 309–31.

[156] E. Solntsev, 'Blagotvoritel'nost'' in *Bol'shaia Sovetskaia Entsiklopediia*, 466–71.

[157] Ibid.

third and final edition (1970) did not contain an entry for 'charity' at all. Less than twenty years later, the idea was everywhere.

Glasnost and perestroika promised to bring about 'socialism with a human face'. This entailed a change in the relationship between individuals and the state. Charitable efforts not associated with the state – which were now often described using the term *miloserdie*, which has religious connotations – came to define the Soviet Union as it collapsed. The *neformaly* ('informals') were in the vanguard of the humanitarian movement in the country.[158] It was they who founded the first charitable organizations. They had widely varying interests and benefited from the striking opportunities for collaboration (e.g. between Chernobyl children activists and film-makers) that resulted from this. The boundaries between 'official' and 'unofficial' became blurred.[159] Informal groups could associate themselves with organizations (such as the Lenin Children's Fund) that were more closely connected to the state if their work seemed genuine and effective.[160] Significant arguments arose inside the independent movements over what form relationships with the state should take; these internal disputes began during perestroika and continued in the post-Soviet era.

In March 1988, almost two years after Chernobyl, the newspaper *Komsomolskaia pravda* observed: 'History progresses. Yesterday the newspapers were full of references to *glasnost* and *perestroika*. Today we meet the word charity [*miloserdie*] increasingly often. In a certain sense the first formed the base, the second the superstructure.'[161] Humanitarianism had indisputably arrived in the Soviet Union now; it would continue to evolve in the years to come, closely intertwined with initiatives from abroad. The Chernobyl children movement is a striking example of this.

---

[158] On the groups initiated by these individuals, see Pilkington, *Russia's Youth*, 160: 'Although generally translated as "informal groups" in fact the groups were not necessarily *in*formal in their organization or activity ... but *non*-formal. What the *neformaly* signified was the existence of a sphere of activity which was outside the realm of the private but was also consciously not incorporated into the state or formal sector' (italics in original).

[159] White, 'Charity', 798.

[160] See Section 3.1. A close interrelationship between the Lenin Children's Fund and the state is suggested, for example, by the fact that six of the leading roles in the fund were included in the official *nomenklatura* system and five seats were reserved for it in the 'public organisations' block in the Congress of People's Deputies. Ibid., 801.

[161] Quoted in ibid., 787.

2

The Disaster and After

Nikolai Fomin, the plant's chief engineer, believes that both man and nature are completely safe. The huge reactor is housed in a concrete silo, and it has environmental protection systems. Even if the incredible should happen, the automatic control and safety systems would shut down the reactor in a matter of seconds.[1]

Two months after these words were printed, the incredible did happen at Nicolai Fomin's plant. It is hard to imagine a crueller historical irony. Only a short time later, he had lost almost everything: his job, his Party membership, and his freedom. Soon after that, he suffered a nervous breakdown in a Kyiv prison and tried to take his own life.[2] It was summer 1986.

Fomin's comments were published in *Soviet Life*, whose glossy pages had been reporting on life in the Soviet Union for readers in the United States since 1956.[3] One of the main themes of the February 1986 issue was 'nuclear power development and management'. The Chernobyl nuclear power station was proudly presented as a shining example. The photographs featured unblemished turbine components, glowing control panels, men and women in white overalls, workers striding toward the

[1] Maxim Rylsky, 'A Town Born of the Atom', *Soviet Life*, no. 2 (353) (1986), 13.
[2] Nikolai Fomin had been working at Chernobyl since 1972 and had taken on the role of chief engineer a year and a half before the accident. He was neither a nuclear physicist nor an engineer; instead, he had been trained as an electrician and taken a few courses in nuclear reactor physics. The ministries responsible assumed that the plant would be run collaboratively – one of Fomin's deputies was a nuclear physicist, another an engineer. However, he did not seek advice from either of them before the test that led to the disaster. Schmid, *Producing Power*, 3, 152–3.
[3] Since 1993, the magazine has been published as *Russian Life*.

camera – images epitomizing progress and triumph over nature. The message throughout was clear: the workers at the power plant and residents of the nearby town 'born of the atom' could be confident of their safety.

When a Ukrainian minister was quoted on the remote chances of a meltdown – once every 10,000 years – he was talking about something utterly beyond the reach of the imagination at the time.[4] The story of nuclear energy was framed as one of unadulterated success. Three leading scientists – including Valerii Legasov – lied that 'in the 30 years since the first Soviet nuclear power plant opened, there has not been a single instance when plant personnel or nearby residents have been seriously threatened; not a single disruption in normal operation occurred that would have resulted in the contamination of the air, water or soil'.[5]

A rosy picture was painted of life in the young, green town of Prypiat. Mayor Vladimir Voloshko aspired for it to be 'as safe and clean as the power plant', adding that there were just a few 'teething problems' that had to be addressed: a baby boom had resulted in the need for more childcare facilities, and more jobs would have to be created for women.[6]

At the end of the 1960s, Soviet planners looking for somewhere to build a new nuclear power station had chosen a site in the backwaters of the Prypiat Marshes in the far north of Ukraine. The Vladimir Ilich Lenin Nuclear Power Plant was built between 1970 and 1983, not far from the BSSR–Ukraine border and around 110 km from the Ukrainian capital, Kyiv.[7] Despite being located close to the planned town of Prypiat,[8] which was founded at the same time, largely as a home for workers at the plant, the power station was known as 'Chernobyl' after the much older town of that name on the River Prypiat. Chornobyl (to use its Ukrainian form) is actually 18 km away and can look back on a history of more than 800

[4] 'The Nuclear Power Industry in the Ukraine', *Soviet Life*, no. 2 (353) (1986), 8.

[5] Valeri Legasov, Lev Feoktistov, Igor Kuzmin, 'Nuclear Power Engineering and International Security', *Soviet Life*, no. 2 (353) (1986), 14–15, here 14. For previous incidents, see 'Spoiling the Fun' in Section 1.4.1.

[6] Rylsky, 'A Town Born of the Atom'. As early as 1980, 195 women had written a joint letter complaining that there were not enough kindergartens (detskie sady). N. Vakulenko, 'Spravka', 26 August 1980, *Z arkhiviv VUChK, GPU, NKVD, KGB* 16, no. 1 (2001), 35–6.

[7] 'Nuclear power plants were sited according to a set of criteria that included proximity to a metropolitan area with high energy demand, … a significant reservoir of water for cooling, existing infrastructure (roads, transmission lines, etc.), and the availability of large contingents of construction workers'; Schmid, *Producing Power*, 35. For more on the background to the Chernobyl plant, see ibid. and Wendland, 'Tschernobyl', in: Arndt (ed.), *Politik und Gesellschaft nach Tschernobyl*.

[8] On Prypiat, see also Section 3.1.1.

years that is of Jewish interest in particular. Since the accident, however, this has all been largely ignored; instead, 'Chernobyl' has become synonymous with the disaster all over the world – and Chornobyl has become an *atomgrad* more than ever. It has served as home for those still working at the nuclear power plant after the evacuation of Prypiat. Scientific laboratories and a hotel have appeared there over the years. In the recent past, increasing numbers of tourists have stayed at the hotel; stag and hen parties have even been held there.[9] It should come as no surprise to find that, in addition to around 150 personnel, 4 outsiders were present in the exclusion zone when Russian troops occupied it together with the power plant during the invasion of Ukraine.[10]

At the power plant, four RBMK-1000-type reactors were intended to produce an electrical output of 1,000 MW each.[11] The first reactor block was completed in 1977. The plant's operators put the ill-fated reactor – the fourth and newest one – into service just three years before the disaster. The graphite-moderated, water-cooled, boiling water RBMK reactor type is a different design from most of those in Western Europe. Specialists were aware from a very early stage that it suffered from a range of problems and vulnerabilities that did not affect other reactor types.[12] There are eight RBMK reactors still in use today – all in the Russian Federation. The last operational reactor in Chernobyl was shut down in December 2000, fourteen years after the accident.

Various other publications have pieced together the events surrounding the accident, right down to the smallest level of detail.[13] The present

---

[9] Florence Wilkinson, 'How the Abandoned Nuclear Wasteland of Chernobyl Became a Bachelorette Party Town', *VICE*, 7 June 2017, https://bit.ly/4fkRUM9 (1 May 2024).

[10] Maxim Kamenev, 'How Russia Took Over Chernobyl', *openDemocracy*, 22 June 2022, https://bit.ly/3ZVlJhM (1 May 2024).

[11] The abbreviation stands for *reaktor bolshoi moshchnosti kanalnyi* – roughly, 'high-output reactor with channels' – followed by the electrical output in megawatts. On the development and operation of the RBMK reactor, see, respectively, Schmid, *Producing Power*, 110; World Nuclear Association, *RBMK Reactors*, February 2022, www.world-nuclear.org/information-library/appendices/rbmk-reactors (1 May 2024).

[12] See, in detail, Schmid, *Producing Power*; V. A. Sidorenko, 'Istoriia RBMK' in V. A. Sidorenko (ed.), *Istoriia atomnoi energetiki Sovetskogo Soiuza i Rossii*, vol. 3 (Moscow: IzdAT, 2003); Sidorenko (ed.), 'Uroki avarii na Chernobyl'skoi AES' in Sidorenko, *Istoriia atomnoi energetiki*, vol. 4 (Moscow: IzdAT, 2002).

[13] E.g. Arndt, *Auswirkungen des Reaktorunfalls*, 33–8; Schmid, *Producing Power*; Paul Laufs, *Reaktorsicherheit für Leistungskernkraftwerke: Die Entwicklung im politischen und technischen Umfeld der Bundesrepublik Deutschland* (Berlin: Springer 2013); Sahm, *Transformation im Schatten von Tschernobyl*; Plokhy, *Chernobyl*. See also Section 1.2.

chapter highlights the key moments and the most important responses, both on the spot and centrally in Moscow, against the background of which the drive to help the Chernobyl children would later unfold. The first part of the chapter focuses on interactions between the Soviet Union and other countries, in particular, the United States. The second part discusses global reactions: both to the accident itself and to the Soviet handling of it. Finally, the third part considers what it means to have to live in and with irradiated landscapes.

2.1 LIKVIDATSIIA: THE ACCIDENT AND ITS AFTERMATH

Even today, the question of what actually caused the explosion during the night of 25/26 April 1986 continues to attract speculation and conspiracy theories. Localized earthquakes, suspicious magnetic fields, US sabotage – many possibilities have been suggested, not all of which have much to do with reality. The literature is filled with contradictory, imprecise, and often downright inaccurate statements. What can be said with certainty is that the explosion took place in the course of a planned safety test at 1.23 a.m. on 26 April. The objective was to check whether the turbine of reactor no. 4 could supply sufficient energy to operate the coolant-circulating pumps in the event of a loss of external power. The accident can be put down to a combination of human error and various technical shortcomings in the reactor design.[14]

During the test, several safety systems designed for emergency situations were intentionally deactivated so as not to interfere with the procedure. The sequence of events culminated in two explosions. As a result, forty identifiable radionuclides were released into the atmosphere; above all, the highly volatile iodine-131 and caesium-137 nuclides formed dangerous aerosols.[15] They were the main components of the radioactive 'cloud', and in some cases were carried thousands of kilometres away before rain removed them from the air. This in turn resulted in the contamination of whole swathes of land far away from the site of the accident. Other radioactive substances, less easily vaporized, were largely released in the form of particles and dispersed primarily in the

---

[14] In particular, the use of graphite 'displacers' on the control rods proved to be a fateful part of the design. For more details on the accident sequence, see Schmid, *Producing Power*, 128–35.

[15] See Iakov Kenigsberg, 'Chernobyl'' i belorusskoe obshchestvo'. *Svobodnaia mysl'*, no. 6 (2008), 145–56, here 145.

vicinity of the reactor. Areas affected by the radioactive fallout included the modern-day Republic of Belarus (70 per cent), Ukraine (5 per cent), and the Russian Federation (0.6 per cent).[16]

Two days after the accident, there were concerning radiation readings at the Forsmark nuclear power plant in Sweden. After the operators had ruled out an incident at their own power station, the meteorological situation led them to suspect that the radioactive particles originated in the Soviet Union, at which point they informed the world community. Only then did the Politburo decide that some kind of public response was needed. Accordingly, on the evening of 28 April, an emotionless television newsreader made it known on *Vremia* that 'an accident has occurred at the Chernobyl nuclear power plant as one of the reactors was damaged. Measures are being taken to eliminate the consequences of the accident. Aid is being given to those affected. A Government commission has been set up.'[17] The following day, the same terse statement approved by the Politburo was published in the lower-right-hand corner of the front page of *Pravda*, the official Soviet newspaper.[18]

2.1.1 On the Frontline

The black-and-white photograph of the wrecked reactor block released by the Soviet authorities – initially in a retouched version without the smoke – has become deeply ingrained in the collective memory of the Global North.[19] It was later joined by shaky, blurred footage of men encumbered by improvised protective gear trying to shovel radioactive debris away from the reactor roof as quickly as possible.[20] These early images have become part of the staple diet for Chernobyl documentaries. They are now readily accessible on YouTube as well.

The initial attempts to put the reactor fire out with water and then to smother it with lead, sand, and other agents failed. Only cooling with nitrogen brought success.[21] From May to November 1986, 'liquidators' – some volunteers, some conscripted – worked on the sarcophagus,

---

[16] Kenigsberg, 'Chernobyl' i belorusskoe obshchestvo', 145.
[17] The translation follows 'Soviet Statements on Nuclear Plant Accident', *New York Times*, 1 May 1986, 10, https://bit.ly/3ZUA268 (1 May 2024).
[18] 'Ot Soveta Ministrov SSSR', *Pravda*, 29 April 1986, 1.
[19] On the visual history of Chernobyl, see Wendland, 'Tschernobyl'.
[20] For details, including radiation exposure and its consequences for the firefighters, see Arndt, *Tschernobylkinder*, 72–3.
[21] For details, see ibid., 73–4.

a concrete structure that was constructed to encase the remains of the reactor building as a provisional protective measure.[22] Only a few years later, it had already begun to develop cracks.[23] Between 2010 and 2016, an international consortium called Novarka built the New Safe Confinement (NSC) in an unprecedented construction project that cost more than two billion euros.[24] The NSC was assembled a short distance away from the reactor before being slid into position over the old and leaking sarcophagus; it is intended to keep the hazards of the radioactive interior at bay for around a century.[25]

Those who took part in clean-up operations and work to make the accident area secure in 1986 and after were exposed not only to radiation but also to a whole range of other dangers. Even today, these seem to be of little interest to researchers and the general public alike; in the worst case, they have become a matter of taboo. Fatal accidents occurred during the 'liquidation' work, for instance[26] – but there were also, as the radio-biologist Natalia Manzurova recalls, risks of a very different nature. Manzurova was sent to Chernobyl because of her background working at the secret Maiak plutonium plant in the Cheliabinsk oblast.[27] Her account evokes a disturbing situation, more akin to wartime than any-thing else, in which morality and the rule of law had, in effect, been put on hold. Excessive consumption of vodka, in the belief it would help minim-ize the effects of radiation, did not improve matters, least of all for the few women in the male-dominated 'zone'. Some of them, Manzurova notes, fell victim to disappearance, rape, or murder.[28]

2.1.2 Away in Moscow

On the morning after the explosion, the Council of Ministers in Moscow established a special commission that got to work at the scene of the

---

[22] The official sources refer to the sarcophagus as the 'Shelter' or 'Cover' construction (*ob"ekt 'Ukrytie'*).

[23] Arndt, *Tschernobylkinder*, 73.

[24] It took almost three more years to seal and complete the structure. SSE ChNPP, *NSC Construction Is Formally Completed*, 3 September 2019, https://bit.ly/4gt4w4O (1 May 2024).

[25] See DSP ChAES, *NBK: Dostupno pro novyi bezpechnyi konfainment*, https://chnpp.gov.ua/nbk/index.html (5 May 2024).

[26] SBU, f. 16, spr. 1114, ark. 146–8.

[27] On Manzurova's biography, see also Brown, *Plutopia*, 282.

[28] Natalia Manzurova, interviewed by Melanie Arndt, 14 March 2016. Paris. More details in Arndt, *Tschernobylkinder*, 75.

accident. It was led by Boris Shcherbina, a Ukrainian and deputy chairman of the Council of Ministers, who also had control over the influential
energy ministries.[29] Three days later, at Gorbachev's suggestion,[30] the
Politburo formed a task force headed by Nikolai Ryzhkov, chairman of
the Council of Ministers, which it endowed with wide-ranging powers.[31]
This group included a number of ministers and Politburo members, including Viktor Chebrikov (head of the KGB) and Anatolii Aleksandrov (president of the Academy of Sciences).[32] The Soviet republics also established
bodies of their own that resembled the Moscow task force;[33] it was,
however, always Ryzhkov's group that called the shots. Most members
of these groups were not suitably qualified – and, when it came to it,
probably not prepared – to understand the true magnitude of what had
happened in Chernobyl and take appropriate measures in response. They
concentrated on doing what they could to get the situation at the wrecked
reactor and in the general public under control. They failed to pass on
everything they knew, and kept some things quiet entirely, under the
pretext of the need to avoid panic at all costs.[34] The reality is that there
was none in the immediate aftermath of the explosion; panic took hold
only because of the ill-judged policy of restricting the flow of information.

The job of keeping a lid on rumours that might spread panic and any
potential negative remarks largely fell to the KGB. It insisted that such
comments were against the law and those responsible had to be held to
account; it was unable to conceive that making remarks of this kind could
amount to anything except 'inciting' or 'encouraging antisocial sentiment'. In spite of glasnost, the KGB was still a long way from being
ready to tolerate critical voices in the Soviet public or take people's
concerns seriously. Instead, the secret service continued to devote itself

---

[29] Boris Shcherbina (1919–90) also chaired a special censorship commission of the State
   Committee for the Utilization of Nuclear Energy (Goskomatom) and later led a disaster
   response commission after the 1988 Armenian earthquake. See Pekka Roisko, *Gralshüter
   eines untergehenden Systems: Zensur in Massenmedien in der UdSSR 1981–1991*
   (Cologne: Böhlau 2013), 255.
[30] See Schmid, *Producing Power*, 134, FN 37.
[31] HA/ACPSS, f. 89, d. 53, op. 2.
[32] The other members were Egor Ligachev and Vladimir Dolgikh (Central Committee
   secretaries – Ligachev for ideology), Vitalii Vorotnikov (chairman of the RSFSR Council
   of Ministers), Sergei Sokolov (minister of defence), Aleksandr Vlasov (minister of internal
   affairs), and, from the third meeting on 1 May 1986 at the latest, Aleksandr Iakovlev
   (Central Committee secretary); HA/ACPSS, f. 89, d. 53, op. 2. Gorbachev first took part in
   a meeting of the task force on 7 May. HA/ACPSS, f. 89, op. 51, d. 20.
[33] Sahm, *Transformation im Schatten von Tschernobyl*, 202.
[34] See ibid., 200.

to tracking down and eliminating potential cracks in the armour of the state.

On 1 May 1986, the International Day of Workers' Solidarity, when the streets were filled with crowds for the usual orchestrated rallies in Kyiv and elsewhere,[35] Chebrikov informed the KGB administrations in the Soviet republics about the accident. His memorandum offered little in the way of new information beyond what its audience had long since gleaned from the press anyway.[36] There was, however, plenty of room for details about the 'propaganda campaign' that was apparently under way in the international mass media. Chebrikov ordered the implementation of familiar intelligence countermeasures – primarily counter-espionage and an information blockade – to safeguard the Soviet Union's reputation. Public health and protecting the environment were of secondary importance.[37] Indeed, the KGB's strategy sometimes made the medical situation worse, as was the case with the requirement that secure channels be used for exchanging information from mid-May 1986 onward.[38] Healthcare facilities had to comply with the directive, but access to secure telephone lines was scarce, hindering rapid communication and ultimately disadvantaging patients.

Chebrikov sought to enlist the help of nuclear power experts so that the security service could conduct its own investigations into the causes of the accident. The plan was that these experts would be enlisted for periods of four months, including one month in the 'zone', for which he intended to provide them with personal protective equipment (PPE). At the end of that time, they would be medically examined and their involvement in the operation would be recorded in their personal files. In actual fact, Chebrikov seems not to have trusted these experts from outside his organization very much at all, as is clear from his instructions that they were all to be accompanied by KGB operatives.[39] Agents were also to engage with workers at scientific institutions in order to 'educate them prophylactically' and remind them of the importance of loyalty and state security, thereby 'protecting' them from doing anything that might lead to prosecution.[40] The KGB did not, however, have sufficient material that it

---

[35] On the 'normalization and standardization' of this ritual, see Yurchak, *Everything Was Forever*, 58.

[36] LYA, f. 46, d. 2448, l. 141–3.

[37] Ibid.

[38] SBU, f. 29, op. 17, no. 9, l. 34.

[39] LYA, f. 46, d. 2448, l. 175–7.

[40] SBU, f. 31, op. 8, no. 2, l. 300–30, here 318. For more on the 'prophylactic' aspect, see Amir Weiner, *KGB: Ruthless Sword, Imperfect Shield* (forthcoming).

could put to use in these efforts. The result of all this was that, while trying to stop information about the accident leaving the country, the KGB was also active in the West, where it was working to get its hands on classified and public-domain material about nuclear power and accidents at nuclear power plants.[41] The Soviet response to the disaster benefited from some of the knowledge that was acquired in this manner.[42]

The insistence on secrecy even affected the ability of the medical commission established by the Politburo on the May Day holiday to respond properly to the disaster.[43] The team at the clinical centre for nuclear accidents at the hospital of the Institute for Biophysics – Hospital No. 6 – had been amassing findings on the diagnosis and treatment of patients with acute radiation sickness for more than two decades. They had even produced a documentary film about acute radiation sickness, complete with graphic images. All this material was classified and thus of no use to other experts. Despite considerable opposition, it was, however, made available to military doctors in the first days after Chernobyl in order to help them prepare for work on the scene.[44]

The KGB was now also starting to take an increasing interest in members of the burgeoning environmental movement. The security service feared that their activities might undermine the Soviet economy and spark social discontent in areas with nuclear power plants. In this context, too, the KGB chose to rely primarily on *profilaktika*, hoping to turn to its own advantage the 'progressive character' that it was clearly prepared to acknowledge in the environmental movement. The plan was to use existing 'operational opportunities' and new personal contacts with leading figures in the environmental movement to explain how the KGB had, for instance, helped to raise safety standards at nuclear power plants. The KGB also wanted to highlight the fight against nuclear terrorism – something that the agency genuinely took very seriously. In order to preempt attacks, it was even planning collaborations with the security services in other countries, to include exchanging information about

---

[41] Cf. SBU, f. 16, spr. 1129, ark. 27–8.

[42] SBU, f. 16, spr. 1128, ark. 36.

[43] The commission was headed by Oleg Shchepin, deputy minister of health (to 26 December 1986) and then minister of health (to 17 February 1987) in the Soviet Union. The commission also included representatives of Hospital No. 6 and the KGB. See Pavel Andreevich Vorob'ev, Andrei Ivanovich Vorob'ev, 'Do i posle Chernobyl'ia', *Nezavisimaia Gazeta*, 28 April 2006, www.ng.ru/health/2006-04-28/8_chernobyl.html (6 November 2019), 42–5.

[44] Vorob'ev, Vorob'ev, 'Do i posle Chernobyl'ia', 49.

significant events and coordinated investigation of actual attacks.[45] The KGB realized that this same counterterrorism work could also be exploited strategically in efforts to placate the environmental movement.

2.1.3 In the Public Eye

A day and a half after the first announcement on *Vremia*, the Council of Ministers released further details: the accident had occurred inside the no. 4 reactor and destroyed part of the building, a 'certain amount' of radiation had been released, the three neighbouring reactors had been shut down, two people had died, the authorities had since taken the necessary measures, the radiation situation at the power plant and in the surrounding area had stabilized, and all those affected were receiving appropriate medical care.[46] Without referring by name to the *atomgrad* on which so much praise had otherwise been heaped, the Council of Ministers simply noted that residents of the nuclear power station settlement (*poselok AES*) and three inhabited locations (*naselennye punkty*) nearby had been evacuated, and that the radiation situation was under constant surveillance.[47]

The coverage of Chernobyl in the Soviet media has been described as a 'first serious test for glasnost'.[48] Contemporary Western commentators at the time certainly saw it as such.[49] The very fact that the accident was being reported on at all was acknowledged – this was, after all, not something to be taken for granted. It was, though, probably more a consequence of the enormity of the catastrophe and the global attention it attracted than an achievement of glasnost itself.[50] In what is probably a remarkable coincidence, *Pravda* printed a four-column article about glasnost on the day of the explosion.[51] This was the latest in a steady stream of such pieces that had been appearing for several months in an effort to promote – and demonstrate – the transformation of the Soviet public sphere. The lack of information about Chernobyl, however, demonstrated that glasnost had not yet been embraced everywhere. The forces of change and the old guard were still vying for the upper hand.

[45] LYA, f. 46, d. 2452, l. 10.

[46] 'Ot Soveta Ministrov SSSR', *Pravda*, 30 April 1986, 2.

[47] HA/ACPSS, f. 89, op. 53, d. 2, l. 4.

[48] Joseph Gibbs, *Gorbachev's Glasnost: The Soviet Media in the First Phase of Perestroika* (College Station: Texas A & M University Press, 1999), 40.

[49] E.g. Marples, *Chernobyl*, 137.

[50] See Gibbs, *Gorbachev's Glasnost*, 40.

[51] V. Kozhemiako, 'O glasnosti. V polnii golos' *Pravda*, 26 April 1986, 2.

Chernobyl-related material was subject to several stages of censorship. Publishing 'information about the accident, information about treatment outcomes', and details of 'the exposure to radiation of personnel involved in liquidating the aftermath of the accident at the Chernobyl nuclear power plant' was prohibited.[52] The general Soviet list of censored topics was becoming ever shorter at the end of the 1980s, but even the considerably pared-down version of 26 June 1987 still required environmental data, including any information relating to radioactive contamination, to be withheld – or, more precisely, it could only be published if the relevant ministry cleared it for release.[53] Various other censorship mechanisms, including self-censorship (whether conscious or not), were also to be reckoned with before or after such approval was given, even outside public contexts (as in the governmental medical commission).

Journalists such as Sergei Kiselev, the Ukraine correspondent of the *Literaturnaia gazeta*, were already trying to bring the policy of secrecy to attention at an early date. In autumn 1986, he wrote a piece entitled 'Radiation and Secrecy', where he took issue above all with the 'censors in white coats'[54] – an obvious reference to forces in the Ministry of Health.[55] It is hardly surprising that those same forces stopped the text from being published. The case of Vladimir Gubarev, science editor at *Pravda*, was different.[56] His drama *Sarkofag*, initially published as a series of extracts, was already being performed on Soviet stages in late 1986.[57] He was even allowed to undertake an international reading tour with his wife.[58] *Sarkofag* gave a largely true-to-life account of the disaster, but Gubarev avoided specific place names; 'Chernobyl' does not appear at all. Similarly, he steered clear of criticizing the state's efforts to restrict the flow of information. The failings of the plant management are at the heart of the drama. Gubarev later recalled that responses to his book had been similar all over the world: 'how can we help?', audiences at the readings had asked. Their ability to draw the missing link with Chernobyl was not lost on the KGB.[59]

---

[52] HA, Russian Subject Collection, Box 30.

[53] Roisko, *Gralshüter eines untergehenden Systems*, 308.

[54] Kiselev, quoted in Roisko, *Gralshüter eines untergehenden Systems*, 257.

[55] See ibid., 255–7.

[56] On *Sarkofag*, see ibid., 257.

[57] For interpretation of the play, see Marples, *Social Impact*, 131.

[58] Gubarev stated that he personally negotiated permission for this tour with the Politburo when he told them about his first visit to the 'zone'. Vladimir S. Gubarev, interviewed by Arndt, 16 October 2015. Moscow.

[59] Ibid.

The KGB probably let Gubarev get away with this because there were no question marks over his loyalty. He was among a group of journalists and documentary film-makers who were given access to the 'zone' early in May 1986. On his return, he summarized his observations in a remarkably critical, practically outraged letter to the Central Committee.[60] He used it to dismantle the infallibility of the Soviet system point by point, describing the precarious situation on the ground and the callous disregard shown for people's lives. His main criticism of the responsible local authorities was that Prypiat had been evacuated far too late. In his report for *Pravda*, however, Gubarev was far more restrained; he and his co-author kept in step with the Politburo's expectations, which stipulated that it was necessary most of all to show that normal life (*zhiznedeiatelnost'*) continued to be possible in the affected areas.[61] The two *Pravda* journalists reported that steps to keep the population safe, including evacuation, had been taken 'very quickly', and that order had been preserved at all times in Prypiat and the surrounding area. They also painted a picture of heroes who had gone above and beyond the call of duty in responding to the explosion; reference was made here to the high temperatures facing the firemen, but the report did not mention the more sinister aspects of the circumstances in which they had to work.[62]

Even in other countries, specifically the United States, Soviet coverage of the accident managed to give the impression that the situation was under some kind of control, at least initially. The alarm bells, however, started to ring in US security circles when even TASS, the Soviet news agency, started admitting the presence of anxiety – particularly among parents worried about their children – while reporting that normal life went on. These concerns only grew worse when it was realized that 'the situation in Kiev was one of disquiet' and that Kyiv schools were closing early for the summer on 8 May (the day before the Victory Day holiday).[63] As well as staff at foreign missions, media correspondents from abroad were also trying to get hold of more information. They were on the

---

[60] HA/ACPSS, f. 89, op. 53, d. 6.

[61] HA/ACPSS, f. 89, op. 51, d. 19, l. 3.

[62] V. S. Gubarev, M. Odinets, 'Stantsiia i vokrug nee: Nashi spetsial'nye korrespondenty peredaiut iz raiona Chernobyl'skoi atomnoi elektrostantsii', *Pravda*, 6 May 1986, 6. The journalists from *Izvestiia* who had visited the 'zone' with Gubarev painted a very similar picture the next day. Cf. Andrei V. Illesh, 'Obstanovka pod kontrolem', *Izvestiia*, 7 May 1986, 6.

[63] RRL, Jack F. Matlock Jr. Files, Box 29, F. 'USSR Nuclear Accident: Chernobyl Apr. 29, 1986 (7/9)'.

lookout for passengers arriving from the south at Moscow airports and stations, for instance, in the hope of speaking to them about what they had experienced.[64] The KGB not only kept foreign journalists under surveillance, it also attempted to influence their reporting by means of staged interviews and similar ploys.[65] For many journalists, the interest of Chernobyl was intertwined with a growing interest in the wider social and political changes that were affecting the Soviet Union.[66]

It was not until 14 May 1986 – almost three weeks after the explosion – that Gorbachev commented on the accident in public.[67] His televised address was split thematically into three parts. Visibly shaken, he began by acknowledging that the Chernobyl 'accident' (*avariia*) was a 'misfortune' (*beda*) that had 'caused public anxiety internationally' as well as affecting the Soviet people. Gorbachev drew attention to the 'extraordinary ... nature' of the accident and underlined this as he went on, while also reassuring his audience that the situation was under control and that everything necessary was being done to 'ensure the safety of the population'. He largely avoided going into detail about the scale of the disaster; it was only the circumstances of his appearance and the emphasis on the unprecedented nature of the accident that might have made his audience start to realize the enormity of what had happened. Gorbachev made clear that 'the most serious consequences have been averted'; 'the top-priority task' now was 'to deal with the effects of the accident' and return 'the area ... to a state that is absolutely safe for the health and normal life of people'.[68]

In the second part of his address, Gorbachev strongly criticized the 'immoral campaign' that he claimed certain Western nations, most of all the United States and West Germany, had been pursuing against the Soviet Union in the wake of the accident. He claimed that the governments and mass media of these countries had been spreading 'dishonest and malicious' lies in order to divert attention away from problems of their own and to avoid engaging with Soviet efforts to pursue nuclear disarmament. Gorbachev used the third and final part of his statement to call for greater

[64] HA/ACPSS, f. 89, op. 53, d. 3, l. 10.

[65] SBU, f. 16, spr. 1114, ark. 259–61. The KGB, for example, probably coached one of the individuals interviewed for Mike Edwards, Steve Raymer, Pierre Mion, 'Chernobyl: One Year After'. *National Geographic* (1987), 632–53.

[66] See, LYA, f. K-1, ap. 46, d. 2450, l. 277–80.

[67] See Arndt, *Auswirkungen des Reaktorunfalls*, 47.

[68] WHO, *TV Address by Mikhail Gorbachev*, 14 May 1986, https://bit.ly/4gi7lG2 (1 May 2024).

international collaboration on nuclear disarmament and the use of nuclear power for peaceful purposes.[69]

Gorbachev's address went down predictably badly in American diplomatic and governmental circles. Its obsession with railing against the West was seen as a mix of 'Soviet embarrassment, historic Soviet self-consciousness, traditional Soviet reaction, and the major role of Soviet propaganda officials'.[70] Gorbachev, it was felt, had 'suffered greatly in the contrast between his "openness" rhetoric and the traditional secretive reaction by Soviet officialdom'.[71]

The first Soviet press conference took place the day after Gorbachev's television address. Remarkably, it was given not just by Andrei Vorobev (a haematologist from Hospital No. 6) and an official from the Soviet Ministry of Health, but also by two Americans: Armand Hammer, the Soviet Union's 'first capitalist',[72] and the bone marrow specialist Robert Peter Gale. Hammer appears to have persuaded Gorbachev that a press conference should be held with Gale,[73] but international audiences were once again sorely disappointed: despite the participation of the Americans, there was still little in the way of new information.[74] A second press conference two weeks later was no different in this respect. It was intended to demonstrate, in particular, the mass heroism that had been displayed and the comprehensive steps that had been taken to ensure the safety and well-being of the population (including medical care, new jobs, and new homes). It also took issue with what the Soviets saw as US propaganda and challenged the notion that the radiation released had caused 'significant environmental and material damage' in various countries.[75] It was around this time that sending children away first began to figure in Soviet plans. All the relevant ministries of the republics and various agencies were to join forces in organizing a summer recuperation campaign for children and time away for workers; the

[69] Ibid.

[70] RRL, Jack F. Matlock Jr. Files, Box 30, F. 'USSR Nuclear Accident: Chernobyl Apr. 29, 1986 (9/9)'.

[71] Ibid.

[72] Edward Jay Epstein, *Dossier: The Secret History of Armand Hammer* (New York: Random House, 1996), 82.

[73] HA/ACPSS, f. 89, op. 53, d.7; Robert Peter Gale, Thomas Hauser, *Final Warning: The Legacy of Chernobyl* (New York: Warner Books, 1988), 91.

[74] See RRL, Jack F. Matlock Jr. Files, Box 30, F. 'USSR Nuclear Accident: Chernobyl Apr. 29, 1986 (9/9)'.

[75] HA/ACPSS, f. 89, op. 51, d. 24, l. 5.

evacuated zones were to be avoided, but otherwise there were no restrictions on travel.[76]

2.2 GLOBAL RESPONSES

Chernobyl gave the security services an opportunity to keep busy as the end of the Cold War was approaching. In the United States, the government and the CIA were both equally troubled by the fact that they had been unaware of the accident at first, even though it had 'occurred right in the heart of NATO's "warning of war" domain'.[77] The United States had only learnt about the incident because of what the Swedish experts had found. The heightened intelligence activity that ensued stood in marked contrast to an increased interest on the part of civil society in what was happening inside the 'evil empire', as Ronald Reagan had called the Soviet Union even as late as 1983.

2.2.1 Political Reactions

Not until three days after the explosion, when the disaster was already being widely reported in the West, did the Central Committee instruct the ambassadors of the Soviet Union to inform the heads of state in its socialist 'brother countries'. Ambassadors to the Western nations were to become active later. In practice, the ambassadors hardly presented anything beyond what had already been released in the Soviet press. Still, they were at least put in the picture sooner than the KGB administrations in the Soviet republics. More detailed information was to be provided to 'friends' (and also, because they were particularly affected by the nuclear fallout, the Scandinavian countries) if deemed necessary,[78] but it is impossible to say on the basis of the sources whether any meaningful exchange actually took place.

2.2.1.1 *Demands for Information*
One day after the ambassadors had contacted the heads of state, Eduard Shevardnadze, the minister of foreign affairs, provided a meeting of the Central Committee with a summary of how the handling of the disaster was

---

[76] Ibid.

[77] RRL, Jack F. Matlock Jr. Files, Box 29, F. 'USSR Nuclear Accident: Chernobyl Apr. 29, 1986'.

[78] HA/ACPSS, f. 89, op. 53, d. 2, l. 4.

being received abroad. The picture was not an encouraging one. Other countries were demanding information about the extent of the accident, what had caused it, and the progress of the 'liquidation work'.[79] The statements released by the Soviet embassies in various countries in an attempt to dispel concern had not been effective. Only a few days after his first report, Shevardnadze had a whole series of other concerning developments to present to the Central Committee. The Romanian press had officially reported unusually high radiation levels, as well as telling people to cover wells and stop letting children play outside. In Warsaw, there were protests on the streets on 1 May 1986, and leaflets denouncing the lack of information and 'poor safety conditions at nuclear reactors' were being distributed.[80]

The recipients of the official Soviet notes soon realized that they were once again dealing with nothing new in terms of information. They sought to get hold of facts by other means.[81] The United States was in close contact with Western European states and organizations, including ones involved in the response to the disaster, through diplomatic channels.[82] Its sources of information were many and varied: embassy workers, intelligence officers, journalists, nationals of other countries in the Soviet Union, and refuseniks and dissidents.[83] Scientists clandestinely collected soil, food, and water samples;[84] the CIA's National Photographic Interpretation Centre supplied images of the accident site.[85] In addition, various agencies gathered information about reactions in East Central and South East Europe. A US embassy officer and military attaché, for instance, observed three extra trains arriving from Kyiv in Moscow on 8 May. Two of the

---

[79] HA/ACPSS, f. 89, op. 53, d. 3, l. 6.

[80] HA/ACPSS, f. 89, op. 53, d. 3, l. 13. For more on the Warsaw protests, see Kacper Szulecki, 'Von Czarnobyl zu Żarnobyl: Die Auswirkungen Tschernobyls auf die grüne Opposition in Polen' in Arndt (ed.), *Politik und Gesellschaft nach Tschernobyl*, 26–52.

[81] RRL, Jack F. Matlock Jr. Files, Box 29, F. 'USSR Nuclear Accident: Chernobyl Apr. 29, 1986 (6/9)'.

[82] Ibid.

[83] See RRL, Jack F. Matlock Jr. Files, Box 29, F. 'USSR Nuclear Accident: Chernobyl Apr. 29, 1986 (8/9)'. 'Refusenik' here is used in the original meaning of Russian *otkaznik*, an unofficial term for Soviet Jews who were denied permission to emigrate.

[84] In early May 1986, for instance, Austrian radiation experts took samples from Zhlobin in the BSSR, where a large number of Austrians were engaged in a Soviet construction project. Representatives of other countries, such as Great Britain, the United States, and West Germany also took measurements covertly. See RRL, Jack F. Matlock Jr. Files, Box 29, F. 'USSR Nuclear Accident: Chernobyl Apr. 29, 1986 (5/9)'.

[85] See RRL, Jack F. Matlock Jr. Files, Box 29, F. 'USSR Nuclear Accident: Chernobyl Apr. 29, 1986 (6/9)'.

trains had stopped in Kyiv; the third had originated there. Most of the passengers on it were women and children, they noticed; the attaché estimated that around 65 per cent of them were 'toddlers, infants and small children'. He found 'the scene ... reminiscent of the evacuation of Finnish children to Sweden during WW II'.[86]

2.2.1.2 *The 'Anti-Soviet Campaign'*
Shevardnadze adopted a familiar approach in his accounts to the Central Committee: rather than discussing the Soviet Union's problems openly, he framed them in terms of an 'anti-Soviet campaign' intended to smear the Soviet Union and its achievements. That went just as much for the Warsaw demonstrations as for US speculation that the Soviet nuclear industry was built on inferior technological foundations and had not even considered the consequences of a possible accident (on the grounds that its reactors were not generally protected by concrete domes).[87]

The Americans had even interpreted the May Day rallies in the context of the accident, Shevardnadze complained. It was, he reported, being alleged that the Soviet leadership had come up with the idea of broadcasting these events, and the Kyiv rally in particular, on television 'in order to make the country's people think nothing bad had happened at the Chernobyl power plant'.[88] It is impossible to say with certainty on the basis of the sources whether Shevardnadze really was that naive or simply trying to make a point to the Central Committee. The same goes for the indignation with which he reacted to the 'claims' in the US media that this was one of the worst accidents in the history of nuclear power, that there were many casualties, and that the Dnipro basin and the breadbaskets of Ukraine had been contaminated with radiation.

The overriding impression given by internal US analyses is one of level-headedness; it is hard to find in them any of the 'anti-Soviet propaganda' that Moscow accused the United States of peddling. Information was pieced together through various channels, and it was not possible to verify each and every detail; consequently, inaccurate accounts did appear and some media reports were indeed exaggerated. Even so, as a rule, rather than taking everything at face value, the analysts and officials at the US embassy in Moscow treated each piece of information with caution,

---

[86] RRL, Jack F. Matlock Jr. Files, Box 29, F. 'USSR Nuclear Accident: Chernobyl Apr. 29, 1986 (7/9)'.
[87] HA/ACPSS, f. 89, op. 53, d. 3, l. 7.
[88] HA/ACPSS, f. 89, op. 53, d. 3, l.14.

questioning it and weighing it up objectively. Likewise, when a Canadian embassy driver reported at the beginning of May that radiation sickness had appeared where his relatives lived – close to the evacuation zone – officials remained cautious and mentioned the possibility that 'psychosomatic factors' were at play.[89]

Shevardnadze was soon able to inform the Central Committee about a new development in the Western discourse. The – from the Soviet perspective – gratuitously overdramatic response had been joined by a move to play down the magnitude of events, reflecting concerns in Western governments and industry that the accident in Ukraine might cast a question mark over nuclear policy in their own countries. It did not take the Soviet minister of foreign affairs long to realize that the West also had an interest in ensuring that the effects of the accident did not cause unnecessary alarm. Even so, Chernobyl had already given a new lease of life to the anti-nuclear movement in several countries, including, among others, the United States, West Germany, and the Netherlands. The most radical activists were demanding an immediate end to the use of nuclear power.[90] Shevardnadze took advantage of the situation and suggested to the Central Committee that Hans Blix, director general of the IAEA, be invited to Moscow in mid-May. Citing the Swede's 'objective approach', he felt that, rather than posing a risk, such a visit would provide an opportunity to legitimize the Soviet response to the disaster and, consequently, make it easier to avoid agreeing to visits by other foreign experts.[91]

2.2.1.3 Assistance in Public and behind the Scenes
Harsh as some of the criticism may have been, many governments had also been quick to express their concern and offer assistance. Even before Shevardnadze's first report on foreign reactions to the Central Committee, Deputy Foreign Minister Aleksandr Bessmertnykh had given Gorbachev the transcript of an oral message delivered on behalf of Ronald Reagan. The US president had communicated his 'deep regret' and offered the Soviet Union wide-ranging 'assistance in dealing with this tragedy'. A specialized group of scientists could, Reagan had informed the Soviet leader, be dispatched at once to 'assist in determining and subsequently coordinating the

---

[89] RRL, Jack F. Matlock Jr. Files, Box 29, F. 'USSR Nuclear Accident: Chernobyl Apr. 29, 1986 (7/9)'.

[90] HA/ACPSS, f. 89. op. 53, d. 3, l. 14.

[91] HA/ACPSS, f. 89. op. 53, d. 3, l. 17.

best use of the resources from the United States nuclear safety and environmental protection programs'.[92] Reagan, an advocate of nuclear power, had even instructed his staff not to 'propagandize' the accident; instead of stoking existing anti-Soviet sentiments, he wanted them to 'keep politics out of it'.[93] The scope of the assistance that he was prepared to send was substantial – from measuring radiation in air, water, and soil, to mapping contamination, to assistance with decontamination, to medical specialists familiar with radiation exposure – and included equipment that the Americans would bring with them.[94] As well as the United States, France (the largest European nuclear power) also offered to send specialists and technical equipment at a very early stage.[95] Great Britain, Canada, Japan, and West Germany reached out in a similar manner.[96]

Even if it was not their sole purpose, such gestures of goodwill were also made with a view to obtaining insights that would benefit the nuclear industry closer to home and could be used to inform disaster management. Genuine offers of assistance could quite easily be treated as an opportunity to extract information at the same time, as is demonstrated by the reaction of Edward J. Markey, chairman of the House Subcommittee on Energy Conservation and Power. Soon after the accident, he too expressed his 'deep regrets' to his Soviet peers and told them he had requested 'that technical and medical assistance, as you deem appropriate, be provided'. At the same time, he urged them to start sharing more information; after all, the accident was 'of enormous scientific and technical interest to all nations employing nuclear power'.[97] Markey also diplomatically pointed to his own country's experience in the case of the Three Mile Island accident, explaining that information about it had been made available to the whole world immediately,[98] and adding that 'lessons continue to be

[92] RRL, European and Soviet Affairs Directorate Files, RAC Box 8, F. 'Chernobyl'; HA/ACPSS, f. 89, op. 53, d. 3.

[93] Jack F. Matlock, *Reagan and Gorbachev: How the Cold War Ended* (New York: Random House, 2004), 188.

[94] RRL, European and Soviet Affairs Directorate Files, RAC Box 8, F 'Chernobyl'; HA/ACPSS, f. 89, op. 53, d. 3.

[95] HA/ACPSS, Volkogonov, Box 27.

[96] HA/ACPSS, f. 89, op. 53, d. 3.

[97] HA, James K. Asselstine, Box 105, F. 'CR-86-50'.

[98] This was not quite true; see the recent Zaretsky, *Radiation Nation*. Three Mile Island activists such as Eugene Stilp, Mary Osborn (Stamos), and Paula Kinney criticized the lack of transparency and openness to local residents following the accident. Eugene Stilp, interviewed by Melanie Arndt, 25 April 2014. Harrisburg, PA; Stamos (previously Osborn), interviewed by Arndt; Paula Kinney, interviewed by Melanie Arndt, 25 April 2014. Mechanicsburg, PA.

shared internationally, as evidenced by the participation of the Japanese in the decontamination of the ... reactor'.[99] Markey called for a 'similar commitment' on the part of the Soviet Union now.[100]

It was, above all, Democratic congressmen and -women who wanted Reagan to share US expertise in contamination and decontamination with the Soviet Union – not least in order to facilitate further international collaboration between the two countries.[101] The Soviet Union, however, officially declined to accept assistance.[102] Countries that offered help were thanked but given to understand that this was not currently required.[103] John Matlock, special assistant to the president for national security affairs, later recalled that the Soviet response to the initial offer of assistance from the United States was 'actually accusatory, as if, in offering to help, we had insulted them'.[104]

For all its insistence that help was not necessary, the Soviet leadership did, in fact, quietly accept technical and medical assistance. Even as early as 29 April 1986, Soviet diplomats were meeting informally with 'individual researchers at FRG nuclear firms and industry organizations'. They were looking for information about, for instance, 'methods of extinguishing graphite fires' and 'manufacturers of nuclear safety equipment'.[105] The Soviet Union ended up purchasing three remote-control vehicles from the Kerntechnischer Hilfsdienst company in Karlsruhe, discreetly mediated by the German Atomic Forum. On 10 and 11 May 1986, German experts showed their Soviet counterparts how to operate the robots in Moscow.[106] Japan contributed expertise in dealing with victims of radiation sickness. The modern equipment that it sent for treating children

---

[99] James Asselstine of the Nuclear Regulatory Commission, though, felt that the implications of the Three Mile Island accident had never been fully addressed in the United States; in his eyes, interest in doing so had dried up, despite an initial drive to improve safety standards in the immediate aftermath of the accident. NRC, 47193/47194.

[100] HA, James K. Asselstine, Box 105, F. 'CR-86-50'.

[101] RRL, WHORM: Subject File, CO165 Soviet Union, Box 201, ID# 392764.

[102] The Soviet embassy in Washington informed the Department of State accordingly on 30 April 1986. NRC, 35994, EPA, 'Soviet Nuclear Accident. A Task Force Report'. 1 May 1986.

[103] HA/ACPSS, f. 89, op. 53, d. 3, l. 8.

[104] Matlock, *Reagan and Gorbachev*, 188. Matlock was US ambassador in Moscow from April 1987 to the fall of the Soviet Union. On collaboration in reactor safety, see Thomas Wellock, *Safe Enough? A History of Nuclear Power and Accident Risk* (Oakland: University of California Press, 2021), 89–109.

[105] RRL, Jack F. Matlock Jr. Files, Box 29, F. 'USSR Nuclear Accident: Chernobyl Apr. 29, 1986 (2/9)'.

[106] Laufs, *Reaktorsicherheit*, 143.

with thyroid cancer, however, soon became unfit for purpose: for one thing, Soviet healthcare workers were unable to operate it properly; for another, the single-use components were rapidly used up and there was no supply of replacements.[107]

The potential implications for perestroika were already being discussed a day after news of the accident reached Washington. 'If there is widespread death, illness, and dislocation', an intelligence paper predicted, 'this event will be a severe psychological blow to the Gorbachev regime and its gospel of optimism, even if the economic effects are limited'. 'In any case,' the analysis continued, 'the system under Gorbachev's new leadership will be put to a politically and psychologically important test: Did it react with the honesty, efficiency, promptness, and public-mindedness he calls for? Or did it manifest the usual sloth, carelessness, evasions, and outright lies?'[108] Faced with the collapsing international confidence in the Soviet Union, Shevardnadze changed his tone at his second meeting with the Central Committee, in early May, and advocated going on the offensive. He argued that the Soviet Union urgently needed to provide more information, 'as specifically as possible', on the effects of the accident. He proposed issuing a daily bulletin, releasing radiation and environmental data with exact figures, and openly admitting uncertainties.[109] He did not, however, succeed in getting his way.

2.2.2 'We All Live in Chernobyl': Reactions in Wider Society

Dear Head of the Kremlin Gorbachev and President Reagan!

We are children from Austria, and we are very concerned about the reactor accident in Chernobyl. We did not know that a nuclear power plant holds within it such sinister dangers with regard to radiation. We therefore have a very big favor to ask of you both: please halt all those nuclear weapons tests and shut down the nuclear power plants as quickly as possible. We would not survive a nuclear war in any case; or are we to spend our entire lives in bunkers underground? ... What are we to live on, once our Earth is contaminated? We can't eat money! We would have to perish miserably of the most varied diseases, such as leukaemia, cancer, AIDS, etc. ...

We are therefore asking both of you in the name of humanity: stop ... nuclear power, invest the money for healthy forests and clean oceans, and

---

[107] Gubarev, interviewed by Arndt.
[108] RRL, Jack F. Matlock Jr. Files, Box 29, F. 'USSR Nuclear Accident: Chernobyl Apr. 29, 1986 (1/9)'.
[109] HA/ACPSS, f. 89, op. 53., d. 3, l. 17.

help the starving people in the Third World; otherwise the whole Earth will soon be a Third World. So much good could be done with this money that perhaps someday all human beings will have enough to eat and drink, and also the PEACE we long for so much will come into the world, and horror, terror, and dread of nuclear war will not prevail, as is the case now.[110]

Thus reads a handwritten letter from schoolchildren in Austria presented to the Department of State by the Austrian embassy in December 1986. The children's distress reflected feelings that were shared by many in the West during the mid-1980s and acquired a new urgency as a result of Chernobyl. Environmental issues had been attracting increasing attention; people were worried about their implications, just as they were about illnesses (AIDS, cancer) that were either new or being seen in a new light. Such concerns were joined by fears of nuclear war and anxiety about the potential effects of the Chernobyl accident.[111]

A detailed response was drawn up. Reagan assured the Austrian children that 'the American people share your interest in preserving the environment', while also explaining the need for nuclear testing. He told them that it helped 'to make nuclear weapons safer' and pointed out that all US tests were now carried out underground in order to avoid causing environmental damage. A world without nuclear power was inconceivable for Reagan; it was, he explained to the children, likely that 'more rather than less' nuclear energy would be needed to replace the dwindling reserves of fossil fuels.[112] His response clearly shows that nuclear euphoria was by no means a thing of the past when it was drafted in 1987.

The Austrian children were not alone in their sentiments. Reagan and Gorbachev alike were inundated with letters and telegrams from people all over the world once the first reports of the accident started appearing in the international media. According to the propaganda section of the Central Committee, most of them expressed concern and solidarity, which led to the conclusion that a not inconsiderable part of Western society was resisting the 'attempts of imperialist propaganda to stoke anti-Soviet sentiment'.[113] Significantly, although the Central Committee recognized that most of those reaching out to Gorbachev displayed empathy and

---

[110] RRL, WHORM Subject File AT 485927. The letter is quoted here in the State Department's internal translation; 'We all live in Chernobyl' in the section title is quoted following Karl Grossman, *Power Crazy: Is LILCO Turning Shoreham into America's Chernobyl?* (New York: Grove Press, 1986), 316.

[111] See, e.g., Biess, 'Sensibilisierung des Subjekts'; Arndt, 'Human Security'.

[112] RRL, WHORM Subject File AT 485927.

[113] HA/ACPSS, f. 89, op. 53, d. 37, l. 1.

a willingness to help, it did not reflect on what this might mean for its own policy of withholding information; instead, it continued to cling to the theory that the West was exploiting the accident for ideological purposes.

The global scale of the accident was often foregrounded in the letters, but the Soviet Union was not condemned out of hand, not even when increased levels of radiation were observed in other countries. Instead, the idea that 'the world was all one place now' was often encountered[114] – an expression of global solidarity in which the language of the Cold War was nowhere to be seen. No differently from many politicians, these letter-writers from civil society expressed their regret by recalling disasters in their own countries; the *Challenger* disaster in January 1986 was mentioned frequently. Three themes figured in almost all the letters: a hope that future technological progress would no longer come at great human cost, a desire for peace, and calls for disarmament. Some of those who wrote seemed to have a clear religious motivation as well.[115] Hardly any of the letters mentioned the plight of children from the area where the explosion occurred. Indeed, the initial media coverage of Chernobyl rarely focused on children at all; the reports and photographs from the elite Artek holiday camp were one of the few exceptions to this.[116]

The impact of the disaster on tourism was surprisingly small. Some trips to the Soviet Union run by Western, specifically US, tour operators were cancelled shortly after the accident, despite written assurances from Intourist (the Soviet state travel agency for foreign tourists) that even travel to the accident area was 'perfectly safe'.[117] Visits were not, however, permanently suspended, and even Kyiv remained on the list of destinations. Such travel to the Soviet Union was controversial in the US context, but this was more for political reasons than because of any concerns about radiation.[118] Even a year after the accident, all foreign visitors were still being placed under KGB surveillance; the agency remained wedded to the mindset that they were only there to gather information.[119] Information flowing in the other direction appears to have been of little concern to the KGB. Some American students who were in Kyiv when the accident occurred, for instance, had told their

[114] HA/ACPSS, f. 89, op. 53, d. 37, l. 2.
[115] HA/ACPSS, f. 89, op. 53, d. 37, l. 3.
[116] Cf. Barringer, 'From Children of Chernobyl'.
[117] Quoted in Sara Rimer, 'For City's Ukrainians, Ordeals of Waiting for Word on Disaster', *New York Times*, 1 May 1986, https://bit.ly/4fjntWI (1 May 2024).
[118] SBU, f. 16, spr. 1118, ark. 65–6.
[119] SBU, f. 16, spr. 1118, ark. 65.

Ukrainian friends about the accident. The response was initially sceptical, demonstrating that faith in technology and trust in the system were still deeply ingrained in Soviet society; only when the Americans started to insist on drinking bottled water did the Ukrainians begin to consider the idea that something was seriously amiss.[120]

Contrary to what is often claimed, the sum total of the information that was circulating – news, official statements, the rumour mill, personal observations – would have been more than enough to give rise to public concern inside the Soviet Union. Significant portions of society, however, simply did not take an active interest in the disaster. The very fact that it was unthinkable clearly prevented many from seeing it for what it was and facing up to what it actually meant. This apathy also needs to be understood against the background of a general lack of trust in Soviet society: as Gubarev put it so well, 'nobody believed anyone anymore'.[121] It is, therefore, an oversimplification to hold functionaries in the state and Party apparatus alone responsible for the silence about the disaster. This does not change the fact that they can and should be criticized for it, or that in many cases people kept quiet because they were afraid of the consequences of speaking out.

2.2.2.1 *Repercussions in the United States*

When around 150 demonstrators gathered near the Shoreham nuclear power plant (Long Island, New York State) on 3 May 1986, they expressed solidarity with those affected in the Soviet Union with a banner that read 'We all live in Chernobyl'.[122] The rally had been organized by the SHAD (Sound–Hudson against Atomic Development) Alliance, an influential coalition of more than twenty anti-nuclear groups in the south of New York State. Joe Paparatto, a representative of the alliance, noted that he was 'encouraged' by the rise in attendance at the event: there were, he said, 'a lot of new faces here today'.[123]

Before the news of Chernobyl spread, the US environmental and anti-nuclear movements had found themselves in a state of stagnation. As can be seen from the Shoreham demonstration, this soon changed once details of the accident started to become known. On 24 May, less than a month

---

[120] RRL, Jack F. Matlock Jr. Files, Box 29, F. 'USSR Nuclear Accident: Chernobyl Apr. 29, 1986 (3/9)'.
[121] Gubarev, interviewed by Arndt.
[122] Grossman, *Power Crazy*, 316–17.
[123] Quoted in ibid., 317.

after the disaster, more than forty organized anti-nuclear protests took place across the nation. Support for the use of nuclear power and the construction of new nuclear power plants plummeted. Opinion polls suggested 'that the majority of Americans oppose the expansion of the US nuclear program'.[124] In an ABC survey, 78 per cent of those questioned were against the use of nuclear power.[125] The efforts of political institutions and the nuclear industry to convince the general public that Chernobyl was a 'typically Soviet' accident, and that the successful handling of the Three Mile Island accident had clearly shown that US reactors were safe, were widely met with scepticism.[126]

In large numbers, Americans began writing to governors, senators, and other elected representatives with their concerns and demands.[127] Dissent sprung up across the land – from the east coast, where it was loudest; to Michigan in the Midwest; to California on the west coast. The protests often began in places like Monroe, Michigan, that felt particularly threatened because there were nuclear power plants nearby.

Evacuation plans and the safety of reactor buildings were the main points of contention in opposition to the use of nuclear power after Chernobyl in the United States. The power plants of Shoreham in New York State and Seabrook in New Hampshire were particularly controversial. Operating licences had not yet been issued, so their future was uncertain. When it became known that even the Soviets had evacuated everyone within a 30 km radius, the evacuation radius of 16 km (10 miles) envisaged in the United States became a matter of bitter dispute. Under pressure from concerned citizens, the governors of Massachusetts, Vermont, Ohio, and New York State refused to agree to evacuation plans that involved their states, thereby delaying or stopping the licensing process for controversial plants.

In spite of the costly media campaigns of nuclear power companies,[128] a survey for *USA Today* revealed that more than half the American population believed that an accident like Chernobyl could 'happen anywhere'.[129]

[124] 'Introduction', *Radical America*, 20, no. 2/3 (1986), 2; Richard Rudolph, Scott Ridley, 'Chernobyl's Challenge to Anti-Nuclear Activism', *Radical America* 20, no. 2/3 (1986), 8–11, here 8.

[125] Alvin M. Weinberg, 'A Nuclear Power Advocate Reflects on Chernobyl', *Bulletin of the Atomic Scientists* 43, no. 1 (1986), 57–60, here 57.

[126] Grossman, *Power Crazy*, 349.

[127] For the public mood, see Stewart Diamond, 'How Chernobyl Alters the Nuclear Equation', *New York Times*, 25 May 1986, https://bit.ly/4eiayDP (1 May 2024).

[128] Cf. the advertisements placed by the Edison Electronic Institute: RRL, WHORM: Subject File AT, Box 4, ID#415695; RRL, WHORM: Subject File AT, Box 5, ID# 501645.

[129] Rudolph, Ridley, 'Chernobyl's Challenge', 8.

This view was shared by James Asselstine, a prominent member of the Nuclear Regulatory Commission (NRC) and its 'most conspicuous dissenter'.[130] Shortly after the explosion in 1986, he distanced himself from the official US position and argued that accidents similar to what had happened in Ukraine were not only possible, but to be expected.[131] His warnings were widely covered in the nation's leading media outlets,[132] but they mostly fell on empty ears among politicians. Asselstine was not reappointed to serve a further term as commissioner in 1987. Even so, his warnings were, at least in part, still taken seriously: they led, not least, to the formation of an international expert commission that also included Soviet nuclear physicists.[133] Supporting nuclear power could have career implications too: just three weeks after the accident in Ukraine, William Carney, a Republican member of the House of Representatives and vice-president of LILCO (the Long Island Lighting Company), which had built the Shoreham nuclear power plant, announced that he would not be standing for re-election, citing pressure from opponents to the Shoreham facility.[134] In the end, Shoreham was the only plant to be dismantled – in 1992, the first such case in the United States. It had never entered commercial operation.[135]

Five years after the Chernobyl disaster – by which time the first Chernobyl children had already been to the United States for respite and acquired a public face in the figure of the pale boy with cancer called Vova[136] – James Edgar, the Republican governor of Illinois, declared 1991 Chernobyl Awareness and Relief Year, with 26 April as Chernobyl Day. He explained this novel initiative with reference to the fact that 'the effects of this catastrophe have been and will continue to be felt in neighboring regions of Europe and other parts of the world, making it a disaster which truly affects all people'. Edgar also noted that 'Americans of Ukrainian and Byelorussian descent and other Illinois residents, many

---

[130] Irvin Molotsky, Warren Weaver, Jr., 'Washington Talk: Briefing; Letters from Asselstine', *New York Times*, 26 October 1986, https://bit.ly/3ZVwbpG (1 May 2024). On Asselstine and his views on safety, see Wellock, *Safe Enough?*, 89–109.

[131] NRC 47193/47194.

[132] Cf., e.g., Molotsky, Weaver Jr., 'Washington Talk'; Matthew L. Wald, 'Retiring US Official Assails Nuclear Plant Safety', *New York Times*, 7 June 1987, https://bit.ly/3V HkPTBf (1 May 2024).

[133] See Wellock, 'Children of Chernobyl'.

[134] Diamond, 'Nuclear Equation'.

[135] See Dennis Hevesi, 'Nora Bredes, Who Fought Long Island Nuclear Plant, Dies at 60', *New York Times*, 22 August 2011, https://bit.ly/4gDQa1o (1 May 2024).

[136] See Prologue.

having friends and relatives in the affected areas, wish to commemorate this fifth anniversary . . . by promoting awareness and assisting in the relief efforts for their brethren'.[137]

2.2.2.2 *The Ukrainian Diaspora*

Alongside the environmental and anti-nuclear movements, there was another segment of North American society that also lost no time in swinging into action: the diasporas of the various Soviet republics, in particular, members of the Ukrainian diaspora. The KGB's North American agents regarded them with suspicion. Even in early May 1986, various Ukrainian diaspora organizations had already held demonstrations and set up a Chernobyl aid fund.[138] A 'Coordinating Committee for Matters Concerning the Disaster in Ukraine' followed in July. The KGB saw this as a nationalist conspiracy, suggesting that the fund's sole purpose was to serve as a cover for murky financial activity and sow anti-Soviet sentiment. The KGB dismissed as unfounded the allegations of the Ukrainian diaspora that the Soviet leadership was pursuing a 'policy of genocide against the Ukrainian people' by having deliberately chosen Ukraine for the 'experiment that was carried out at the Chernobyl power station'.[139] The KGB was equally indignant at the suggestion that Chernobyl was an attempt to induce another human-made famine akin to the Holodomor of the 1930s, this time by contaminating the fields with radiation.[140]

What is true is that members of the Ukrainian diaspora gathered and published information about the accident and the situation in their 'homeland';[141] they made at least one documentary film about Chernobyl and used the first anniversary of the explosion to organize, among other things, a series of symposiums, a 'funeral procession' through Washington, and a memorial service 'in memory of the victims of Chernobyl'.[142] A few years later, the Ukrainian diaspora organizations were to be among the first to invite Chernobyl children to North America.

---

[137] Private Archive Pankrats, Jim Edgar, Governor of the State of Illinois, 'Proclamation', 11 April 1991.

[138] SBU, f. 16, spr. 1114, ark. 44; 'Ukrainians in Canada Try to Organize Relief Efforts', *The Ukrainian Weekly*, LIV, no. 18 (1986), 5 May 1986, 2, 15.

[139] SBU, f. 16, spr. 1114, ark. 188, 233–5.

[140] SBU, f. 16, spr. 1114, ark. 234. The Holodomor did provide a point of reference for members of the Ukrainian diaspora trying to make sense of the accident, even very early on. See Rimer, 'For City's Ukrainians'.

[141] SBU, f. 16, spr. 1114, ark. 218.

[142] SBU, f. 16, spr. 1116, ark. 4–7. Cf. also *The Ukrainian Weekly*, LV, no. 17 (1987), 26 April 1987.

2.2.3 Between Two Worlds: Gale and His Compatriots

At the end of 1986, the Ukrainian KGB concluded that the 'anti-Soviet campaign' launched by the West with regard to Chernobyl had clearly had a negative impact on academic exchange. The campaign – other explanations, such as the KGB's own obsession with information control or concerns about radiation exposure, were apparently not considered – was blamed for a marked drop in the number of foreign researchers coming to the Soviet Union. For a start, 4 large and 'particularly important' conferences at which up to 400 foreign delegates had been expected had to be cancelled because of a lack of registrations.[143] That was not all, however: the KGB also felt that the 'anti-Soviet campaign' it never tired of invoking had made its own espionage activities more difficult. It had, apparently, become much harder to plant agents in Western research institutions: seven researchers acting as KGB agents, for instance, were unable to undertake research visits in 1986 because the host countries delayed their visas.[144] This state of affairs hardly supported the usual view that the West was doing everything it could to get hold of information about the accident – not that the contradiction seems to have mattered much to the KGB.

By this point, those scientists who did go to the Soviet Union in spite of the circumstances were generally specialists acquainted with the effects of radiation on humans and the environment or matters such as reactor technology and decontamination methods. These visitors sought information from first-hand sources, formed impressions of their own, and tried out medicines and procedures.[145] Their appearances were generally low-key; the Soviet government had, after all, officially refused to accept external assistance. Decisions about who was granted entry and who was not often came down not so much to specialist expertise as to connections in the Soviet apparatus and potential usefulness to the Soviets in projecting a positive image abroad. It is otherwise hard to explain why Robert Peter Gale, a California-based specialist in bone marrow transplants with no experience of treating victims of radiation exposure, got to come to the Soviet Union, whereas the world-famous French oncologist and immunologist Georges Mathé did not. Both had offered their help.[146] In Gale's case, the Politburo's decision was connected with the intervention of the

---

[143] SBU, f. 31, op. 8, no. 2, l. 304.
[144] Ibid., 327.
[145] On this, see also Brown, *Manual for Survival*.
[146] HA/ACPSS, Volkogonov, Box 27.

legendary US businessman and networker Armand Hammer (already aged eighty-eight at the time). He had contacted Gorbachev personally in a letter of 29 April 1986, and asked him to bring Gale to the Soviet Union.[147] Hammer himself had gone to the Soviet Union in the 1920s to help children suffering from typhus;[148] since then, he had shuttled back and forth between Los Angeles and Vladivostok, between Washington and Moscow – between two worlds, in other words – and was involved in the oil and health sectors, as well as the art market, in various positions; he was chairman of Occidental Petroleum until his death in 1990.[149]

2.2.3.1 *Privileged Access: Robert Peter Gale*

Gale's work in Hospital No. 6 has been written about elsewhere.[150] Treating seriously ill patients was not the main reason that he was allowed to see 'inside this holiest of all holy secrets' (*v sviataia sviatykh sekret-nosti*), as Vorobev, head of the clinical centre for nuclear accident victims there, rightly observed.[151] That much was clear enough to everyone who had anything to do with Gale's involvement. Gale and his team – he was subsequently joined by four more doctors – likewise knew that they had been invited primarily 'to lend credibility' to the Soviet response to the accident.[152] Gale, who was forty-one at the time, had the potential to become a reliable hero – and he knew how to make the most of it.[153] The

---

[147] A copy of the letter is printed in Gale, Hauser, *Final Warning*, 41.

[148] Gale reports that Lenin had rebuked Hammer for this on the grounds that the Soviet Union already had enough doctors; what they really needed, Lenin apparently said, were tractors; shortly after that, Hammer introduced Ford tractors to the Soviet Union. Robert Peter Gale, interviewed by Melanie Arndt, 17 May 2014. Online.

[149] See Steve Weinberg, 'Armand Hammer's Unique Diplomacy', *Bulletin of the Atomic Scientists* 42, no. 7 (1986), 50–2, and biographies and autobiographies such as Armand Hammer, Neil Lyndon, *Hammer: Witness to History* (Sevenoaks: HarperCollins Distribution Services, 1988); Armand Hammer, Neil Lyndon, Gerda Bean, *Mein Leben*, 5th ed. (Bern: Scherz, 1989); Carl Blumay, Henry Edwards, *The Dark Side of Power: The Real Armand Hammer* (New York: Simon & Schuster, 1991); Steve Weinberg, *Armand Hammer: The Untold Story* (Boston: Little, Brown and Co., 1989).

[150] Marples, *Chernobyl*, 137–40; Marples, *Social Impact*, 33; Nataliia P. Baranovs'ka, *Ukraina – Chornobyl' – Svit: Chornobyl's'ka problema u mizhnarodnomu vymiri 1986–1999* (Kyiv: Nika-Tsentr, 1999), 36–7; 43–4; 236–47; Petryna, *Life Exposed*, 44–8. Most recently: Plokhy, *Chernobyl*, 243–6; Higginbotham, *Midnight in Chernobyl*; Brown, *Manual for Survival*, 13–25.

[151] Vorob'ev, Vorob'ev, 'Do i posle Chernobyl'ia', 40.

[152] Gale, interviewed by Arndt.

[153] Gale's account of his experiences was published two years after his first visit to the Soviet Union: Gale, Hauser, *Final Warning*. The book was turned into a film in a co-production with Soiuzkinoservis, released just in time for the fifth anniversary of the disaster in 1991: Anthony Page, *Chernobyl: The Final Warning*. USA, 1991: Carolco Pictures.

demonstration of his loyalty to the Soviet Union and the nuclear industry culminated in July 1986, when – with considerable media attention – he went to Kyiv with his then-wife and three children 'to help calm the population'.[154]

It is not always easy to pin down where Gale and Hammer stood in a web of competing interests. Almost 300 Chernobyl workers and emergency responders had been brought to the specialist unit in Moscow;[155] their condition was in some cases so horrific that it would have been very hard not to draw the right conclusions about how bad the disaster really was. Gale, however, was very careful about what he did with what he learnt in the Soviet Union, and diplomatically understood that some things were better not disclosed to the international press.[156] His portrayal of the Soviet response to the disaster was neutral, at times even positive, which won him the reputation of pursuing 'reactor diplomacy'.[157] His treatment methods remained controversial.[158]

Gale's invitation to the Soviet Union was a public sensation in the United States. The media, the government, and the CIA all took a strong interest in his activities there. At the same time, Gale recalls, they kept a 'decent distance' from him and did not demand that he reveal everything he knew.[159] In a meeting with George Shultz, the US secretary of state, in May 1986, Gale is recorded as saying that Soviet doctors were 'well-read

---

[154] Robert Peter Gale, Eric Lax, *Radiation: What It Is, What You Need to Know* (New York: Alfred A. Knopf, 2013), 238.

[155] Most of the 299 individuals admitted to Hospital No. 6 in the first 48 hours after the explosion were men aged between 25 and 35. Apart from the two exceptions – both local residents who had been in the vicinity of the power plant when the accident took place – they were all plant workers or firemen. Of those admitted, 89 were discharged because they were deemed healthy. Of those who remained in hospital, 26 had died of radiation sickness and burns by mid-July. Michael McCally, 'Hospital Number Six: A First-hand Report', *Bulletin of the Atomic Scientists* 42, no. 7 (1986), 10–12.

[156] Gale, interviewed by Arndt.

[157] Felicity Barringer, 'Reactor Diplomacy', *New York Times*, 31 July 1988, www.nytimes .com/1988/07/31/books/reactor-diplomacy.html (1 May 2024).

[158] The approach involved administering GM-CSF, a drug from the Swiss pharmaceutical company Sandoz, that had not been approved at the time and had only been tested on monkeys and, at the last minute, by Gale and Vorobev on themselves. See Petryna, *Life Exposed*, 46–8. For Gale's and Vorobev's own accounts, cf. their autobiographies: Gale, Hauser, *Final Warning*; Vorob'ev, Vorob'ev, 'Do i posle Chernobyl'ia'. Cf. further Robert P. Gale, Andrej I. Vorob'ev, 'First Use of Myeloid Colony-stimulating Factors in Humans', *Bone Marrow Transplantation* 48, no. 1358 (2013); Alexandr. E. Baranov et al., 'Hematopoietic Recovery after 10-Gy Acute Total Body Radiation', *Blood* 83, no. 2 (1994), 596–9. See also Brown, *Manual for Survival*, 18–20.

[159] Gale, interviewed by Arndt.

and knowledgeable', and that 'perhaps fifty to a hundred thousand were potentially at risk over the years to come';[160] he was later to adopt a much more optimistic line. Shultz was full of praise for Gale and his team, telling them that they had 'given the world an example of the common bond of humanity that links its peoples'.[161]

2.2.3.2 *Coming to Realize 'They Are Humans, Too'*

Before Gale took his family to the Soviet Union, he went to his daughters' school in Los Angeles to tell the fourth-graders there about his first visit. His appearance resulted in a bundle of children's letters and drawings that Hammer sent to the White House. Gorbachev received a similar consignment; in fact, his staff pointed out to Reagan, he had already replied to it. Random House even expressed an interest in publishing the letters and the responses of the two leaders.[162]

The Reagan administration received more than fifty letters from the school through Hammer. On one level, they are documents of children's creativity and imagination; on another level, they give an indication of what Gale and the two teachers who hosted him wanted the children to take away from his visit: US superiority, but clearly without the rhetoric of the 'evil empire'; the emphasis lay instead on a shared humanity. The formulation 'they are humans, too' appeared, with minor variations, in almost all the letters. Several boys and girls expressed concern that a nuclear accident might happen in the United States as well. Dr Gale had told them that US reactors were safe, the letters explained, but they were still afraid. They called on Reagan to check US nuclear power plants and to make sure that better equipment was on hand to deal with an emergency than had been available in Chernobyl. One letter said that no more houses should be built near nuclear power plants. If an accident did occur in the United States, one pupil suggested, the Soviet Union should be asked for help – after all, the United States was helping the Soviet Union now. The children took a marked interest in the everyday life of the 'Russians': what they wore, what they ate, what feelings they had. Another idea was to convene an emergency session of the United Nations to alleviate the human suffering that Chernobyl had caused. 'I don't want any Russians to die', wrote Michele; she promptly followed this up by

---

[160] RRL, Jack F. Matlock Jr. Files, Box 30, F. 'USSR Nuclear Accident: Chernobyl Apr. 29, 1986 (9/9)'.
[161] Ibid.
[162] RRL, WHORM Subject File, DI001, Box 2, ID#40743. The publication never materialized.

suggesting: 'Why don't you bring all the Russians to the US?' The children wanted the president to make sure the Soviet Union knew that the thoughts of Americans were with those in the accident area. It was also suggested that Reagan should send more doctors like Gale to 'teach the Russians' – and not just bone marrow transplantation. Gale had made a deep impression. 'I want to be just like him', Chong, aged nine, declared.

2.2.3.3 *Other Western Visitors*

Gale was merely the most prominent Western doctor and scientist who went to the Soviet Union; there were also many other, less vocal experts. At the beginning of June 1986, Michael McCally, a Chicago doctor and the treasurer of International Physicians for the Prevention of Nuclear War, visited Moscow to see the medical care that the radiation victims were receiving. Following his return, McCally reported that the sight of the patients in Hospital No. 6 had reminded him of pictures of Hiroshima: 'Now I have seen nuclear war. This could happen to us', he thought. McCally wrote that he was impressed by how 'the psychological needs of these terribly injured people had been considered' by the Soviet medical teams. 'Families were in attendance and the nurses were attentive. The patients were concerned to tell us the histories of their injuries. They had clearly been encouraged to "work through" their experiences.' The American delegation was allowed to photograph patients with their consent. Some of the pictures were printed in the *Bulletin of the Atomic Scientists*.[163] The journal itself had a narrow readership, but more and more people encountered the images because they also found their way into presentations given by activist movements.

Encouraged by the success of his advocacy for Gale, Hammer also asked Gorbachev to agree to a visit by Floyd Culler. Culler was president of the Electric Power Research Institute (EPRI) in Palo Alto, California, and, as Hammer put it, 'one of America's most eminent nuclear power authorities'.[164] Hammer reminded Gorbachev of the Three Mile Island accident and the setback it had represented for the US nuclear power industry; he also observed that 'local and world populations' were 'very frightened as a result of two serious accidents in seven years'. The unspoken message to Gorbachev – that these fears would have to be addressed if the Soviet nuclear industry were to be spared the same

---

[163] McCally, 'Hospital Number Six'.
[164] HA/ACPSS, f. 89, op. 53, d. 7. Culler had been deputy director of Oak Ridge National Laboratory from 1970 to 1977.

repercussions – is clear. Culler, a scientist and member of the nuclear lobby who was mindful of the need for discretion, seemed the ideal choice to Hammer. The idea was to make it look like a private trip: Hammer, who would be arriving in Moscow on 13 May for the opening of his new art exhibition, proposed that Culler fly with him. Culler would then go on to meet Soviet officials in secret. Hammer had, he added, 'already received State Department approval as this a [*sic*] private visit'.[165] The scheme played out exactly as Hammer intended.[166]

In June 1987, the *EPRI Journal* published an issue with the theme 'Chernobyl in Perspective'. Culler is quoted in it, but nowhere is the background to his information – the fact that he visited Moscow in person soon after the disaster – revealed. The journal also quotes remarks by Gale at an EPRI briefing in which he endorsed the delayed evacuation of Prypiat, arguing that it had been better to wait than risk moving residents to locations that might actually have been more contaminated.[167] The visual presence of children in the issue is noteworthy. One photograph, taking up half a page, depicts happy and healthy-looking Prypiat children at a Black Sea holiday camp in their Pioneer uniforms. Another picture shows a teenage evacuee accompanied by a counsellor; they seem to be facing the future positively together.[168] The bottom line is that the journal presented the Soviet response to the disaster as measured and professional. Hammer had, it would appear, delivered.

All the same, the Soviet leadership's confidence in Gale and Hammer – and even more so in its own citizens – had its limits. When the two Americans asked that Soviet experts be allowed to travel to join a meeting of international radiation experts in Los Angeles in July 1986, the Central Committee refused. It clearly still felt that there was too great a risk of crucial information entering the public domain in the West.[169]

2.3 IRRADIATED LANDSCAPES

At the time of the disaster, there were around seven million people – including around three million children – living in contaminated

[165] Ibid.
[166] RRL, Jack F. Matlock Jr. Files, Box 29, F. 'USSR Nuclear Accident: Chernobyl Apr. 29, 1986 (8/9)'.
[167] 'Chernobyl and Its Legacy: A Special Report', *EPRI Journal* 12, no. 4 (June) (1987), 5–21, here 16.
[168] Ibid., 13.
[169] HA/ACPSS, f. 89, op. 11, d. 51, l. 1.

landscapes. Between 200,000 and 350,000 people were relocated or left on their own initiative in several waves. Today, more than five million people in Belarus, Ukraine, and the Russian Federation are still living on land that was contaminated as a result of the accident.[170]

During the period we are concerned with, the 'zone' (singular) was actually an umbrella term for various – generally four – zones that were distinguished according to levels of ground contamination: the evacuation (or alienation) zone (*zona otchuzhdeniia*), the obligatory resettlement zone (*zona bezuslovnogo otseleniia*), the subsequent/voluntary resettlement zone (*zona posleduiushchego/dobrovolnogo otseleniia*), and the radiological control zone (*zona radiologicheskogo kontrolia*).[171]

According to official estimates, people living in contaminated areas were exposed to an average effective dose of 10–30 mSv. According to Iakov Kenigsberg, a Belarusian medical expert, this figure could be as high as 50 or even several hundred mSv in the radiological 'control zone'. He quickly puts this in perspective, though, by remarking that the average dose was less than the 100–200 mSv that people in areas with a high level of natural background radiation in India, Brazil, and China, for instance, would be exposed to in the course of twenty years. In the case of the Chernobyl accident, he writes, most of the affected population received an effective dose of less than 1 mSv per year.[172]

When the explosion occurred, around 130,000 people were living in the immediate vicinity of the reactor. Almost 50,000 of them were in Prypiat and around 14,000 in Chornobyl; the rest were scattered across 76 villages and smallholdings. The evacuation of a 10 km radius around the reactor began at 2 p.m. on 27 April 1986 – 36 hours after the accident. This was extended to a 30 km radius on 4 May. Further locations outside the exclusion zone were included later. The deciding factor was the level of ground contamination with caesium-137. Residents had the right to resettle if they so wished at contamination

---

[170] Kenigsberg, 'Chernobyl' i belorusskoe obshchestvo', 148.

[171] The names vary slightly during the period covered by this study, but the underlying principle behind the divisions (whether resettlement was compulsory, optional but permitted, or not permitted at all) remained the same. Zones with a 10 km and 30 km radius around the power plant are still distinguished within the alienation zone today. On the evolution of the zones, see, in particular, the very helpful tables in Sahm, *Transformation im Schatten von Tschernobyl*, 261–97, 403–5.

[172] Kenigsberg, 'Chernobyl' i belorusskoe obshchestvo', 149.

levels above 5 Ci/km². In Ukraine, compulsory evacuations were carried out above 15 Ci/km²; in Russia and the BSSR, the threshold was 40 Ci/km².[173]

2.3.1 New Homes

By the end of 1986, 116,000 people from 188 locations were having to get used to life in a new home.[174] This figure was to double in the years to come. Prypiat was not even two decades old when, toward the end of 1986, work began on another planned town to replace it. Located about 50 km east of the Chernobyl power plant, Slavutych was, like its predecessor, named after a river.[175] It has been connected by rail with the power plant since 1987.

Workers from the power plant were not overly enthusiastic about moving to the new town. Many people who had been relocated to Kyiv were not keen to leave the capital and move again, especially to somewhere that was relatively isolated. They felt that the standard of housing in Slavutych was unacceptable and were concerned about the lack of prospects for their families there. To make matters worse, it was rumoured that the town and the surrounding areas were contaminated. Some workers threatened to take their own radiation readings; they even said that they were prepared to make the results public and report them to the IAEA if the Ministry of Health did not explicitly guarantee that there was no risk to human health in Slavutych.[176]

In 2000, the two reactors that were still in service were shut down. Up until that point, most residents of Slavutych had been working in the immediate environment of the accident reactor; the number has been steadily decreasing since then. Nonetheless, even today, more than 100 residents are engaged in safety, clean-up, and oversight work there. Just like Prypiat in its time, Slavutych has one of the youngest populations and the highest birth rates in Ukraine. In 2016, thirty years after the disaster, it had around 25,000 inhabitants. Many people have left as the number of jobs at the power plant and in the surrounding area has

---

[173] Arndt, *Auswirkungen des Reaktorunfalls*, 40–3. The resettlements and the provisions behind them are covered in detail in Sahm, *Transformation im Schatten von Tschernobyl*, 261–97.

[174] Kenigsberg, 'Chernobyl' i belorusskoe obshchestvo', 147.

[175] Slavutych is an ancient name for the Dnipro.

[176] SBU, f. 16, spr. 1119, ark. 25–26.

decreased. It might seem that, much like Prypiat before it, Slavutych too has had its day[177] – except that Prypiat later came back to life in the virtual space of the Internet. Former residents created a website for the town and used it to amass memories, stories, and photographs, share information, and discuss films and books about Chernobyl. The site also provided an opportunity to reconnect: friends, schoolmates, and colleagues whose paths diverged after the disaster could make contact again there.[178]

Even as late as the beginning of 1991, the Central Committee counted 46,000 people who were still living in areas where radiation contamination was high enough to pose a health risk. As the case of Slavutych has already indicated, the realities of resettlement were usually very different from what was depicted in the photographs in the Soviet press and more than a few Western outlets.[179] Those who were required to relocate, or who chose to do so, complained about the poor quality of their new accommodation, in particular, the heating. They also expressed dissatisfaction with the lack of community spaces and cultural venues. Furniture and consumer goods were in short supply as well.[180] Early in 1991, Aleksandr Vlasov – the former Soviet minister of internal affairs, no less – criticized the Soviet media, which he felt were exacerbating social tensions by papering over the true state of affairs.[181] Criticism of the disaster response was now being voiced more and more frequently inside the Party – for instance, at Party meetings in the worst-affected areas and at meetings of the Communist Party in the BSSR and Ukraine.

2.3.2 'Living in Prognosis'

In May 1991, forty-nine women from the children's polyclinic in the small Russian town of Novozybkov, near the border with Ukraine, wrote a lengthy letter to Gorbachev and other officials in which they described

[177] See Daniel Wechlin, 'Die sterbende Stadt der Liquidatoren', *Neue Zürcher Zeitung*, 22 April 2016, www.nzz.ch/international/europa/die-sterbende-stadt-der-liquidatoren-ld.15314 (1 May 2024).

[178] Pripyat.com, http://web.archive.org/web/20220313094141/http://pripyat.com/ (1 May 2024).

[179] See: Edwards, Raymer, Mion, *Chernobyl*, and the photographs in Aliaksandr Dalhouski, *Tschernobyl in Belarus: Ökologische Krise und sozialer Kompromiss (1986–1996)* (Wiesbaden: Harrassowitz, 2015).

[180] HA/ACPSS, f. 89, op. 23, d. 21, l. 12.

[181] Ibid., l. 13.

life inside the radiological control zone.[182] The woman were weary of
their plight, and of being reduced to objects of study. The anger and
desperation in their words are plain to see. The letter attracted considerable attention in California, where it became known through the efforts of
the Russian sociologist Vladimir Lupandin.[183] It is a perfect example of
what Olga Kuchinskaya has called 'the work of living with it': dealing
with the consequences of the disaster on a daily basis at a time when
economic and social difficulties only made matters worse.[184]

These doctors and nurses felt that they and the children in their care were
being exploited and abandoned. A steady stream of commissions – there
had been thirty in 1988 alone, they wrote – had come to Novozybkov in
order to carry out medical studies. For the first few years, they had still
believed in the commissions' work. Now, however, the women had reached
the disillusioned conclusion that the commissions had done nothing for
them apart from adding to the demands on their time and giving them more
things to worry about. They had, the women felt, simply come to get data
that they needed for their research, without giving anything back to the
polyclinic or the children and their families. Should the region's children be
immunized with the usual whooping cough–diphtheria–tetanus vaccine?
Or would it be better to use the other one that immunologists from Moscow
had tested on children in Novozybkov more than two years earlier? The
doctors at the polyclinic didn't know, and their efforts to find out met with
no response. The results of the trial had been withheld from them. They
were equally unsure about what to tell new mothers: should they breastfeed
their babies, even though their own milk was probably contaminated, or
give them formula milk instead? They had had high hopes of Viacheslav
Kalinin, the new RSFSR (Russian Soviet Federative Socialist Republic)
minister of health, who had visited Novozybkov shortly after taking office,
only to be let down again. His appearance had been an empty gesture, they
felt, as indicated by the fact that he had responded to their concerns by
explaining that living on Novaia Zemlia had not done him any harm.[185]

Such comparisons inevitably seemed more than cynical to the women
from the polyclinic. They had to grapple with very real challenges on an

---

[182] GARF, f. R8009, op. 51, d. 5138, l. 259–72. The heading 'Living in Prognosis' is
borrowed from chapter 1 of Lochlann Jain's cultural history of cancer in the United
States. Jain, *Malignant*.
[183] HA, Francis U. Macy, Box 16.
[184] Kuchinskaya, *Politics of Invisibility*, 39–64.
[185] The twin islands of Novaia Zemlia in the Arctic Ocean had been a nuclear weapons test
site from 1955 to 1990.

everyday basis, and had been left to face a whole range of uncertainties for which there was no ready solution. Not least among them was the lack of resources, especially where vitamins were concerned. Staff at the clinic were told they should give the children juice – but nobody could tell them where they were supposed to get it. Previously, the children had eaten fruit grown at home and berries from the woods – but that was a thing of the past now. In 1990 every child was allocated 3 kg of oranges and the same amount of grapefruit; those aged 3 or younger received 1 kg of raisins. The authors of the letter acknowledged the fact that vitamin supplements were provided, but they doubted whether they did much good – not least because they caused allergic reactions among the children. The letter reflected the reality facing most of those who lived in contaminated areas: contrary to what was claimed in official statements and coverage in the mass media, they did not have 'clean' food. They consumed local produce and what they were able to grow at home. The women from the polyclinic had long since heard about new medicines – above all, sorbents that were meant to help the body excrete radionuclides ingested with food.[186] But all they could find in their pharmacies was a drug that had only been tested on animals. Could they really prescribe that to children, the doctors asked?

The authors of the letter described the summer holidays as an 'utter nightmare' because they did not know where to send the children. Opportunities for recuperation inside the Soviet Union had narrowed rapidly as the state collapsed;[187] according to the letter, the number of Novozybkov children able to recuperate somewhere else in 1991 was a third to a quarter of what it had been. Those children who had to remain in the town had little choice, the authors added, but to 'swallow the radioactive dust'. The dust – which might or might not really have been radioactive – came from the largely unsurfaced roads in small towns like Novozybkov. The central streets and squares were routinely cleaned up when a commission was due to visit, the women from Novozybkov complained, but that was it. The problems with air quality were exacerbated by the fact that the town did not have one of the district heating

---

[186] Shortly after the accident, the KGB's military medical service and the Ukrainian Academy of Sciences had collaborated to begin testing the effectiveness of sorbents with 'liquidators', with promising results. However, there were insufficient laboratories and equipment to produce sorbents in large quantities. SBU, f. 16, spr. 1126, ark. 50–1. Sorbents were also used to the clean the radiation-contaminated cooling water at the accident site. Igor Belitskii, interviewed by Melanie Arndt, 5 December 2015. Tübingen.

[187] See Section 3.1.2.

systems that were common in the Soviet Union: stoves were often used instead, and people fed them with wood from the surrounding forests – every family thus had its own 'little reactor', as the letter cynically put it. Radiation levels as high as 40 Ci at some locations in Novozybkov led the authors of the letter to conclude that something was wrong with the RSFSR's resettlement policies compared to the BSSR and Ukraine: 'It is sad to think that room [for people to relocate to] could be found even in those republics but there is no help to be had anywhere in mighty Russia!'

The women actively took issue with the claims of radiophobia that were appearing more and more frequently in the official discourse. 'We admit', they wrote, 'that many of us have radiophobia, that our distress and ailments come from that.' The same could, however, hardly be the case with the children, they contended. 'Where', they asked, 'do headaches, nose-bleeds in 4- to 5-year-old children come from, [why] are dysbacteriosis, dysmetabolic nephropathy on the rise, ... lymphocytes behaving in this strange manner? Why are our children drowsy, exhausted, every second child suffering from dizziness, stomach, joint, and bone pain?' Steps taken by the state to protect the children became a warning sign that seemed to confirm their suspicions: 'If they are healthy, does that mean their class-room hours have been reduced ... their breaks and holidays extended, for no good reason?', they pointedly asked.

The authors of the letter also complained about the shortage of doctors. More and more doctors were leaving the contaminated areas. Novozybkov was no exception: in particular, there were not enough specialists for the town's 9,000 children. The women were particularly worried about a potential rise in leukaemia cases. They had been assured that the incidence had not increased, but also asked 'what will we do if it suddenly starts to go up rapidly?' They reported that Moscow paediatricians had recently started making a new diagnosis: 'no specific indication of being affected by radiation' (*radioaktivnoe vozdeistvie ne utochneno*). The women wanted to know what this was supposed to mean: who was going to identify anything specific, and when? And if even specialists were unable to make a precise diagnosis, they argued, was it not high time to consider resettlement? As a final warning shot, the women added a postscript with their own cancer statistics: 'According to our data, there was one cancer case in 1989; this had already risen to four in 1990. Up to 1986 there had been one to two cases among a large number of children.'[188]

---

[188] GARF, f. R8009, op. 51, d. 5138, l. 259–72.

If the residents of Novozybkov were not resettled, the letter demanded, a radiological centre planned for Briansk (the oblast capital, 200 km away) should be built in Novozybkov instead:

What more can we do to accommodate those who want to study us? Why can't they come to us and see things for themselves? It seems to us that plenty of dissertations have been written in recent years, plenty of decorations and degrees awarded; but there haven't been any selfless doctors, and there still aren't any.

The women looked abroad to underline their demands: in Japan, for instance, they pointed out, there were centres for victims of radiation exposure in Hiroshima and Nagasaki themselves, not in Tokyo. They also cited United Nations resolutions.

The letter did not go unnoticed. On 18 July, Larisa Baleva, chief paediatrician (*glavnyi pediatr*) at the RSFSR Ministry of Health, went to Novozybkov to meet the women from the polyclinic. Baleva composed a report on her visit for the RSFSR's Chernobyl committee, confirming the picture drawn in the letter: paediatric and gynaecological care was affected by a shortage of doctors and modern instruments; provisions for getting uncontaminated food, and vitamins overall, were inadequate. The mitigations suggested by Baleva reflect just how powerless the distant centre was. All she could promise in Novozybkov was to have the children examined annually in Moscow, 600 km away; to send groups of specialists there for at least two weeks at a time; and to arrange for doctors to hold medical surgeries at the polyclinic every month.[189] The polyclinic staff are unlikely to have been impressed by these commitments.

A copy of the letter is present in the papers of Francis Macy, who facilitated exchange between Soviet and US experts from the early 1980s onward, initially in the field of humanistic psychology and later in the context of the environmental movement.[190] Macy was provided with this copy by the Moscow sociologist Vladimir Lupandin, who was trying to persuade Western governments that an international radiological centre should be built in Novozybkov. The idea was that Macy would bring the letter to attention outside the Soviet Union. Lupandin drove home his argument by presenting Macy with a picture of panic and mass exodus to Western Europe from contaminated areas if details of a rise in thyroid cancer cases in Belarus were to reach Russia. The only way to stop such

---

[189] GARF, f. 8009, op. 51, d. 5138, l. 261–3.
[190] HA, Francis U. Macy, Box 16.

a situation from developing, he argued, was to bring overseas pressure and funding to bear.[191] Nothing came of any of this, however: the centre was never built – not in Novozybkov, and not in Briansk either.

Macy already had connections with a non-state Chernobyl organization in the city of Briansk. The following year, he visited Novozybkov with his wife Joanna and Adolf Kharash, a social psychologist from Moscow. They held workshops there on coping psychologically with the legacy of the disaster, as they had done in Ukraine and Belarus on the previous stages of their journey. Joanna Macy wrote that Novozybkov had shown them how mistaken they had been to include 'post-' in the title of the workshops 'Building a Strong Post-Chernobyl Culture'. 'Our workshops, we soon realized, were not so much to help people recover from a catastrophe as live with an ongoing one', she later recalled.[192]

As well as doctors, skilled workers of all kinds were (and still are) generally few and far between in places affected by the nuclear fallout: they had been the first to leave, and were now giving anywhere affected a wide berth. The effects of this could be felt not least in schools and kindergartens: children in contaminated areas received a worse education than their counterparts elsewhere. Weakened immune systems and constant illness took their toll on the children's performance as well. What the children really needed was dedicated support for their educational needs[193] – something that was, in most cases, never going to happen.

Anyone who had anything to do with the contaminated landscapes would often find themselves 'living in prognosis' now. Their existence was defined by statistics and the reduction of their bodies to objects of study, yet they did not have access to the underlying data. To make matters worse, the statistics – which rendered the individual human beings behind the numbers invisible[194] – seemed to contradict what they themselves felt and observed. It was in protest at this state of affairs that the women of Novozybkov wrote their letter.

---

[191] HA, Francis U. Macy, Box 16.

[192] Joanna Macy, *Widening Circles: A Memoir* (Gabriola Island, BC: New Society Publishers, 2007), 262.

[193] See Sahm, *Transformation im Schatten von Tschernobyl*, 252.

[194] As Lochlann Jain writes regarding cancer statistics, 'statistical aggregations provide a logic through which bodies become interchangeable numbers for which nothing need be felt, neither guilt, nor pleasure, nor horror'. Jain, *Malignant*, 35.

CONCLUSION

In July 1987, Chief Engineer Fomin was sentenced to ten years in a labour camp, together with the nuclear physicist Anatolii Diatlov (deputy chief engineer at Chernobyl) and Viktor Briukhanov (plant director). In 1988, Fomin was admitted to a psychiatric hospital. Two years later, he was discharged as recovered and simultaneously granted early release from imprisonment. He worked at the Kalinin nuclear power plant, around 260 km north-west of Moscow, until his retirement.

The trial that held these three men responsible for the explosion in reactor no. 4 was politically motivated.[195] It was more convenient to blame individuals for getting things wrong than it was to hold an entire political system and one of its technological prestige projects to account. In the eyes of the world, however, the Soviet Union was very much on trial.

Rather than simply condemning the Soviets in the usual Cold War manner, responses to the disaster around the world involved a level of interest and readiness to help that amounted to much more than just 'techno-diplomacy'.[196]

Almost instinctively, the Soviet Union withdrew into a defensive position for the first two years. On the other side of the Iron Curtain, by contrast, a whole range of individuals and institutions were quick to offer assistance. It was not long before the curtain was being lifted, often discreetly, to let some of them through to help. These men – and they generally *were* men – got to set foot in the Soviet Union, even including the sensitive spots in and around Chernobyl. The Soviet leadership used them in a calculated effort to present its response to the disaster in a favourable light.

The Party and the KGB kept a very close eye on developments outside the Soviet Union, but they often continued to interpret them in a manner that seemed more indebted to the old Soviet tradition than representative of the age of glasnost and perestroika. Characterizing Chernobyl as 'first and foremost a media event', as the historian David Marples did from a contemporary perspective,[197] is not enough to tell the whole story because it almost entirely misconstrues the reality of the Soviet media prior to 1989. Media coverage of the disaster was certainly one reason for the deep-seated insecurity that took hold in those parts of the population

---

[195] On the trial and its use for political and ideological purposes, see Schmid, *Producing Power*, 2–14.

[196] Glenn Schweitzer, quoted in Petryna, *Life Exposed*, 44.

[197] Marples, *Social Impact*, 125.

that were affected or believed they were affected – or both. There was, however, also a more general feeling that people were not being treated in the way they deserved. This sense of being ignored was particularly pronounced among those who had been – or *felt* that they had been – directly affected by radiation exposure. Their 'multiple voices of pain'[198] fell on empty ears in the wider discourse – not that there were sufficient means or resources to help the inhabitants of contaminated areas anyway.

In mid-July 1986, although the resolution in question was kept a closely guarded secret at the time, the Central Committee accepted as fact what it had dismissed as an anti-Soviet fabrication at the end of April. It followed the governmental commission's assessment of the accident as 'one of the worst in the history of nuclear power', one that had resulted in 'human casualties and the release of a significant amount of radioactive matter' as well as causing 'serious damage to morale and the economy'. Even the 'enormous concern' to which the accident had given rise 'throughout the Soviet Union and in the world community' was registered.[199] The Central Committee not only saw the accident as a critical lesson with far-reaching implications that should be recognized by all concerned in the economy and state apparatus; it also went so far as to say outright that it had 'exposed serious shortcomings in the development of nuclear power, the selection of sites for nuclear power plants, and the measures taken to ensure their safety'.[200]

---

[198] David B. Morris, *Illness and Culture in the Postmodern Age* (Berkeley: University of California Press 1998), 128.

[199] HA/ACPSS, f. 89, op. 53, d. 12, ZK KPSS, l. 2.

[200] Ibid., l. 6.

3

The Soviet Chernobyl Children

3.1 THE STATE AND ITS SHORTCOMINGS

The myth of 'happy Soviet childhood' was one of the casualties of perestroika and the revelations that came with it.[1] More than anything else, conditions in homes for orphans and children with disabilities could be shocking at times – but even healthcare provisions for children who grew up with their families were lagging behind those in Western Europe at the end of the 1980s.[2] Action was urgently needed in several areas to address the impact of years of neglect on provisions for children's welfare in the Soviet Union. By this point, however, the state was hardly in a position to do anything about it. The result was a fundamental shift in where responsibility for looking after children's interests lay. Just as had happened in other aspects of Soviet social life during perestroika, the place of the state began to be taken by society as a whole, by independent initiatives, and by the family. In summer 1987, the author Albert Likhanov wrote in *Pravda* that one of the greatest misconceptions of socialism was the idea that being raised by the state was best for children. Likhanov, who a short time later was to chair the Lenin Children's Fund (Detskii fond imeni V. I. Lenina) founded on his initiative,[3] was, in effect, calling for

---

[1] See Chapter 1.

[2] See A. A. Baranov, 'Maternal and Child Health Problems in the USSR', *Archives of Disease in Childhood* 66, no. 4 (1991), 542–5; Elena R. Iarskaia-Smirnova, Pavel V. Romanov, *Sovetskaia sotsial'naia politika: Stseny i deistvuiushchie litsa, 1940–1985* (Moscow: TsSPGI, 2008); P. Romanov et al. (eds.), *Sotsiologicheskoe issledovanie problem invalidnosti i reabilitatsii invalidov v Rossiiskoi Federatsii: Analiz osnovnykh rezul'tatov issledovaniia* (Moscow: n. p., 2008).

[3] Strictly speaking, this was a revival of an earlier initiative rather than an entirely new one: a Lenin Children's Fund had previously existed from 1926 to 1938. It has hardly been studied at all. See White, 'Charity', 788.

a bourgeois-conservative 'cult of the family' in the Soviet Union, with a particular focus on children: 'Let us bring about all our other transformations – in the fields of the economy, the restructuring of society, the environment, culture, art – by asking right from the outset what we are doing in the process for children, for their minds and souls, rather than leaving that to last.'[4]

The Chernobyl children were coming to the attention of state and society more and more often now as well. Likhanov addressed this only indirectly in his *Pravda* article, but the simple fact that he encouraged donations to the state-run Chernobyl relief fund showed that their plight did not have to remain out of sight and out of mind. The growing presence of the Chernobyl children in the public consciousness added to a general awareness that provisions for children's welfare and well-being in the Soviet Union were in need of a fundamental overhaul.

This chapter centres on the Soviet response to the disaster in the first four years after the accident – which also turned out to be the final years of the Soviet Union. In order to map out the possibilities open to those involved, it sets efforts to help the Chernobyl children in the political, economic, and social contexts of the time. A growing number of environmental crises came to light during this period; the chapter examines how, against this background, the disaster was handled in such a way that Party and state progressively lost their legitimacy in a population that was gradually being mobilized 'in the name of the children'. The chapter also retraces how the practice of sending Chernobyl children away for recuperation began. Introduced and funded by the state, it was from the outset considered a particularly effective measure by the Party and state apparatus and the general public alike. For now, though, those who went away in this context were still very much *Soviet* Chernobyl children: evacuation and recuperation took place inside the borders of the Soviet Union, and the organizational processes and experience remained distinctively Soviet in nature.

3.1.1 From the 'Zone' to Artek: The First Chernobyl Children

3.1.1.1 'Amid Enchanting Pines'
When the accident occurred, Prypiat had 49,614 inhabitants, including around 17,000 children.[5] Like all the other much-vaunted 'nuclear towns'

---

[4] Al'bert Likhanov, 'Obernut'sia k detstvu: Net zaboty vazhnee', *Pravda*, 13 May 1987, 3, 6.
[5] Sahm, *Transformation im Schatten von Tschernobyl*, 207.

or 'energy-worker towns' (*goroda energetikov*) with power stations nearby, Prypiat also made much of being an unusually modern, pleasant, and family-friendly town surrounded by pristine nature: in this case, the Prypiat Marshes, the largest wetland area in Europe.[6] Such propositions were designed to attract a well-trained workforce to serve in nuclear power stations and live in *atomgrady*.[7] The desired effect was certainly achieved in Prypiat. Its inhabitants had an average age of twenty-six, and there were numerous facilities for children and young people, including thirty-five playgrounds, fifteen kindergartens, five middle schools, one vocational school, and one art school. A cinema, a house of culture, and a park provided recreational opportunities for residents. There was also a hospital to take care of the sick.[8]

Alena Oginets belonged to a model *atomgrad* family. With the help of icebreaker and nuclear-submarine experience, her father had gone on to work in nuclear power plants, first in central Russia and then in Prypiat. Alena was four when her family moved into their brand-new apartment block.[9] Like many others who once lived in Prypiat, she remembers it as a 'really beautiful town'. Her building was surrounded by a little forest; she could even feed squirrels from the balcony. She later paid tribute to the place she had left behind in a poem called 'And the Town Is Waiting':

> Suddenly, amid enchanting pines, a town appeared.
> So beautiful and young, like a pristine spring.
> The air there was clear, flowers everywhere.
> Hope had made its home there. It was the town of dreams.[10]

The topos of nature – and, in particular, the 'beautiful, enchanting, unforgettable forest'[11] – is everywhere in the memories of the people of

---

[6] The same goes for the small secret town of Chornobyl-2, which is still largely unmentioned in accounts for specialist and wider audiences alike. The Duga-1 over-the-horizon radar installation is located there. See Sergei Paskevich, *Chernobyl'-2: Sekretnyi dvoinik goroda Chernobyl'*, http://chornobyl.in.ua/chernobyl-2.html (1 May 2024).

[7] See, e.g., Josephson, *Red Atom*; Wendland, 'Tschernobyl'.

[8] See Higginbotham, *Midnight in Chernobyl*, 17.

[9] Oginets, interviewed by Arndt, 27 March 2017.

[10] 'Sred' sosen volshebnykh vdrug gorod voznik, / Prekrasnyi i iunyi, kak chistyi rodnik. / V nem vozdukh byl svezhii, povsiudu tsvety, / Zhila v nem nadezhda. Byl gorod mechty.' The poem is part of *Nine Grown-Up Days in a Typical Childhood*, a play that Oginets wrote as a young adult in an effort to come to terms with her experience of the disaster. Alena Oginets, *Deviat' vzroslykh dnei obychnogo detstva*, unpub. manuscript, Author's own records, n. d.

[11] Quote from Sergei Shabanov, who was ten when the town was evacuated. Quoted in Aleksandr Cheban, 'Chernobyl' 30 let spustia. Evakuatsiia glazami detei', *LiveJournal*, 27 April 2016, https://alexcheban.livejournal.com/290807.html (1 May 2024).

Prypiat. It pushes the industrial landscape of the power station – and with it, one suspects, the memory of the accident – away into the background.

The Last Weekend in Prypiat

Various events for children were planned for the weekend of 26/27 April 1986. Party and state officials in the town were unsettled by the initial reports concerning the power plant, but arrangements for the Saturday were not changed in any meaningful way. School took place as usual, just as it did on any other Saturday in the Soviet Union. Eyewitnesses report that teachers covered the windows with damp cloths and pupils were not allowed to go outside between lessons,[12] but children were left to their own devices once school was over. Alena Oginets recalls that the popular bathing area on the river was 'packed' on that 'unusually hot' day. Some other children in her apartment block who had not gone to the river were playing in the courtyard instead. Alena watched them from behind the windows of her apartment: her father had told her to stay at home and keep them shut.[13]

For the most part, however, even those who must have known there had been an accident – even if not how bad it really was – seemed untroubled. Aneliia Perkovskaia, secretary of the Prypiat Komsomol committee, remembers the child of a senior engineer at reactor no. 4 having fun in a playground sandpit as if nothing had happened – and that the committee had been shocked by this. The faith in technology appears to have been so great that it was simply inconceivable that there might be cause for concern. Parents were still out in town enjoying ice cream with their children an hour and a half before the evacuation.[14] One teacher had been planning a running event for children; when the town's Party committee (*gorodskoi komitet*, or *gorkom*) advised her that it would not be able to take place because Prypiat was being evacuated, she apparently replied indignantly 'What evacuation? Honestly, people! We've got the health run today!'[15] Similar parallel realities are to be found everywhere in the sources (official records, eyewitness recollections, literary and artistic works).

---

[12] Shcherbak, *Chernobyl'. Dokumental'noe povestvovanie* (Moscow: Sovetskii pisatel', 1991), 90.
[13] Oginets, interviewed by Arndt.
[14] Shcherbak, *Chernobyl'*, 90.
[15] Quoted in ibid.

By mid-May 1986, 89,360 people had been evacuated from the 'danger zone' (*opasnaia zona*), as the exclusion zone was initially called in internal documents. This included 62,746 people – among them 27,400 children – from the towns of Prypiat, Chornobyl, and Chornobyl-2. According to official records, all 26,900 evacuated schoolchildren were taken to holiday facilities 'for a break'. In most cases, they remained there for several months, until the start of the new school year in September. Pregnant women and mothers with babies and toddlers were placed in sanatoriums or guest houses in the Kyiv oblast.[16]

Before the children were sent away, most evacuees stayed with families in villages outside the 'zone'. Alena Oginets, who was twelve at the time, found herself among them; she had a particularly hard time because she had to leave Prypiat without her parents. Her mother was visiting Alena's grandmother in Voronezh in Russia, and her father had been summoned to the stricken power station. As a result, she was evacuated with her neighbours rather than her parents. They were billeted together in the village of Sukachi, 40 km from Prypiat. Oginets recalls that everything seemed like an adventure at first, but that she was very unsettled when the local children at the village school taunted her that she would never see Prypiat again. The village was close to the road between Kyiv and Prypiat, and she remembers how the never-ending lines of lorries there added further to her unease. As it began to dawn on her how serious the situation really was, her thoughts turned to her father: 'I cried at night. I was seriously worried about my father. A kind of inner panic that I would never see him again.'[17] It was a good week or so before Oginets's mother came to pick her up; they went back to her grandmother in Russia together. According to Oginets, the silence in the media and the absence of other means of staying up to date meant it had taken her mother days to learn what had happened in Prypiat.

After taking her to Voronezh, Alena's mother returned to the 'zone'. She had, Alena recalls, seen doing so as a moral duty that left her no alternative. It was at this time that Alena was sent to the Artek Pioneer camp in Crimea. She recounts how having to move repeatedly – being sent to new places with different groups of children each time, never knowing what was going to happen to her next – took its toll on her.[18] She

---

[16] Shcherbitskii, 'Informatsiia', 22 May 1986, in Natalia P. Baranovs'ka (ed.), *Chornobyl's'ka trahediia: Dokumenty i materialy* (Kyiv: Naukova Dumka, 1996), 152–3.
[17] Oginets, interviewed by Arndt.
[18] Ibid.

remembers that she had badly wanted at least to be together with her friends in the same holiday camp, but that this had clearly been considered undesirable. To begin with, she thought the authorities were trying to avoid drawing attention to the accident by including only a few Chernobyl children in each group. She no longer felt that way when interviewed, hoping instead that the intention was ultimately a good one and that groups were mixed in an effort to give the children a 'normal life' as far as possible. After all, she pointed out, the evacuee children at Artek were – apart from a few medical examinations – treated just like all the other children there.

Alena recognized herself in footage recorded in Artek by the amateur film-maker Mikhail Nazarenko when it was uploaded to YouTube some years ago.[19] Looking back, she recalls that the children had all been trying to put on a brave face for the camera – which did not take all that much effort: being visited by two such notable figures from Prypiat (Nazarenko was accompanied by a popular musician and DJ) had genuinely cheered them up. The children, she remembers, had been only too eager to believe the two men when they said that everything would be fine and they would get to see their parents again soon. Equally, she acknowledges that the children had been 'properly depressed' at the start of their time in Artek, and that this mood of dejection had returned after Nazarenko and his companion departed: 'After all, we didn't know then where we were going to live. Nobody knew that then yet, did they? Where? Where to? Who?' She notes that this uncertainty affected all of them deeply.[20]

Nazarenko's footage includes undertones of disquiet that will not be encountered in any of the official accounts. The *Pionerskaia pravda* newspaper for children, for instance, painted an untroubled picture of normality in June 1986: a short report framed by two photographs described how 110 children from Prypiat were recuperating at Artek, going on excursions with others at the camp, taking part in sports competitions and work assignments, and swimming. The accident was not mentioned once (Fig. 3.1).[21]

Foreign reporting gave a rather different impression: the American journalist Felicity Barringer from the *New York Times*, who got to talk

---

[19] For Nazarenko's recordings, see the Prologue.
[20] Ibid.
[21] *Pionerskaia Pravda*, 10 June 1986. I would like to thank Kathleen Beger for drawing my attention to this source.

У подножия Аюдага

110 ребят из города Припяти Киевской области отдыхают сейчас в «Артеке». Они вместе с остальными артековцами ездили в Севастополь, участвовали в спортивных соревнованиях «Весёлые старты», в трудовых десантах, готовили вечера. Каждый день наполнен самыми разнообразными делами! Все купаются в море! А если погода бывает холодной — бассейн с морской водой ждёт ребят. Вон как весело октябрятам из 13-го отряда «Морского» лагеря в бассейне! Да и не только здесь. Над этими ребятами взял шефство восьмой отряд. Пионеры помогают маленьким: дежурить по палате Лиле Ящук или Яне Кузьминой, не опоздать на зарядку Игорю Коновалу, всем вместе подготовиться к конкурсу «живых газет».

И сразу видно, что очень хорошо Богдановой Юле и Рязановой Оле с вожатой пятого отряда Олей Клитней. Крепко подружились ребята!

Фото В. САШИНА.

FIGURE 3.1 A report in the children's newspaper, *Pionerskaia pravda*, explicitly stated that 110 children from Prypiat were staying in Artek, but did not mention the accident at all. Instead, it focused on leisure activities.

to children and doctors at Artek, was met with a picture that included shades of anxiety and concern – and she wrote about it.[22]

Not all the evacuated children got to go away like Alena Oginets did. Age limits were strictly enforced, even among those at school. It was primarily schoolchildren from second to ninth class (age 7–16) who were eligible. Parents were just as unhappy about the fact that children were discriminated against for being too young or too old as they were about some of the destinations. These ranged from prestigious flagship camps in sought-after tourist areas – such as Artek and Molodaia Gvardiia on the Black Sea and Zubrenok on the popular Lake Narach in the BSSR – to relatively obscure holiday camps where conditions were generally more spartan.[23] The showcase venues did at least allow for schooling to be continued, as well as offering a wider range of leisure activities; but, even there, it can hardly be said that the evacuated children were able to lead a 'normal life'.

3.1.2 Provisions for Sending Children Away for Recuperation

Even the initial measures described here presented the authorities with ethical dilemmas when it came to determining who was a Chernobyl child and, accordingly, was entitled to spend time away recuperating. These challenges only got worse as the years went by. A document called a *putevka* was generally required to travel to holiday facilities in the Soviet Union; this also applied to the Chernobyl children. Issuing offices and applicants alike often displayed typically Soviet behaviours when it came to these *putevki*. Right from the start, the rules were deliberately circumvented for a variety of reasons.[24] People who had connections in the right places (known as 'vitamin B': taking advantage of personal relationships or acquaintances in order to obtain coveted goods, including scarce goods) or who were able to procure commodities in short supply – of which there were many at this point in time – had better chances of being approved. Parents and grandparents were prepared to cross ethical boundaries to secure one of the coveted *putevki* for their children or get their names on the list of those travelling. There were also acts of

---

[22] Barringer, 'Children of Chernobyl'.

[23] GARF, f. 8009, op. 51, d. 3559, ll. 8–11; Shcherbak, *Chernobyl'*, 90.

[24] On the problems of ad hoc decision-making in humanitarian crises, see Elizabeth C. Dunn, 'The Chaos of Humanitarian Aid: Adhocracy in the Republic of Georgia', *Humanity* 3, no. 1 (2012), 1–23.

compassion, including fabricating paperwork so that a child could be included. Places at the popular elite camps were particularly sought after. Foreign children also went there for holidays (e.g. in Artek or Molodaia Gvardiia), and parents saw this as evidence of the standards there. On the other hand, a child could be sent back home from a holiday camp if it came to light that they were not entitled to travel as a Chernobyl child;[25] the sources do not permit clear conclusions to be drawn about whether such cases were isolated incidents or more widespread. At any event, suspicions that the system in general was being manipulated were present from the beginning and only increased as the years went by.[26]

Socialist competition was another inescapable fact of Soviet life that affected the issuing of *putevki*. On paper, it was primarily exemplary 'top workers' who received vacation *putevki*, particularly for free or subsidized trips. Individuals who got noticed for the wrong reasons would find the authorities withholding *putevki* from them as a mark of disapproval.[27] Much the same was true when it came to the selection of Chernobyl children; this became particularly apparent in the early 1990s, when children (or, as the case may be, their parents) began competing for places abroad. At least in part, the distribution of *putevki* for Chernobyl-related travel served as a means of encouraging desirable behaviour – a state of affairs that remains the case today.

By the end of May 1986, parents were so keen to have their children participate in these trips that restrictions had to be announced on the radio. Only children who could be documented as registered residents of Prypiat would now be entitled to recuperation.[28] Kyiv parents, however, were not prepared to accept this. Public pressure increased to such an extent that the leadership of the Ukrainian Soviet Socialist Republic, under the chairman of its Supreme Soviet, Valentyna Shevchenko, unilaterally decided to evacuate schoolchildren from the city and its surroundings, and the oblasts of Zhytomyr and Chernihiv, as a temporary

---

[25] Shcherbak, *Chernobyl'*, 232.

[26] The sociologist Vladimir Lupandin also expressed this view, albeit not until December 1991. Cf. the interview in Mikhail Podgoronikov, 'My ot prirody ne sposobny k demokratii? Sostoianie genofonda v SSSR vnushaet trevogu: No mozhno li govorit' o "debilizatsii" naseleniia?', *Literaturnaia Gazeta*, 11 December 1991, 7.

[27] See Christian Noack, 'Von "wilden" und anderen Touristen: Zur Geschichte des Massentourismus in der UdSSR', *Werkstatt Geschichte* 36 (2004), 24–41, here 26; Henningsen, *Der Freizeit- und Fremdenverkehr*, 55.

[28] Shcherbak, *Chernobyl'*, 232.

measure.[29] If the accompanying adults are included, capacity had to be found for around 700,000 people to spend 4 months away.[30] The move was criticized by Gorbachev and Oleg Shchepin (at this point, deputy health minister on the Union level).[31] However, only two months later, in July, a group of geneticists from the Soviet Academy of Sciences retrospectively endorsed the Ukrainian decision and advocated temporary evacuation.[32]

The school holidays had begun early for children from Prypiat and the rest of the evacuation zone, but the temporary evacuation of children from the city and oblast of Kyiv coincided at least in part with the usual start of the holidays in May.[33] Some children had already been allocated places to stay, even if not for the whole summer, as part of the usual arrangements prior to Chernobyl. This took off some of the pressure, at least. The school holidays also meant that there was no need to arrange teaching for the children in the immediate future. Nonetheless, a comparison of Shevchenko's figure of 526,000 children with the demand anticipated before the disaster indicates that only about a fifth of the places needed were actually available.[34] The situation was exacerbated by the fact that thousands of accompanying adults and mothers with babies had to be accommodated as well. Most facilities were not remotely suitable for them, for the simple reason that babies and toddlers did not generally go to Soviet holiday camps.[35] Hot water was not available around the clock, and there were inadequate provisions for preparing baby milk, no beds or bathrooms suitable for children, and no play areas. The fear of radiation, however, was such that many mothers took their children to such places

---

[29] Lina Kushnir, 'Valentyna Shevchenko. "Provesti demonstratsiiu 1 travnia 1986-ho nakazaly z Moskvy', *Istorychna pravda*, 25 April 2011, www.istpravda.com.ua/articles/2011/04/25/36971/ (1 May 2024).

[30] Kushnir, 'Valentyna Shevchenko'. Statistics or other sources on this are lacking, so the figures cannot be verified; at any event, several hundred thousand people must have been involved.

[31] Shchepin, 'Protokol', 14 May 1986, in Baranovs'ka (ed.), *Chornobyl's'ka trahediia*, 132–4.

[32] 'Spravka', 19 July 1986, *Z arkhiviv VUChK, GPU, NKVD, KGB* 16, no. 1 (2001), 123.

[33] El'chenko, 'Dovidka', 23 May 1986, in Baranovs'ka (ed.), *Chornobyl's'ka trahediia*, 162–4.

[34] See ibid.

[35] Families had only been able to vacation together at state-run facilities since the late 1970s and early 1980s. From 1981, some facilities offered places for families with children over five; the age limit had previously been seven years, or even higher where boat trips and the like were involved. See Aleksei K. Abukov, *Turizm na novom etape: Sotsial'nye aspekty razvitiia turizma v SSSR* (Moscow: Profizdat, 1983), 92.

anyway and put up with the shortcomings.[36] As a rule, mothers (fathers were out of the question) were only permitted to accompany their children if they were under three years of age. Children over three were accompanied by women who would look after them instead.[37] Pregnant women were given additional leave – the costs of which were covered by the Red Cross – so that they could avail themselves of the hastily arranged trips.[38]

The Chernobyl accident also affected school attendance – quite considerably so. In Kyiv, only 70–75 per cent of schoolchildren continued going to lessons; a fifth of eighth- to tenth-class children (age 13–15) were not at school. In addition, the city committee noted that visits to theatres, concert halls, and cinemas had fallen by as much as around 40 per cent in mid-May. The committee also observed, however, that 'political work with the general public', together with the Ukrainian health minister's assurance that there was no risk to health, had had an effect towards the end of the month: use of cultural venues and school attendance had stabilized back at the levels seen before the accident.[39]

When it came to the BSSR, the executive committee of Homel (Ru. Gomel) oblast was the first to act. In a 'small meeting' (*suzhennoe zasedanie*) on 6 May 1986, three days after the decision to evacuate the 30 km zone, the committee determined that all children, pregnant women, and women with pre-school children in the contaminated raions of Brahin (Ru. Bragin), Khoiniki, and Narouvlia (Ru. Narovlia) should be taken to stay at facilities outside the Homel oblast. Once there, they were all to be medically examined and schooling for the children would continue.[40] Earlier, on 1 May, 680 people, in particular pregnant women and children, had already been evacuated from 25 locations.[41] The authorities now evacuated a further 27,700 people.[42]

---

[36] Solohub, 'Zapyska', 30 May 1986, in Baranovs'ka (ed.), *Chornobyl's'ka trahediia*, 204–6.

[37] NARB, f. 4p, op. 157, d. 168, ll. 15–19.

[38] NARB, f. 4p, op. 157, d. 168, l. 9. The distribution of state-subsidized *putevki* corresponded to the usual practices of holiday organization in the Soviet Union before Chernobyl. See Christian Noack, 'Tourismus in Russland und der UdSSR als Gegenstand historischer Forschung: Ein Werkstattbericht', *Archiv für Sozialgeschichte* 45 (2005), 477–98, here 489.

[39] El'chenko, 'Dovidka', 163.

[40] Grakhovskii, 'Reshenie No. 176-14', 6 May 1986, in V. I. Adamushko et al. (eds.), *Chernobyl': 26 aprelia 1986 – dekabr' 1991. Dokumenty i materialy* (Minsk: NARB, 2006), 39–40.

[41] Grishagin, 'O rabote', 12 October 1989, in ibid., 300–12, here 303.

[42] Ibid., 304.

A good two weeks later, on 23 May, the BSSR Council of Ministers followed suit and decided on sweeping measures to send children away temporarily.[43] Roughly 114,000 children spent the summer of 1986 in various recuperation facilities and sanatoriums; around 16,300 of them were accompanied by their mothers. Places were found for around a quarter (about 31,000) of children in the BSSR, where they were largely accommodated in 82 Pioneer camps and labour and rest camps (*lageria truda i otdykha*, LTOs). LTOs existed not only to provide young people with recreation opportunities but also to put them to work. Unlike other holiday facilities, they often consisted of makeshift accommodation that was rented on a temporary basis or provided by the enterprise that needed the work.[44]

Most of those sent away to recuperate – around two-thirds of them – were accommodated in facilities outside the BSSR. Thirty different Soviet regions took in children and young people: 64,200 in Pioneer camps and 9,800 in LTOs.[45] More than 5,300 children from the Homel and Mahileu (Ru. Mogilev) oblasts spent at least some time in the elite Pioneer camps of Artek, Orlenok, Okean, and Zubrenok.[46]

For 1987, the BSSR decided that recuperation would be 'more differentiated', without specifying in greater detail what that meant in practice. At any event, the number of children sent away dropped by more than a third that year.[47] The numbers went down even in areas where the Soviet Health Ministry had confirmed that conditions could 'not be described as normal'.[48] Of those children who did get to go away for recuperation, more than 60 per cent were sent to locations within the BSSR. The ministries responsible even planned not to send any children outside the republic in 1988; it was anticipated that only just under 32,000 children would be provided with recuperation or rehabilitation.[49]

---

[43] 'O merakh po organizatsii v 1986 godu letnego otdykha i ozdorovleniia detei i podrostok Homel'skoi oblasti', 23 May 1986. The decision is mentioned in NARB, f. 4p, op. 157, d. 168, ll. 15–19.

[44] See Henningsen, *Der Freizeit- und Fremdenverkehr*, 68. Borisenko, 'Iz zvitu instruktora', 16 June 1986, in Baranovs'ka (ed.), *Chornobyl'ska trahediia*, 247–8.

[45] Sovet Ministrov BSSR, 'Informatsiia', 5 June 1987, in Adamushko et al. (eds.), *Chernobyl'*, 118–28, here 123.

[46] NARB, f. 4p, op. 157, d. 168, ll. 15–19, here l. 17.

[47] Only 82,000 children went away, 20,700 of them together with their mothers. This was far fewer than planned, even under the new arrangements for that year (131,980). Ibid.; Sovet Ministrov BSSR, 'Informatsiia', 5 June 1987, 127.

[48] GARF, f. 8009, op. 51, d. 3559, l. 5.

[49] NARB, f. 4p, op. 157, d. 168, ll. 15–19; Sovet Ministrov BSSR, 'Informatsiia', 5 June 1987, 123.

These stays were primarily paid for by enterprises and trade union organizations, which also operated many of the facilities.[50] In addition, a number of ministries, including the Ministry of Health and Education, as well as the Komsomol and various agencies, not only allowed Chernobyl children to stay at facilities they ran but also footed the bill.[51] At this point, it was the health departments of oblast executive committees that decided who could spend time away as a Chernobyl child and for how long.[52] Chronically ill children were to be sent away for 'no fewer than forty-five days', healthy children from places under 'radio-logical control' for 'no fewer than twenty-four days'.[53] The latter figure corresponded to the usual length of visits to children's rest and recuperation facilities in the Soviet Union (twenty-four or twenty-five days).[54]

In the BSSR, it was above all children from the Homel and Mahileu oblasts – the closest to Chernobyl – who were considered Chernobyl children in the first two years after the accident. Some parts of the Brest oblast were also heavily contaminated, but only from around 1988 do the roughly 48,000 children from there appear in the statistics.[55] The recognition of the Brest oblast as a disaster area only added to the problems of finding sufficient capacity in the system. Official sources indicate that it proved possible to accommodate all children who were entitled to recuperation or rehabilitation – but there was nonetheless a 'glaring lack' of accommodation for small children and mothers with children.[56]

The situation grew even worse when regional institutions realized that a brief period of rehabilitation over the summer was not going to resolve health issues or remove the risk of long-term harm. Children who were required to attend periodical check-ups organized by the state (*dispanserizatsiia*) because of their potential exposure to radiation needed year-round rehabilitation care in specialist sanatoriums.[57]

[50] NARB, f. 4p, op. 157, d. 168, ll. 15–19, ll. 11–12, 17; see also Noack, *Geschichte des Massentourismus*. The trade unions were responsible for much of the Soviet holiday system.

[51] See Henningsen, *Der Freizeit- und Fremdenverkehr*, 42; Gennadii P. Dolzhenko, *Istoriia turizma v dorevoliutsionnoi Rossii i SSSR* (Rostov na Donu: Izdatel'stvo Rostovskogo universiteta, 1988), 146.

[52] NARB, f. 4p, op. 157, d. 168, l. 9.

[53] Ibid.

[54] Henningsen, *Der Freizeit- und Fremdenverkehr*, 68.

[55] Bokach, 'Spravka', 4 October 1991, in Adamushko et al. (eds.), *Chernobyl'*, 405–11, here 408.

[56] Ibid.

[57] Nevdakh, 'Predlozheniia', 1990, in Adamushko et al. (eds.), *Chernobyl'*, 364–6, here 365.

One reason for the shortage of sanatoriums lay in the priorities that the Soviet leadership had set for investment in tourism during the 1970s. Funding for expensive and workforce-intensive forms of holiday accommodation centred on health benefits, such as sanatoriums, had been rolled back in favour of a new focus on simpler, more economical accommodation for healthy guests with minimal needs. Numerous 'rest bases' and 'tourist bases' appeared, often of a very spartan nature.[58] Young people who were sent to LTOs due to a lack of capacity in holiday camps often found that the promised recuperation failed to materialize. Two hundred and seventy pupils from the vocational school in Poliske in Ukraine, which was close to Chernobyl and later permanently abandoned, were sent to Berdiansk on the Black Sea, where they got a break from radiation exposure – while working for six to seven hours a day. Those responsible for looking after them at the LTO clearly took little interest in their plight. They failed to arrange medical check-ups or psychological support, and the food on offer was woefully inadequate. Ultimately, some of the group refused to work; four individuals fled the camp. A shocked Komsomol representative expressed his concern about conditions there to the Central Committee.[59]

Despite such occurrences, many parents initially seemed grateful for the efforts being made by Party and state bodies. One letter, written by parents from the 30 km exclusion zone to the Central Committee of the Belorussian Communist Party, said that 'the Party and the government had done everything they can to provide the children with moral and material support'.[60] It is likely that this was not simply a genuine expression of gratitude but was also meant to demonstrate political loyalty in the hope of garnering further support. This kind of ritualized communication was, furthermore, something reassuringly familiar that parents could fall back on in a disaster situation where there were so many unknowns.[61] The fact is that, on the whole, the facilities available lacked both the specialists and the know-how to look after the children properly. What paediatricians there were had no training in how to deal with patients who might have been injured by radiation. In medical terms, there was not much they could do: they were able to establish the presence of high levels of

---

[58] See Noack, *Geschichte des Massentourismus*, 27.
[59] Borisenko, 'Iz zvitu instruktora', 16 June 1986.
[60] Quoted in Dalhouski, *Tschernobyl in Belarus*, 71.
[61] On ritualized speech before and during perestroika, see, respectively, Yurchak, *Everything Was Forever*; Ries, *Russian Talk*.

radiation in the thyroid, or note children's lethargy and fatigue, but they were not in a position to understand the effects of radiation exposure properly or treat them accordingly.[62]

3.1.3 Return

In most cases, the situation even at substandard facilities was still better than what awaited the children when they returned from recuperation that first summer after Chernobyl. Many went back to places that had not been properly decontaminated. They found themselves facing new constraints on how they went about their lives, and had to modify their eating and drinking habits and change what they did in their spare time. Locally produced food and drink (e.g. milk) could no longer be consumed; contaminated sites were out of bounds. December 1986 saw 12 of the 107 locations in the BSSR oblasts of Homel and Mahileu (24,700 residents) that had been evacuated in May, June, and August being reoccupied. In Ukraine, 14 of the 75 locations in the Kyiv and Zhytomyr area that had been evacuated by August (90,784 people) were reoccupied.[63] Submissions to state authorities show that the public did not simply accept going back as a matter of course. The number of complaints increased when the children began to return.[64] It was very often women – and not just directly affected mothers either – who took their concerns to state and Party institutions.[65]

The new school year commenced as usual on 1 September in contaminated areas, just like everywhere else. It was primarily authorities on a local level that made a significant effort to decontaminate kindergartens and school buildings and protect them from further accumulations of radioactive material. Contaminated soil layers were removed and school playgrounds paved over because any suspicious dust could then be removed more easily than from bare earth. Even so, while significant amounts of money were spent on such measures, they remained superficial and had little long-term health impact. Accordingly, one deputy head teacher spoke of 'cosmetic decontamination'; she was of the view that resettlement would have been the better course of action – both in

---

[62] Sokol, 'Dopovidna zapyska', 18 June 1986, in Baranovs'ka (ed.), *Chornobyl'ska trahediia*, 258–9.

[63] Sahm, *Transformation im Schatten von Tschernobyl*, 207.

[64] See Dalhouski, *Tschernobyl in Belarus*, 77.

[65] Aliaksandr Dalhouski came to the same conclusion; see ibid., 138.

financial terms and for the children's health.[66] Those making the decisions, however, arrived at a different conclusion: according to internal
calculations, decontamination measures could reduce contamination
levels of 30 Ci/m² or more by half – but, even if this was the case, the
children would still be being exposed to high levels of radiation.

Children were given free footwear to change into so as not to bring
radioactive dust into the classroom from outside. Schools and kindergartens in contaminated oblasts also improved their catering and provided
children with three or four free meals per day, respectively.[67] Some were
given new fridges to help keep everything fresh now that there was so
much more to be stored. The quality of the offerings, however, remained
underwhelming. Good sources of vitamins – or indeed, uncontaminated
food and drink of any kind – were lacking. Baby food was in particularly
short supply.[68] Even the BSSR Ministry of Health criticized the canteen
meals as 'insipid and low in vitamins'.[69]

Other risks were the result of carelessness, insufficient knowledge, or
an inexcusable lack of awareness. Even in early 1989, men in the southern
BSSR raion of Slauharad (Ru. Slavgorod) were cleaning contaminated
agricultural machinery in the immediate vicinity of a kindergarten 'without any qualms and in full knowledge of their supervisor'. The cleaning
should have been done at special facilities – which did exist.[70] It was such
hazards, as well as the lack of healthy and 'clean' food, that led many
parents to try to resist moves to bring children back to contaminated areas
once their recuperation was over.[71]

In early April 1988, together with the BSSR Ministry of Education and the
secretary of the BSSR Council of Trade Unions, the BSSR Ministry of Health
agreed to continue sending children away for recuperation. Even two years
after the accident, they recognized that there was still a 'very pressing' need to
get children out of the affected areas, at least for a time. They based their

---

[66] Quoted in ibid., 139.
[67] For various reasons, however, there were many cases where this could not be put into
practice successfully. See ibid., 85.
[68] See Ignatenko, Kras'ko, 'Informatsiia', 18 November 1986, in Adamushko et al. (eds.),
Chernobyl', 96–9, here 97; Petrov et al., 'O rabote', 23 October 1986, in ibid., 85–91, here
86.
[69] NARB, f. 4p, op. 157, d. 168, ll. 15–19, here l. 18.
[70] Tkachev, 'Otchet', 24 March 1989, in Adamushko et al. (eds.), *Chernobyl'*, 179.
[71] On this, see also Dalhouski's analysis of representations to the BSSR Council of Ministers
in the second half of 1986. He identifies the return of children to contaminated locations as
one of the most common issues that was raised. Dalhouski, *Tschernobyl in Belarus*, 89.

reasoning on medical grounds.[72] Bodies on the Union level and in the republics alike saw such trips as beneficial for children. In August 1986, a group of experts from the Ukrainian Academy of Sciences had already argued that recuperation measures were medically desirable. They calculated that sending children away temporarily had the potential to reduce incorporated doses in the first year after the accident by a factor of three to five. Despite the recuperation trips, they had also found that the scope and quality of the steps being taken to safeguard children were 'not satisfactory'.[73]

The Soviet Ministry of Health had likewise endorsed sending children away as a 'very effective measure for reducing exposure to radiation'. It was considered one of the most successful responses to the disaster of all. Other mitigations, including checking food and agricultural products for radiation, shipping 'clean' food in from outside, and efforts to minimize ground contamination in agriculture, were rated much less highly.[74] These assessments were based on an increasingly extensive medical surveillance of children. It was not until May 1987, one year after the accident, however, that the Soviet Ministry of Health mandated extensive centrally coordinated check-ups (*dispanserizatsiia*).[75] Some regions had already introduced such measures and finished examining the target population – the Homel oblast, for instance, did so in December 1986.[76] *Dispanserizatsiia* results increasingly became the basis for deciding which children were to be given time away.[77] Properly functioning dispensaries, however, were far from a given.[78] For children in many areas, whole-body radiation measurements in particular necessitated a trip to the capital of their republic. (For an example of a whole-body radiation measurement, see Fig. 3.2.) Parents and children were often abandoned after these tests had been completed.[79] Parents generally lacked the necessary knowledge

---

[72] NARB, f. 4p, op. 157, d. 168, ll. 15–19, here l. 16.

[73] Nauchnyi sovet pri prezidiume AN USSR, 26 August 1986, in Baranovs'ka (ed.), *Chornobyl'ska trahediia*, 327–36, here 334.

[74] NARB, f. 4p, op. 156, d. 393, ll. 94–96.

[75] NARB, f. 4p, op. 157, d. 168, ll. 15–19, here l. 18.

[76] Grinkevich, 'Spravka', 8 December 1986, in Adamushko et al. (eds.), *Chernobyl'*, 101.

[77] Khusainov, 'Zapiska', 18 May 1988, in Adamushko et al. (eds.), *Chernobyl'*, 145–53, here 149.

[78] Dispensaries were a form of healthcare centre that was introduced following the October Revolution to combat particular illnesses (e.g. tuberculosis or sexually transmitted diseases). They had their own in-patient facilities and worked in collaboration with polyclinics, hospitals, and sanatoriums. See Irene Uhlmann, Irene Klemm, Günther Liebig (eds.), *Kleine Enzyklopädie Gesundheit*, 3rd ed. (Leipzig: Bibliographisches Institut, 1957), 450.

[79] Ermak, 'Pis'mo' in Adamushko et al. (eds.), *Chernobyl'*, 245.

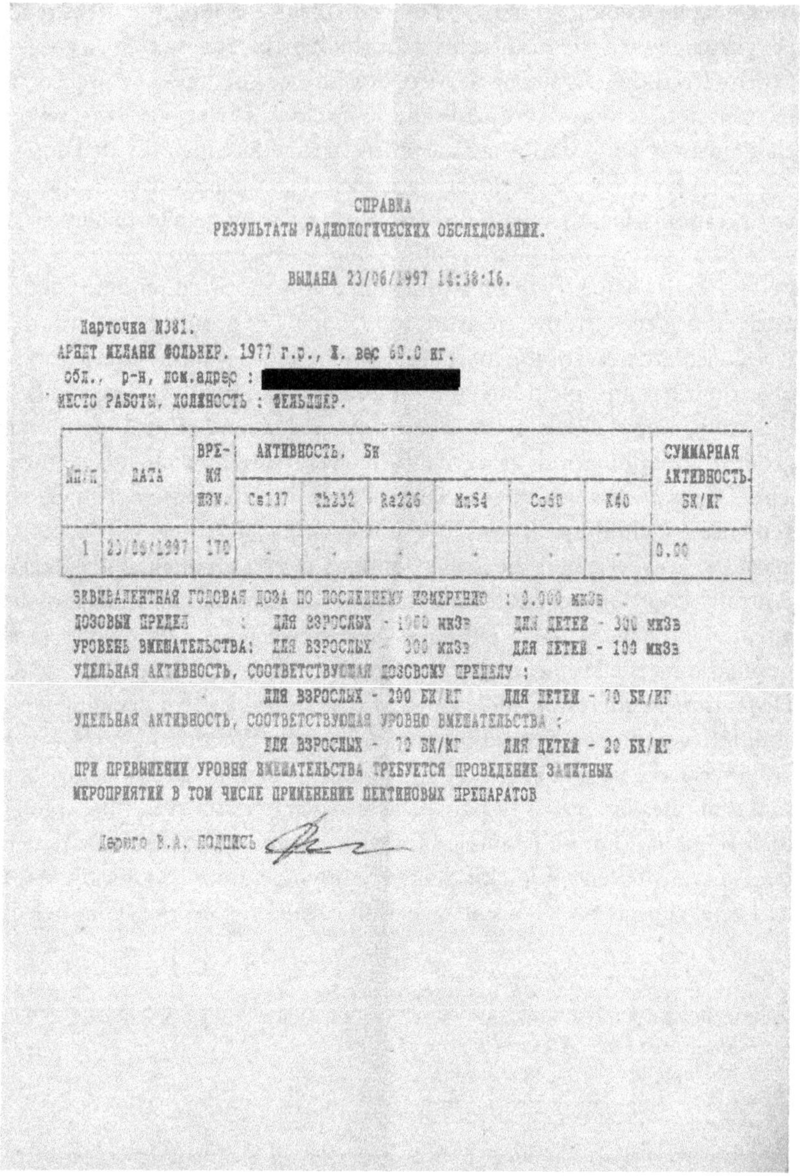

FIGURE 3.2 Whole-body radiation measurement results for the author, Minsk, 1996; elevated radiation levels were not found. The measurement was carried out at the independent Belrad Radiation Institute. The results sheet contains information on the assessment of the measured radiation and recommendations for action. These were missing for measurements in state facilities.

to interpret the results and, in most cases, they were not given any support to help them make sense of them. The upshot was that the tests simply became even more anxiety-inducing than they were already. Parents had begun to see being placed on the *dispanserizatsiia* list or getting a *putevka* as a sign that a child's health was at risk. If the radiation situation was nothing to worry about, a work collective in Kyiv was wondering as early as May 1986, why was teaching time being reduced for some children? The collective was also suspicious because other children were meant to go to school as usual.[80] The fact that some children were being sent for recuperation while others were not resulted in a growing number of direct and indirect questions about the scale of the disaster in letters to the Central Committee of the Communist Party of Ukraine.[81] The obvious deterioration in the health of many children only added to the general public's mistrust of what the state was telling them.

From the very beginning, some voices were calling for foreign assistance to be enlisted. They even reinforced their demands by threatening to turn to international organizations. In July 1986, a group of women who had been temporarily evacuated from Narouvlia with their small children and sent to the north-west of the BSSR made a representation to the chair of the Committee of Soviet Women, warning that they would turn to the IAEA if their own government proved unable to help them and their children.[82]

3.2 'IN THE NAME OF OUR CHILDREN': PUBLIC CONCERN, DISCONTENT, AND MOBILIZATION

From the moment news of the 'accident' (*avariia*) reached the Soviet public, concerned citizens were writing to state and Party institutions with offers of assistance. The effort to send children away was one reason why the idea of a containable accident gave way to that of a full-scale disaster in the minds of the general public. People were involved in the trips in some way, heard rumours about them, tried to interpret the few official statements about them. The offers of help initially conformed to the usual expectations about what kind of things could be said in the Soviet system. They did not even begin to call the Soviet handling of the

[80] Liakhov, 'Informatsiia', 12 May 1986, in Baranovs'ka (ed.), *Chornobyl'ska trahediia*, 126–8.

[81] Various documents in Baranovs'ka (ed.), *Chornobyl'ska trahediia*, 125.

[82] Dalhouski, *Tschernobyl in Belarus*, 95.

disaster into question, let alone criticize the political system. Specific mention was made of children. Correspondents who wanted to help often cited suffering they themselves had been through or witnessed as a reason – experiences during the Second World War were a common point of reference here. People made donations or expressed a readiness to take in affected individuals so that they would not have to suffer in the same way.[83] Independent actions of this nature were, however, not politically desirable in the early years after the accident: they might have suggested that the state was not capable of solving the problems itself.[84] Not until four years later did the Party reconsider its stance and turn to civil society with an injunction to help, as an 'act of genuine compassion and shared humanity with people who have met with great suffering through no fault of their own'.[85]

The context changed fundamentally on 8 February 1989, when the BSSR daily *Zviazda* finally broke step with the policy of secrecy regarding the true scale of the disaster and published radiation maps for the first time.[86] It was followed a day later by *Sovetskaia Belorussiia*, the highest-circulation BSSR daily.[87] Ukraine followed suit on 1 March, when its largest daily, *Radianskaia Ukraina*, likewise published radiation maps for the first time.[88] Almost three weeks later, on 20 March, just in time for the elections to the Congress of People's Deputies, more detailed maps of radioactive contamination were printed in the Soviet daily, *Pravda*.[89] It is true that all these maps were still of a very primitive nature and gave only a rough indication of how badly contaminated individual regions were. Nonetheless, the very fact that they were published at all was astonishing – and it released irreversible forces that played a crucial role in the collapse of the Soviet Union. As has been noted elsewhere,[90] the publication of the maps initiated an information boom and, in effect, dismantled the Soviet

[83] Cf., e.g., Shevchenko, 'Informatsiia', 2 June 1986, in Baranovs'ka (ed), *Chornobyl'ska trahediia*, 222–3, and further sources in Baranovs'ka (ed.), *Chornobyl'ska trahediia*; Dalhouski, *Tschernobyl in Belarus*, 88.

[84] HA/ACPSS, f. 89, op. 53, d. 5.

[85] NARB, f. 4p, op. 156, d. 757, ll. 110–3.

[86] The BSSR Council of Ministers had already decided (4 July 1988) to make information about the radiation situation public in the press and on the radio. See ibid., 86; Sahm, *Transformation im Schatten von Tschernobyl*, 212.

[87] *Sovetskaia Belorussiia* was the mouthpiece of the Central Committee of the Communist Party of the BSSR and the BSSR Council of Ministers and Supreme Soviet.

[88] Ibid.

[89] I. Izrael, 'Chernobyl': Proshloe i prognoz na budushchee', *Pravda*, 20 March 1989, 4.

[90] See: Marples, *Belarus*; Sahm, *Transformation im Schatten von Tschernobyl*.

utopia for good.[91] Even Chernobyl, once a state secret, had now fallen to glasnost – an enormous boost to the opposition movement. Getting at the 'truth' about the disaster became a key driving force in mobilizing wide swathes of the population. It is hard to think of a better example of Yurchak's widely quoted dictum that 'everything was forever, until it was no more'.[92]

On the day that *Pravda* published its maps, Gorbachev had set off on a visit to Ukraine that would include the accident site and Slavutych, Prypiat's successor town. He was accompanied by the KGB and members of the press. Nikolai Golushko, head of the Ukrainian KGB, seemed pleased when he later noted that the journalists had seen the Soviet leader's appearance as marking the 'beginning of the second stage of perestroika' because the emphasis had been on practical questions rather than emotions.[93] At the same time, his notes also reflect the ineffectiveness of the government's limited efforts to quell the growing sense of discontent. Workers at the Chernobyl plant reacted with indignation to a television programme involving the two most important figures when it came to radiation safety in the Soviet Union: Boris Shcherbina (a deputy chairman of the Council of Ministers) and Evgenii Velikhov (director of the Kurchatov Institute in Moscow). The workers felt that the two men had represented the interests of the nuclear industry and had no time for anyone with different views. The criticism was telling: Shcherbina, who headed the first Union-level Chernobyl commission, was at this time also chairman of the State Committee for the Utilization of Nuclear Energy's (Goskomatom) special censorship commission and had thus been closely involved in the policy of secrecy.[94] The appearance of Iurii Izrael, chairman of the State Committee on Hydrometeorology (Goskomgidromet), was also a source of particular resentment. Together with Leonid Ilin, he had been responsible for almost all the key measures to minimize radiation exposure after Chernobyl, and had just released the *Pravda* radiation maps. His remarks were now being dismissed in Slavutych as 'total disinformation'. The plant workers felt let down because he had talked only about gamma radiation and not the consequences of other kinds of radiation. His assurance that there was no radiation in the forest did not

[91] I am alluding here to Scott Shane's *Dismantling Utopia*, which unfortunately does not consider Chernobyl. Scott Shane, *Dismantling Utopia: How Information Ended the Soviet Union* (Chicago: Ivan R. Dee, 1994).

[92] Yurchak, *Everything Was Forever*.

[93] SBU, f. 16, spr. 1122, ark. 68–73, here 70.

[94] Roisko, *Gralshüter eines untergehenden Systems*, 255.

stand up to critical scrutiny in their eyes.[95] For now, such criticism was all state representatives had to worry about, but that would change: a year later, the Ukrainian environmental organization Zeleny Svit declared Shcherbina, Izrael, and even, for a time, Gorbachev personae non gratae in Kyiv.[96]

Once *Pravda* had published its radiation maps, it was to be a further two months before the news blackout on Chernobyl-related topics was completely lifted. During the run-up to the 26 March 1989 elections (the first in Soviet history where there was a choice between candidates), opposition movements drew attention to the fact that information had been kept secret for three years. It was only on 24 May, a day before the first sitting of the Congress of People's Deputies, however, that the Soviet governmental commission decided to release almost all available data about the disaster.[97] The newly elected representatives were able to discuss the accident in public – the first time that this had been possible.

3.2.1 The Congress of People's Deputies and the First Public Reckoning with Chernobyl

All across the land, the atmosphere was tense and people were on the edge of their seats while the first Congress of People's Deputies was in session.[98] Millions followed the debates, which were broadcast live on television (a first in the Soviet Union). The congress amounted to a reckoning with the shortcomings of the Soviet system. Most speakers, though, wanted the problems to be resolved within the context of that system. The events of 9 April 1989, when twenty people died during the bloody dispersal of demonstrators in Tbilisi, were the main point of contention. Environmental issues – above all, nuclear energy and hydroelectric power – joined the recent massacre in Georgia as being among the most hotly debated topics at the congress.

Gorbachev described the Chernobyl accident as an 'elemental disaster' (*stichiinoe bedstvie*).[99] Efrem Sokolov, first secretary of the Belorussian

[95] SBU, f. 16, spr. 1122, ark. 68–73, here 72.

[96] Sahm, *Transformation im Schatten von Tschernobyl*, 218.

[97] Astrid Sahm, *Die weissrussische Nationalbewegung nach der Katastrophe von Tschernobyl: 1986–1991* (Münster: Lit, 1994), 87.

[98] On the Congress of People's Deputies, see Helmut Altrichter, *Russland 1989: Der Untergang des sowjetischen Imperiums* (Munich: Beck, 2009), 123–204.

[99] Verkhovnii Sovet SSSR, *Pervyi s"ezd narodnykh deputatov SSSR: 25 maia–9 iiunia 1989g. Stenograficheskii otchet*, 6 vols. (Moscow: 1989), 33 (vol. III), 439 (vol. I).

Communist Party, criticized the Soviet evacuation policy: people should not, he insisted, be living in places where locally grown food was too heavily contaminated with radiation to be consumed.[100] Borys Oliinyk, secretary of the Ukrainian Union of Writers, called both for an end to nuclear power and for a national programme to safeguard children.[101] It was also argued that this programme should facilitate the establishment of sanatoriums and similar facilities for mothers with children from the affected areas.[102] Z. N. Tkacheva, a Belorussian paediatrician from Slauharad in the Mahileu oblast, also wanted the number of recuperation places available to be increased. She presented deputies with medical data and observations from the affected areas: there were increasing numbers of children with congenital cataracts, vision issues, weak immune systems, anaemia, and heart attacks. Experts from outside, she complained, came up with all manner of explanations for these ailments – 'nitrates, poor nutrition, inadequate breastfeeding' – just not the 'radiation factor'.[103] She had had enough of the ignorance of Soviet specialists and the empty promises of medicine, equipment, personnel, and 'clean' food. Instead, she proposed asking foreign experts for help. Chernobyl was very much an 'elemental catastrophe' in her eyes: 'We live in a world where clean earth, water, air, the forests and meadows have been taken from us. Without them, humans don't live; they just exist.'[104]

Albert Likhanov, chairman of the Lenin Children's Fund, spoke about its Chernobyl children programme and the precarious state of Soviet provisions for children's well-being. He pointed out that the Soviet Union took a 'shameful lead' (*pozornoe liderstvo*) when it came to child mortality in developed countries.[105] Likhanov also mentioned the high number of children who were dying of cancer (6,000 every year). He predicted that this number was only going to get worse if the number of beds available for young cancer patients was not increased from 800 to 3,000 as a matter of urgency.[106] Likhanov also appealed for donations and wanted civil society to take action to support the children of the Soviet Union. He reminded his audience of the anonymous children 'chosen for their happiness' who, to

[100] Ibid., 476 (vol. I).
[101] Verkhovnii Sovet SSSR, *Pervyi s"ezd narodnykh deputatov SSSR*, 469–72 (vol. VI).
[102] Ibid., 38 (vol. II).
[103] Ibid., 314.
[104] Verkhovnii Sovet SSSR, *Pervyi s"ezd narodnykh deputatov SSSR*, 314 (vol. II).
[105] Ibid., 144. On child mortality: Zatravkin, Vishlenkova, 'Kluby' i 'getto' sovetskogo zdravookhraneniia, 218–21.
[106] Verkhovnii Sovet SSSR, *Pervyi s"ezd narodnykh deputatov SSSR* (vol. II), 143.

'touching applause', ran toward the leaders of the nation clutching bouquets of flowers at the annual rallies outside the Lenin Mausoleum on Red Square in Moscow. It would be better, he suggested, if 'children marked by suffering' were to take their place. Rather than handing out *konfety* (sweets), he proposed, deputies might ask the children for their addresses and send them a card every six months.[107]

After this first session of the Congress of People's Deputies, social mobilization began to gather increasing momentum. The main towns of the worst affected areas became focal points of mobilization and protest. On 1 June 1989 – Children's Day – more than 1,000 people gathered in the BSSR town of Cherykau (Ru. Cherikov) in the Mahileu oblast to call for 'children to be protected'. The demonstrators wanted social justice: they felt that the Chernobyl accident had deprived them of a full life and that, unlike the victims of other disasters, they had not even been compensated for this.[108]

This need to compete with those affected by other disasters was a by-product of the fact that such events were being handled with growing transparency. The support that the victims of the 1988 Armenian earthquake found at home and abroad became a central point of reference for critics of the (lack of) assistance that was forthcoming after Chernobyl. Television and radio were quick to broadcast in-depth coverage of the earthquake, leading some Chernobyl victims to feel that they had been passed over.[109] To make matters worse, survivors of the earthquake received aid from the West – a first of its kind. It was against this background that the residents of Cherykau demanded not only recompense in the form of financial payments and a reduced retirement age, but also recuperation trips of three months for children in pre-school and school education, as well as for mothers with children up to the age of three.[110]

Almost as soon as the disaster became known, some in the BSSR felt that they were not being treated in the same way as their counterparts in Ukraine. This impression was not always factually correct. The

---

[107] Ibid., 142.

[108] 'Protokol', 1 May 1989, in Adamushko et al. (eds.), *Chernobyl'*, 248–50, here 248.

[109] On the Armenian earthquake, see Katja Doose, *Tektonik der Perestroika: Das Erdbeben und die Neuordnung Armeniens, 1985–1998* (Vienna: Böhlau 2019); Doose, 'Green Nationalism? The Transformation of Environmentalism in Soviet Armenia, 1969–1991', *Ab Imperio* 1 (2019), 181–205; Nigel A. Raab, *All Shook Up: The Shifting Soviet Response to Catastrophes, 1917–1991* (Montreal: McGill-Queen's University Press, 2017).

[110] 'Protokol' in Adamushko et al. (eds.), *Chernobyl'*, 249.

anonymous author of a June 1986 representation, for instance, complained that the Ukrainian government was evacuating those at risk whereas the BSSR government was not doing anything.[111] This was simply not true – evacuations began at roughly the same time in the two republics.

3.2.1.1 *Growing Dissent*

The first strikes in the BSSR took place in the town of Narouvlia almost a month after the first session of the Congress of People's Deputies ended in June 1989. Aliaksandr Dalhouski has rightly characterized these strikes as a 'milestone in the history of the republic'.[112] These developments were hardly noticed further afield because the mass media did not report on them. Nonetheless, the consequences on a regional level cannot be overstated: the strikes were followed by numerous similar protests in the south of the republic, where the radioactive fallout had been at its worst.[113]

Events began with work collectives in Narouvlia. Early in the summer of 1989, the collectives decided to strike in order to express their discontent at the ongoing failure, after almost three months, to meet demands they had made at a rally on the third anniversary of the disaster and subsequently presented to various representatives of the Party and state apparatus. That rally – demanding 'health and a future for the children of Narouvlia' – had been prompted by a newspaper report on 'concerning deteriorations in health', particularly among the region's children.[114] The workers in Narouvlia, though, wanted *putevki* not just for children but for entire families.[115] The strike was also directed at an international audience. An IAEA delegation had visited contaminated areas on 22 June, but Narouvlia had not been part of its itinerary. The workers, therefore, wanted the IAEA and the WHO to inspect their region too.[116]

As was the case elsewhere, numerous women were involved in Narouvlia, as can be seen from a photograph of an unapproved demonstration held in September 1989 to protest at the lack of consultation on the response to the legacy of Chernobyl. The front rows are packed with women. Children's health was – once again – among the demonstrators'

---

[111] Dalhouski, *Tschernobyl in Belarus*, 94.
[112] Ibid., 161.
[113] Sahm, *Die weissrussische Nationalbewegung*, 94.
[114] Quoted in Dalhouski, *Tschernobyl in Belarus*, 159.
[115] 'Rezoliutsiia', 26 April 1989, in Adamushko et al. (eds.), *Chernobyl'*, 238–41, here 240; Baluev, 'Informatsiia', 22 June 1989, ibid., 272–3.
[116] Baluev, 'Informatsiia', 22 June 1989, ibid., 272.

concerns. The protestors also included representatives of the Belarusian Popular Front (Belaruski Narodny Front, BNF), which had been founded three months previously. The BNF played a crucial role in making Chernobyl a political issue.[117] Only a month later, state representatives took part in another rally (this time, an approved one) – alongside 3,000 other demonstrators. They even went so far as to endorse the resolution that was drawn up at the event – a far cry from September, when political demands had still been viewed with trepidation.[118]

The results of a survey by a regional BSSR newspaper called *Maiak Palessia* demonstrate the level of feeling in the population that accompanied the lifting of censorship on Chernobyl and the first public protests concerning the accident. In June 1989, the newspaper wanted to find out what residents of the Brahin (Ru. Bragin) raion (almost 7,000 people) thought about the Soviet handling of the disaster. It transpired that 90 per cent of those questioned were not satisfied and felt that the town was a dangerous place to live because of elevated levels of radiation exposure there.[119] People – on the streets and in letters of protest – became increasingly outspoken with the fourth anniversary of the accident in April 1990. The correspondence included forceful articulations of the connection between childhood and the experience of nature. 'Our children have been robbed of the chance to discover the natural world of their homeland: the forest, the river, the earth, everything around us, without which childhood and a fulfilling life are impossible', as one letter put it. Its authors were saddened that their children had become 'hostage[s] of this most dreadful environmental catastrophe'.[120]

3.2.2 The Role of the Trade Unions

From the outset, sending Chernobyl children away for recuperation had the backing of a substantial part of the population. The children benefited above all from the support of the BSSR's trade unions, which were heavily involved in funding Chernobyl recuperation and rehabilitation trips for children and adults alike. The unions also acted as conduits for concern and discontent, and thereby exerted political pressure. As a result, state organs rightly considered them a barometer of the underlying mood in

---

[117] See Section 4.2.1.
[118] Cf. the cover photograph of Dalhouski, *Tschernobyl in Belarus*; see also ibid., 161–3.
[119] Ibid., 142.
[120] 'Obrashchenie', n. d., in Adamushko et al. (eds.), *Chernobyl'*, 331–4, here 331.

society. The trade unions did more than just criticize the actions of the state (or the lack thereof): they also called on those concerned to act proactively on their own initiative. Work collectives in enterprises, for example, were enjoined to make their holiday facilities available to the Chernobyl children without bureaucratic hurdles.[121]

The shortfall in accommodation capacity was a crucial factor in allowing the strikes that had begun in Narouvlia in June 1989 to spread throughout the south of the BSSR in 1990. In March that year, work collectives at two production facilities in contaminated locations in Homel sent Ryzhkov a telegram setting out their demands 'in the name of our children and grandchildren'. They felt that the state had a duty to do something, and called for the cultivation of food for human and animal consumption on contaminated land to be prohibited, for the public to be supplied with 'clean food' and dosimeters for personal use, and for all the region's children to be able to spend the summer holidays in 'clean' locations.[122]

On 6 and 7 July 1990, the southern Belorussian trade unions organized a 'march for survival' (*marsh za vyzhivanie*). The route took the marchers on foot from Homel in the BSSR to Briansk in Russia, and then on to Moscow, where they drew attention to the plight of the *Chernobyltsy* while the Twenty-Eighth Party Congress was under way.[123] The picture was one that displayed striking similarities with global forms of protest. Demonstrators in 'alternative' fashion sported T-shirts professionally printed with 'Chernobyl: our pain' (*Chernobyl' nasha bol'*); many of them had headbands. They expressed their views with banners that, in their graphic design and outspokenness, were entirely different from the usual placards at orchestrated Soviet demonstrations. One read: 'The Belorussians: a people without a future'[124] – a total antithesis to the bright socialist future that Soviet rallies had been propagating only a few months previously. The partnership with local media was also new: a radio and television team from Homel accompanied the marchers all the way to Moscow.[125]

The protest brought results. Although Ryzhkov refused to invite the marchers to the Kremlin, a government commission appeared in Homel

---

[121] GARF, f. R5446, op. 162, d. 1968, ll. 33–6.
[122] Ibid.
[123] The BNF and various women's organizations were also involved in organizing the march. See Dalhouski, *Tschernobyl in Belarus*, 171.
[124] Cf. the photographs in ibid., 171.
[125] See ibid., 170–2.

two weeks later. Together with trade union representatives, it drew up a long list of measures in which the protestors at least managed to get their socio-economic demands recognized.[126] The central points on the list included medical examinations for all children in the contaminated area by 1 September 1991, and more rehabilitation opportunities for children, young people, and pregnant women. Enabling chronically ill children to visit sanatoriums and similar facilities was to be prioritized; the Soviet Council of Ministers was to make facilities used by the workers of state agencies available for this purpose.[127]

3.3 CHERNOBYL POLICY BETWEEN CENTRE AND PERIPHERY DURING KATASTROIKA

Once the radiation maps had been published, a return to censorship was impossible. The public were becoming more and more vocal in their insistence on being given the facts; they were also making clear that isolated measures would no longer be enough to allay their concerns. It was in the context of this atmosphere of discontent that the BSSR leadership decided in March 1989 to draw up a 1990–5 Chernobyl programme for the republic. The draft was presented to the republic's Supreme Soviet for approval as soon as July – a short time by Soviet decision-making standards. It was also released for public discussion in *Sovetskaia Belorussiia* – an entirely unprecedented move.[128] The '35 rem concept' attracted particularly heavy criticism and became the main point of contention between representatives of the Union-level bureaucracy and the republic's scientists.[129] The BSSR Supreme Soviet eventually passed an

---

[126] Doguzhiev et al., 'Protokol', 25 July 1990, in Adamushko et al. (eds.), *Chernobyl'*, 350–61.

[127] Ibid., 352–3.

[128] 'Postanovlenie', 29 July 1989, ibid., 291–5.

[129] The '35 rem concept' was the brainchild of the Soviet Ministry of Health. It was based on the assumption that a lifetime dose (set at seventy years) of up to 35 rem (350 mSv) would not have an adverse effect on health; the Soviet Standards of Radiation Safety from 1976 had similarly envisaged an annual dose of up to 0.5 rem (5 mSv) for people living in close proximity to nuclear power plants. Members of the Academies of Sciences in Ukraine and the BSSR were always critical of the 35 rem maximum, which they felt was far too high. They argued that even low levels of radiation, in particular, in combination with other harmful environmental factors, are damaging to health. They were of the view that humans should absorb no more than 7 rem (70 mSv) in the course of seventy years, and that the inhabitants of areas where an annual dose of 0.1 mSv (0.01 rem) might be exceeded should be resettled. By way of comparison, occupational exposure to radiation in the United States is limited to 50 mSv (5 rem) per year. See Sahm, *Transformation im Schatten von Tschernobyl*, 200–35, 408–13.

extensively revised version of the programme on 26 October 1989.[130] It would be going too far to see the BSSR as acting independently here, given that its leaders ran the various drafts past Union-level organs. The initiative, however, had passed to the republic nonetheless. The text that was ultimately signed off retained the '35 rem concept', but also included a compromise regarding the right to resettlement and extended the provisions for recuperation.[131] The approval of the Chernobyl programme was followed by a new wave of evacuations in the early 1990s, as a result of which capacity issues reared their head once more. Entire schools (children and teachers) were pre-evacuated, as it were, by being sent to facilities outside the 'zone' until families were able to relocate, at which point the children joined them again.[132]

The Ukrainian authorities drew up a Chernobyl programme for their republic in summer 1989 as well; it too covered the years 1990–5, but also looked ahead as far as 2000. The document was meant to be discussed in the Ukrainian Supreme Soviet in October 1989, but this did not come to pass; other challenges were clearly considered more pressing. The Supreme Soviet did not engage intensively with the draft programme and the legacy of Chernobyl again until February 1990, in the context of wider debate on the environment. The Ukrainian programme was largely similar to its Belarussian forerunner, but there were some respects in which it differed from the provisions in the neighbouring republic to the north. For instance, there was a right to resettlement as soon as ground contamination with caesium-137 reached levels of 10 Ci/m^2 or above, or whenever there was a 'risk to life'. Criteria for establishing where such a risk was present were not, however, laid out in the programme; in practice, decisions continued to be made on an ad hoc basis instead. The Ukrainian SSR clearly adopted this vagueness in order not to pre-empt decisions that might be made by Union-level organs.[133]

3.3.1 The Unified Union Programme: A Late Attempt at (Re)centralization

Approval by the Supreme Soviets of the two republics did not mean that the Ukrainian and Belorussian programmes came into effect there and

[130] 'Postanovlenie "O Gosudarstvennoi programme po likvidatsii v Respublike Belarus' posledstvii avarii na ChAES na 1990–1995 gody"', No. 2916-XI. 26 October 1989, *SZ BSSR*, 1989, 31, 313.
[131] Ibid.
[132] GARF, f. R5446, op. 162, d. 1968, ll. 33–6.
[133] Sahm, *Transformation im Schatten von Tschernobyl*, 226.

then. The republics' plans were examined by no fewer than six expert groups under Gosplan, the Soviet State Planning Committee, before they were ultimately drawn together in a unified Union programme. The costs were the main object of scrutiny. The BSSR government accused the Soviet leadership of wanting to use the programme to fund general infrastructural measures, and threatened to suspend payment of its contributions if Moscow failed to fund the whole package set out in the programme. Loyalty to Moscow decreased even more quickly when funds failed to arrive in summer 1991. In Ukraine, on the other hand, there was suspicion from the outset that payments from the Union were inequitable because the Russian and Belorussian programmes were receiving markedly more funding. The impression of unfair treatment gave a further impetus to calls for Ukrainian independence.[134]

The unified Union programme was approved by the Supreme Soviet of the Soviet Union on the eve of the fourth anniversary of the accident. It ushered in a paradigm shift. Until this point, it had been standard practice in Union-level documents to refer to an 'accident' (*avariia*); the unified Chernobyl programme now spoke of the 'worst disaster of the modern age, a source of suffering that has afflicted the entire nation and has implications for the fate of millions of people living over a vast area'.[135] The programme, furthermore, determined that the measures taken so far were largely inadequate, resulting in what it described as an 'extremely critical social and political situation' in the affected areas. The programme put this down to contradictory scientific recommendations on radiation safety, delays in implementing the necessary measures, and, ultimately, a loss of trust in local and central authorities among some of the general public.[136] This amounted to an admission that public pressure was the only reason that a programme had been drawn up in the first place.

This new Chernobyl programme envisaged seeking support and practical assistance on an international level. Elected representatives and international organizations all over the world were to be contacted proactively and enlisted in the drive to alleviate the damage that this 'truly global disaster' had caused. The Soviet Council of Ministers was to work together with 'public organizations' (*obshchestvennye organizatsii*) in

[134] Ibid., 222, footnote 425.
[135] A. Luk'ianov, 'Postanovlenie "Ob edinoi programme po likvidatsii posledstvii avarii na Chernobyl'skoi AES i situatsii, sviazannoi s etoi avarii"', 1452–1, 25 April 1990, in Valerii Stanislavovich Levonevskii, *Zakonodatel'stvo*, http://pravo.levonevsky.org/baz a/soviet/sssr0933.htm (2 May 2024).
[136] Ibid.

laying the necessary groundwork for international cooperation and coordinating foreign assistance.[137]

The programme initially meant that, in spring 1990, the interests of the republics could still be reconciled with those of Moscow. But this concord was not to last for long. Once the newly elected Supreme Soviets in the republics got down to business and set up their own Chernobyl committees, they often diverged from the Union-level norms. The social protection laws passed by Ukraine and Belarus in February and November 1991, respectively, were based on radiation safety concepts of their own, for instance.[138] The debates about this legislation reflected the state of competition between the two republics on the one hand and between the centre and the republics on the other. All were striving to regain moral credibility in the eyes of the general public. The result was a labyrinth of rules in which Union law made more generous provisions in some areas and the laws of the republics were more generous in others. Overall, people in Belarus had access to fewer benefits than those in Ukraine in several respects. Nonetheless, the provisions – which included salary supplements and income-tax exemptions – did make a difference to families in the worst-affected areas during this time of economic crisis. The prospect of becoming ineligible for such assistance after resettlement could be a crucial factor in deciding whether to relocate or not, not least because those living in the 'zone' were also entitled to material benefits such as free meals for children and students at public educational institutions, the right to a free recuperation trip of one or two months every year, and heavily subsidized (or even free) medicine.[139] The cornerstones of state Chernobyl policy – resettlement and social protection – were weighted differently in Ukraine and Belarus. Up to 1995, Belarus invested primarily in resettlement and housing for people who had (or had been) resettled. In Ukraine, on the other hand, the emphasis was always on social protection. Around a fifth of the Belarusian population was entitled to a range of benefits under the republic's Chernobyl social protection legislation; the equivalent figure in Ukraine was only 6 per cent.[140]

A new development in Chernobyl policy in the early 1990s was the idea of involving the general public in various aspects of radiation safety. Dosimeters were gradually being distributed over a wide area, which

[137] Ibid.
[138] See Sahm, *Transformation im Schatten von Tschernobyl*, 228, 412–13.
[139] Ibid., 233, 247–50.
[140] Ibid., 234–50.

meant that people were now in a position to make their own risk assessments – for what they were worth – and decide for themselves whether they wanted to leave a particular area. Having one's own radiation monitor, however, was by no means a guarantee of staying healthy. The devices could be used only under certain conditions, and there was a lack of instruction and training in their use. They provided little relief from the fundamental insecurity caused by conflicting expert views on the risks of radiation exposure, and were entirely indifferent to emotional attachment to one's home and (perceived or real) social duties and obligations. *Rodina* – a concept akin to the German *Heimat*, which involves the sense of being inseparably bound to the earth of one's dead ancestors – was a recurring topos in people's explanations for why they could not bring themselves to abandon contaminated areas.[141] In many cases, older family members in particular felt unable to face the emotional and social consequences of resettlement. Whole families sometimes chose to stay with them because of this. Some even went back because they felt they owed it to those who had not wanted (or been able) to relocate.[142] A sense of social responsibility, the pressure of traditional expectations, and a reluctance to face up to the risks associated with living on contaminated ground also limited the options that many people felt were open to them.

3.3.1.1 *A Siberian Disneyland: The Lenin Children's Fund*

Pravda's official call for the plight of children in the Soviet Union to receive more attention, including the establishment of a fund devoted to them, was followed by the foundation of the Lenin Children's Fund in October 1987, with Albert Likhanov as its chairman. Its successor, the Russian Children's Fund, makes much of having been the first independent aid organization for children in the late Soviet Union. This self-image, however, is not quite in line with reality. It is true that the fund was the first of its kind in the Soviet Union, but as a 'public organization' it clearly stood – and continues to do so today – in the tradition of Soviet mass organizations, whose activities were closely intertwined with the organs of Party and state in staffing, structural, and financial terms. Only later did true non-governmental organizations that broke with the norm of combining social activism and political indoctrination emerge. The influence

---

[141] On this, see also Arndt, 'Babushkas of Chernobyl'.
[142] Cf., e.g., the interview with Olga Lipskaia and Nikolai Lipski in Melanie Arndt, Margarethe Steinhausen (eds.), *'Wir mussten völlig neu anfangen': Opfer der Tschernobylkatastrophe berichten* (Bielefeld: Luther-Verlag, 2011), 26.

of the Party and state apparatus on the creation of the fund is obvious: it had been approved by the Central Committee of the Communist Party of the Soviet Union, which was also in all likelihood behind the move. Even so, the fund did try to reach out to the general public, encouraging them to help set its priorities. Likhanov was keen to point out that the fund and state complemented each other: 'The efforts of the fund, and that means of the people, should not replace the state; instead, they should support, should help the state . . . After all, how can we, the people, not support and help, not least where childhood is involved?'[143]

The arguments used to support the establishment of the fund stand out for their open criticism of the socialist maxims of the scientific-technological revolution. Rather than worshipping progress in the old style, the founding members criticized industry, holding it responsible for the alarming environmental conditions that, in their eyes, explained the omnipresent risks to health reflected in rising child mortality rates. Shalva Amonashvili, a Georgian psychologist and the pioneering figure in Soviet humane pedagogy, condemned scientific and technological progress for coming between children and their parents and dehumanizing teachers. The artist Rolan Bykov felt that creating a children's fund amounted to a 'revolutionary new beginning'.[144] The heavily industrialized Kuznetsk Basin in western Siberia seemed to epitomize everything that had gone wrong. As chairman, Likhanov proposed using the fund's resources to build a 'children's town' modelled on Disneyland 'somewhere in Siberia'.[145] The idea was positively received, particularly in the Kuznetsk Basin. One miners' brigade leader thought it should be chosen because the children were particularly deserving of it.[146]

The very idea of building a 'Disneyland' indicates that the West was constantly present in people's minds – and that it no longer functioned as a negative point of comparison but as a positive point of reference. It was clear from the presentations at the fund's inaugural congress that the ideal of a 'happy Soviet childhood' had all but fallen apart at this point. The event gave voice to a criticism of modernity that was not dissimilar to the discussions that social scientists and psychologists had been having in Western

[143] Al'bert Likhanov, 'Obernut'sia k detstvu'.
[144] 'Sovetskii detskii fond imeni V.I. Lenina sozdan!', *Smena* 23, no. 1453 (Dekabr') (1987), 1–6, http://smena-online.ru/stories/sovetskii-detskii-fond-imeni-vi-lenina-sozdan (1 May 2024), here 1.
[145] Al'bert Likhanov, 'Obernut'sia k detstvu'.
[146] 'Sovetskii detskii fond imeni V.I. Lenina sozdan!', 1.

Europe and the United States since the mid-1970s.[147] Soviet experts were, in a sense, now joining them in predicting a 'disappearance of childhood'.[148]

The Lenin Children's Fund focused on children's well-being, attaching particular importance to children in need – above all, orphans and children with disabilities. This was something new in the Soviet approach to child welfare, and was clearly to be understood as a criticism of the practice of shutting even children with minor disabilities away in homes. Institutions were now enjoined to support children's individual needs instead. The fund's founding members envisaged a wide range of ways in which it could work toward this goal. They wanted to build accommodation where children with physical disabilities could go for holidays, a centre for children with intellectual disabilities, and specialist clinics for paediatric oncology, cardiology, and ophthalmology. These facilities were to be established not only in major towns and cities, but also on the neglected periphery. Some of these plans became a reality, too, even if this did not always happen at once. The fund also awarded scholarships to artistically gifted children and set up a network of children's video libraries and a book bank, as well as publishing two magazines: *Deti* (*Children*) and *My i raduga* (*The Rainbow and Us*).[149]

Concurrently with the fund's head office in Moscow, sections were established in the regions of the RSFSR and the capitals of the other Soviet republics. In the case of the BSSR, the Belorussian Children's Fund was chaired from its inception by the author Vladimir Lipskii. He 'described his organization as a "child of *perestroika*". While it received some funding from Moscow, it depended on local and international support ... As an "official" state agency, it was also free to relate directly to international agencies throughout the world with a minimum of restrictions', even in the late 1980s.[150] Likhanov in Moscow and Lipskii in Minsk both had a knack for lobbying and promoting their cause; they proved to be particularly successful outside the Soviet Union.[151] Initially, however, they sought to generate resources at home. As the funding

[147] Cf. Hartmut von Hentig's foreword to Philippe Ariès, *Geschichte der Kindheit* (Munich: Hanser 1975), 38; similarly, Dieter Spanhel, Stefanos Hotamanidis (eds.), *Die Zukunft der Kindheit: Die Verantwortung der Erwachsenen für das Kind in einer unheilen Welt* (Weinheim: Deutscher Studien-Verlag, 1988); Hans-Günter Rolff, Peter Zimmermann, *Kindheit im Wandel: Eine Einführung in die Sozialisation im Kindesalter* (Weinheim: Beltz 1985). See also Chapter 1 in this book.
[148] Neil Postman, *The Disappearance of Childhood* (New York: Delacorte, 1982).
[149] Al'bert Likhanov, 'Obernut'sia k detstvu', 6.
[150] Carter, Christensen, *Children of Chernobyl*, 16.
[151] See Chapter 4.

situation became increasingly tight, Lipskii persuaded the BSSR Council of Ministers and the USSR Ministry of Atomic Energy and Industry (Ministerstvo atomnoi energetiki i promyshlennosti, or Minatomenergoprom) to encourage donations to the Belorussian Children's Fund.[152]

The Belorussian Children's Fund launched a Chernobyl children programme of its own in 1988, in the context of which it arranged medical treatment and recuperation, including periods of time away from home, for children. That same year, the BSSR Ministry of Health set up a 'radiation protection centre' for children with the fund's help; the centre is said to have been equipped with medical equipment worth US$300,000.[153]

3.3.2 The State Chernobyl Children Programme

The first – and last – Union-level Chernobyl children programme was drawn up with input from the Lenin Children's Fund at the end of 1990. Its starting point was the unified Union programme for 'liquidating' the consequences of the disaster, which identified medical care, access to 'clean' food, and recuperation provisions for people from contaminated areas, 'particularly the children', as inadequate. The text also called for a 'special "Chernobyl Children"' programme to be developed by September 1990 in order to prevent or minimize any 'impact of unwelcome aspects of the Chernobyl disaster on the generation that is currently growing up, taking all the specific socio-economic and medical aspects of this challenge into account'.[154] Although this was framed as a matter of urgency, it was eight months before the Soviet government signed off the Chernobyl children programme for 1991–5.[155]

In the meantime, in July 1990, the Twenty-Eighth Congress of the Communist Party of the Soviet Union had decided to make 500 million roubles of Party funds available 'for implementing recuperation measures for those parts of the population that live on contaminated land'.[156] The

[152] 'O posledstviiakh avarii na Chernobyl'skoi AES v 1986 g. dlia Belorusskoi SSR i ikh preodolenii', 5 June 1990, in Adamushko et al. (eds.), *Chernobyl'*, 339–46, here 346.
[153] Rossiiskii Detskii Fond, *Deti Chernobylia*', www.detfond.org/about/stranicy-istorii/det i-Chernobylia/ (6 November 2019).
[154] 'Postanovlenie "Ob edinoi programme"'.
[155] GARF, f. 8009, op. 51, d. 5138, ll. 127–8.
[156] XXVIII s"ezd KPSS, 'Obsuzhdenie i priniatie rezoliutsii XXVIII s"ezda KPSS "O politicheskoi otsenke katastrofy na Chernobyl'skoi AES i khoda rabot po likvidatsii ee

commission behind the resolution presented at the congress was clearly not unanimous regarding the amount of funding required. Some of its members returned to these differences of opinion at the congress. Grigorii E. Omelchenko, who taught at the police academy in Kyiv, called for a billion roubles instead. Only then, he said, would 'all children living in zones they shouldn't be living in, children living in zones where the situation is critical and other contaminated zones', be able to go for recuperation that year. He also pointed out that, although mothers were being allowed to take time off work, they were having to pick up the costs, as a result of which many were unable to afford recuperation with their children. 'These are our children',[157] Omelchenko implored his comrades. But he was unable to get his way.[158]

Half of the Party funds released – 250 million roubles – went to the BSSR. The relevant Central Committee group of the Belorussian Communist Party proposed using 172 million roubles (almost 70 per cent of the total) to build recuperation facilities and sanatoriums, purchase medical and computer technology, and develop modern radiation-monitoring equipment. Its rationale was that the money had to be used effectively – but also in such a way that it created a favourable impression at home and abroad. The priorities were set not just according to what was best for the children but also – even at this late stage – with an eye to 'the great significance of this political effort for the Party's authority'.[159] The remaining roubles were to be divided between Homel, Mahileu, Hrodna (Ru. Grodno), Brest, Minsk, and Vitsebsk (Ru. Vitebsk) – the republic's six oblasts, all of which were now classed as affected – where they were to be used to purchase medication and baby food.[160]

Some very senior government officials considered the resolution passed at the Twenty-Eighth Party Congress as one of the few cases of real progress in the handling of the disaster. Aleksandr Vlasov, head of the Central Committee's Economic and Social Policy Department and former minister of internal affairs, was among them. He was thus all the more concerned when he complained in February 1991 about how long it had taken to finalize the Chernobyl children programme and the ongoing

posledstvii"' in Institut teorii i istorii sotsialisma TsK KPSS (ed.), *XXVIII s''ezd Kommunisticheskoi partii Sovetskogo Soiuza: 2–13 iiulia 1990 goda*; Stenograficheskii otchet (Moscow: Izdatel'stvo politicheskoi literatury, 1991), 585–98, 586.

[157] Ibid., 594.
[158] Ibid., 596.
[159] NARB, f. 4p, op. 156, d. 889, l. 4.
[160] Ibid.; NARB, f. 4p, op. 156, d. 889, l. 14.

delays in its implementation.[161] The details of the Union-wide Chernobyl children programme had been worked out not only by the Lenin Children's Fund but, above all, by Union-level institutions and ministries: the Ministry of Health, the Academy of Medical Sciences, the Academy of Sciences, the State Committee for People's Education, and the State Committee for Labour and Social Questions. The Councils of Ministers were the only bodies from the three most affected republics to be involved in the planning process.[162]

Crucial aspects of the scientific expertise behind the programme took shape in close collaboration with foreign partners, above all from the United States. Soviet scientists consulted publications by their foreign counterparts on debacles such as the Love Canal toxic waste scandal in the late 1970s and the Three Mile Island accident in 1979. They even carried out field studies together with US specialists. The Russian sociologist Vladimir Lupandin visited contaminated areas with a group of Americans, including the nuclear physicist and molecular biologist John Gofman, to undertake research there. Gofman was well known in the United States for his insistence that the effects of low-level radiation on humans should not be underestimated. The group's main sceptic when it came to nuclear power, though, was Judith Johnsrud, a geographer and anti-nuclear activist. The findings of the trip had a direct influence on the drafting of the Chernobyl children programme. According to its introduction:

The Chernobyl accident is the worst nuclear disaster ever; it has affected over four million people, a quarter of them children. The accident has fundamentally changed the already precarious environmental situation in a number of raions in Russia, Ukraine, and Belorussia, and posed a real risk of deterioration in children's health, given that they are the most vulnerable to poor environmental conditions.[163]

Although targeted measures between 1986 and 1989 had made it possible to reduce radiation doses for the general public by more than half, the introduction noted, there were still 'worrying trends in the behaviour of certain demographic markers and in children's health'. The birth rate and actual population growth had decreased, it reported, and illnesses among children and complications during pregnancy and childbirth were being observed more often. The 'question of children's health now and in the

---

[161] HA/ACPSS, f. 89, op. 23, d. 21, l. 9.
[162] GARF, f. 8009, op. 51, d. 5138, ll. 127–8.
[163] HA, Francis U. Macy, Box 16.

years to come' was thus identified as particularly pressing in the introduction to the programme.[164]

The programme acknowledged medical consequences that had previously been denied, as well as glaring gaps in knowledge. The need for the proposals was even framed in these terms: practical measures to safeguard children, the programme said, should be developed and implemented without delay because the effects of low-level radiation on them were not fully understood. The move to develop a specific programme for children, separate from the general programme for dealing with the disaster, was derived from the 'overriding right of children to life and health irrespective of environmental disasters and socioecological situations that may arise in society'.[165]

The programme included an extensive medical research section, listing nine research fields in which the effect of ionizing radiation – not simply alone but in combination with other influences (regionally specific factors, psychoemotional stress) – on the developmental cycle of children from the foetus to the teenage years was to be examined. The priority, though, was 'to achieve means of diagnosis, prophylaxis, and cure'. The results of the research were not to be of scientific interest alone, but to have 'both a medical and a social impact' that lent itself to practical applications.[166] The effect of long-term low-level radiation exposure on molecules, cells, and organs was another focal point.[167]

Research findings were to be entered into various databases (e.g. for clinical, dosimetric, and pathomorphological information) so that it would be easier to make use of them. The programme also envisaged setting up general registries of cancer cases and birth defects, as well as specific registries for the groups of children being studied, including the children of 'liquidators'. Up to this point, the psychological effects of the disaster had generally been dismissed as 'radiophobia', but they too were acknowledged in the research envisaged in the programme. It included, for instance, analysis of children's responses to the chronic psychoemotional stress caused by the limitations that had been imposed on their lives.

---

[164] The archival source is a draft of the programme, but it can be assumed that it was also passed in this form. GARF, f. R8009, op. 51, d. 5138, ll. 129–96.

[165] Ibid., ll. 130–1.

[166] 'Spravka', 8 April 1991, in Adamushko et al. (eds.), *Chernobyl'*, 374–8, here 377.

[167] The programme document noted an increase in the following illnesses: malignancies such as neuroblastoma or leukaemia, bone marrow disorders such as myelodysplastic syndrome, secondary immunodeficiency disorders, congenital abnormalities and their complications, and intellectual disabilities and chromosomal changes.

The programme made clear that international standards were to be followed on all levels of healthcare provision. It also responded to calls for more specialists to be on hand in contaminated areas and for cadres and medical personnel to be better qualified in general. Rehabilitation and recuperation for mothers and children who had been adversely affected by the legacy of the accident formed an integral part of the medical programme. Annual research reports were to be made publicly accessible in order to put an end to the confusion of the preceding years where access to and provision of information was concerned.[168]

These were ambitious goals. It never proved possible to realize them in their entirety. The foundations for the Chernobyl children programme were laid by the unified Union programme, which was approved in April 1990. Seismic shifts took place in the internal and international political situation during the months that followed, and the political and social climate in the Soviet Union changed profoundly. The RSFSR declared sovereignty in June 1990; Ukraine and the BSSR followed suit in July. The Twenty-Eighth Congress of the Communist Party of the Soviet Union that same month, which was also its last, was in effect a reckoning with a passing system. All of this had significant implications for the drafting and implementation of the Chernobyl children programme.

The simple truth is that, when the programme finally got going early in 1991, it had already been overtaken by events. The republics were already pursuing their own Chernobyl policies with some degree of independence; this was, for obvious reasons, not conducive to the successful implementation of a centrally organized programme.[169] In mid-February, the USSR Ministry of Health still did not know what funds it would have to put the programme into practice. Aleksandr Baranov, a paediatrician and first deputy minister of health, tried to do something about the situation. He expressed his displeasure to the State Commission for Emergency Situations, criticizing the behaviour of organs in the republics which were clearly starting to organize medical measures without involving the centre. Baranov urged the commission to discuss practicalities with all

---

[168] GARF, f. R8009, op. 51, d. 5138, ll. 129–96.

[169] Strictly speaking, this was the 'Union–Republican Programme for Immediate Measures to Liquidate the Consequences of the Accident at the Chernobyl Nuclear Power Station for 1990–1992' ('Gosudarstvennaia soiuzno-respublikanskaia programma neotlozhnykh mer na 1990–1992') – a subprogramme that was approved with the Union programme. GARF, f. 8009, op. 51, d. 5138. ll. 205– 6.

relevant representatives of the republics as soon as possible so that the programme could be put into practice.[170]

In February 1991, the USSR Ministry of Health decided that it and the Ministries of Health in the Soviet republics should set up expert panels, and quickly: the ministers of health in the republics were given five days to get the panels in place. Moscow also instructed the ministers of health in the republics to draw up concrete plans for implementing the Chernobyl children programme.[171] The Union-level expert panel, chaired by Baranov, coordinated activities in the republics, as well as setting work and funding priorities; it also assessed the relevance of research findings and decided how international collaboration should figure in the context of the programme.[172] The panel, which was composed of twenty-nine members, was dominated by representatives of Union-level bodies and medicine; there was only one representative of a 'public organization': Robert Tilles, the first deputy chairman of Chernobyl Help (Chernobyl' pomoshch).[173] The Lenin Children's Fund was represented by its medical advisor, who also held a powerful position in the Soviet Ministry of Health.[174]

But it was all too late. There was little the expert panel could do at this point. Even putting the research programme into action in the institutions of the republics turned out to be difficult. Hardly any of the laboratory equipment needed for experiments was available, and the funding situation for the future was uncertain. What money that was available was not even enough to cover the most important needs: comprehensive check-ups and improvements to specialist endocrinological and oncological facilities.[175] The expert panel found itself facing an impossible task. Representatives of the republics complained of receiving anticipated funding from Moscow late, if at all.[176] They, in turn, withheld their contributions to the central budget. By the end of 1991, neither the BSSR nor the RSFSR had paid their contributions for that year (2.08 million and 3.07 million roubles, respectively).[177]

---

[170] GARF, f. 8009, op. 51, d. 5138, ll. 205–6.
[171] GARF, f. 8009, op. 51, d. 5138, ll. 207–10, here ll. 207–8.
[172] Ibid., 208–9.
[173] Thus the name in the source. The organization being referred to is probably a liquidators' association called the Soius Chernobyl'tsev.
[174] Ibid., 209–10.
[175] GARF, f. R8009, op. 51, d. 5138, l. 250.
[176] 'Spravka', 8 April 1991, in Adamushko et al. (eds.), *Chernobyl'*, 377.
[177] GARF, f. 8009, op. 51, d. 5138, l. 286; GARF, f. 8009, op. 51, d. 5138, ll. 288–9, here 288.

The problems came to a definitive head with the Ukrainian and Belarusian declarations of independence in August 1991.[178] Two months later, in November, Larisa Shchepliagina, deputy chair of the Union-level expert panel, retroactively informed the Ministries of Health in the Russian, Ukrainian, and Belarusian republics that Moscow would only cover the costs of their research institutes up to September 1991; the republics, which were now formally independent, would have to find the means to fund the facilities themselves from then on.[179] This move put the main Belarusian research institute, the Institute of Radiation Medicine in Minsk, in a particularly challenging position because the BSSR government had already removed funding for the institute from the state budget in April.[180] It was now facing a completely uncertain financial situation, despite being supposed to play a leading role in implementing the Chernobyl children programme.

Thus, in Belarus it was unclear both how the planned research was going to be made possible and how the rights that the Chernobyl children now had in writing were going to be met. The official line was that there was sufficient capacity in the system to meet recuperation needs; there were, however, not enough of the better facilities, specifically sanatoriums. In 1991, only 41 per cent of demand among children aged 7–14 could be met. This was still better than the situation for older children, though: for 15- to 17-year-olds, the figure was as low as 14 per cent; their accommodation was provided by the Union Ministry of Health. The trips were also being affected by practical problems. In many cases, *putevki* did not arrive until two or three days before the departure date. It was difficult to distribute them in time; parents and children hardly had any time to get ready. Construction of the Pioneer camps, sanatoriums, and similar facilities for children that had been planned did not commence in 1991;[181] in 1992, the Belarusian government then dropped the idea of building new recuperation facilities, entirely for financial reasons; it refurbished and modernized the existing accommodation instead. A 1993 report by the Belarusian Chernobyl committee emphasized that it would not be possible to make up for the shortfall in accommodation in this manner. The

[178] The RSFSR was the only member of the Soviet Union that did not make a declaration of independence.

[179] Ibid.; GARF, f. 8009, op. 51, d. 5138, l. 286.

[180] 'Spravka', 8 April 1991, in Adamushko et al. (eds.), *Chernobyl'*, 377.

[181] NARB, f. 4p, op. 156, d. 757, ll. 107–8.

Belarusian government had reckoned with needing 300 million roubles for 1992; only a third of this sum had actually been found.[182]

CONCLUSION

The steps taken to support the Chernobyl children in the immediate aftermath of the disaster and the early years of the response (1986–91) were the result of decisions made by Moscow and its institutions – but also by Party and state officials further away from the centre: in Minsk and Kyiv, and in the regional centres of the areas worst hit by the radioactive fallout. Almost all those involved acknowledged that one group – the children – were the most vulnerable and most in need of protection, even if this did not always result in the necessary steps actually being taken to safeguard them. This period was characterized by improvisation, and by sometimes contradictory courses of action that were embarked on before the Party had resolved what to do, or that even subverted its decisions. Particularly on a regional level, on occasions bureaucratic structures displayed a remarkable degree of flexibility during the confusion that immediately followed the disaster. There was often no alternative, for the simple reason that the system did not have solutions with which to bring order to this kind of chaos.

Faced with these uncertainties about who was – and who was suitably qualified to be – responsible for what, as well as a lack of resources and fading loyalties, the central leadership turned to typically Soviet solutions in the form of bureaucracy and secrecy. It was therefore not long before it had largely succeeded in closing its own window of opportunity. Trade unions and non-state actors were starting to question the old system and its ideas. In the republics, a variety of forces, often directly linked to Chernobyl protest movements, were seeking national sovereignty, making the situation even more volatile.

Unilateral actions such as the evacuation of children from Kyiv early in May 1986 remained the exception rather than the norm. Nonetheless, they indicate that, even before people took to the streets to demand independence, leaders in the republics knew that they had some scope to act independently of Moscow. They made full use of this when it started to become clear that the Communist Party and the Soviet state were losing their legitimacy. The publication of the first radiation maps put the ball firmly in the court of the leaders and Party organs in the republics: it was

---

[182] NARB, f. 507, op. 1, d. 32, ll. 68–71.

now on them to show that they could do a better job than the Soviet centre in Moscow.

Children always figured prominently in damage limitation efforts in the aftermath of Chernobyl. Medical and administrative workers, for instance, made a point of collecting separate data for those children they considered 'affected individuals' (*postradavshie*). Yet there was a lack of tangible long-term plans that were designed around the specific needs of the Chernobyl children rather than simply with research purposes in mind. The Soviet leadership introduced (often makeshift) measures to provide protection from radioactive dust and organized permanent evacuations. Their preferred tool, though, was sending children away on a temporary basis – a concept that was subsequently to be widely adopted and adapted, as well as transformed into a global symbol.[183]

It was only late in 1990 that the Soviet Union made an attempt to systematically address the children's plight on a relatively long-term basis. These efforts on the part of the state – in the guise of its Chernobyl children programme – were genuinely ambitious, but they soon encountered financial and structural hurdles. As a result, many of the planned measures never became a reality. The programme came too late – not just for the children and their families, but also for the political ends it was meant to pursue. The Union-level organs never regained their legitimacy, and it never proved possible to bring back the old political order. The worst-affected areas saw the emergence of increasing awareness and processes of mobilization in the general public that spread and grew into protest movements of unprecedented dimensions. These processes were irreversible. The Soviet leadership and the governments of the republics tried to stop such developments from taking hold more widely by prohibiting coverage of them in the mass media. That may still have worked in the case of the Chernobyl demonstrations in Minsk late in 1989, but the strategy did not succeed in the worst-affected areas. With its merciless exposure of the weak points in the system, the disaster became a social catalyst.[184] Change was inevitable. The catastrophic environmental situation in the Soviet Union was a central theme at the Congress of People's

---

[183] The concept of the global in this book is admittedly a very limited one; with the exception of Cuba and India, it primarily refers to the Global North. This does, however, reflect the fact that it was primarily the nations of the highly industrialized North and Japan that were in a position to – and did – participate in efforts to help the Chernobyl children.

[184] On this concept, see, e.g., García-Acosta, 'Historical Disaster Research', 50; Annelies Heijmans, 'From Vulnerability to Empowerment' in Bankoff et al. (eds.), *Mapping Vulnerability*, 115–27; Thomas A. Birkland, 'Natural Disasters as Focusing Events:

Deputies in late May/early June 1989. For the first time, elected representatives and the wider public were able to engage critically with the consequences of the explosion, which was now being called a 'disaster'. The
situation of those affected only grew more challenging with independence:
not just because of a lack of funding everywhere, but also because of the
collapse of production chains for medicines and medical instruments that
were managed on the Soviet level.

Chernobyl children and their families were largely left to face everyday
life and their insecurities on their own during this period. Hundreds of
thousands of children were sent away so that they could escape from the
radiation for a time, at least. The extent to which their physical and
psychological needs were addressed during these trips, however,
depended on the commitment of the people looking after them – and
how qualified they were to do so. Hardly anyone had been properly
trained or prepared for the task. It is, therefore, hardly surprising that
concern for the children lay at the centre of the protest movements. The
majority of the population were no longer prepared to 'liquidate' the
disaster in Soviet style and sacrifice the children in the process. From the
very beginning, they looked abroad for support, be it as a means of lending
clout to their demands or as an additional source of legitimacy.

The Soviet Union was, quite simply, structurally incapable of dealing
with disasters. Its response to the Armenian earthquake clearly demonstrated this again two years after Chernobyl.[185] At the same time,
Chernobyl was more than just a typically Soviet catastrophe: the
response displays various features that are typical of disaster situations
more generally – be it the separation of parents and children,[186] information blockades, or poor decision-making grounded in ignorance,
fear, and complacency.

Policy Communities and Political Response', *International Journal of Mass Emergencies
and Disasters* 14, no. 2 (1996), 221–43.

[185] See Doose, *Tektonik der Perestroika*; Raab, *All Shook Up*.
[186] See Lucy Bonnerjea, 'Disasters, Family Tracing and Children's Rights: Some Questions
about the Best Interests of Separated Children', *Disasters* 18, no. 3 (1994), 277–83.

4

The Chernobyl Children as 'Children of the Whole Planet'

At a press conference in Moscow shortly after the fifth anniversary of the disaster, the sociologist Vladimir Lupandin read out a statement from a group of US–Soviet researchers and activists.[1] They had made a ten-day journey through Ukraine and Belarus, including the 30 km exclusion zone, in order to see the legacy of the accident for themselves. The press conference reflected how the dynamics of power had changed in the late Soviet Union: it was organized by the country's largest independent environmental organization, the Socio-Ecological Union of the USSR (Sotsialno-ekologicheskii Soiuz SSSR), together with the Soviet Sociological Association (Sovetskii sociologicheskii Soiuz) and the periodical *Vrach* (*The Doctor*).

The group was led by Francis U. Macy – an expert on Russia and a facilitator of international exchange – and also included the ecophilosopher and deep-ecology activist Joanna Macy,[2] the geographer and anti-nuclear activist Judith Johnsrud, and Adolf Kharash, a pioneering figure in Soviet humanistic psychology. Lupandin informed the representatives of the press that the group had concluded that Chernobyl was still a 'bleeding wound'. Far from getting smaller, he said, they had found that the effects of the accident were taking on dimensions of increasing

---

[1] The quotation in the chapter title is from HA, Francis U. Macy, Box 4, F. 'Chernobyl Trip Reports '91'.

[2] Joanna Macy, interviewed by Melanie Arndt, 2 December 2013. Berkeley, CA. For her 'Despair and Empowerment Work' and 'Work that Reconnects', cf. Macy, *Despair and Personal Power in the Nuclear Age* (Philadelphia: New Society Publishers, 1993); Macy, Molly Young Brown, *Coming Back to Life: Practices to Reconnect Our Lives, Our World* (Philadelphia: New Society Publishers, 1999).

concern.³ Lupandin also said that children's illnesses were among the most serious consequences: thyroid and blood cancer, as well as non-specific symptoms such as constant headaches or persistent exhaustion.

Lupandin explained to his Soviet audience how the Americans had not previously been aware of the psychological problems that had appeared in the wake of the disaster. 'People', he said, 'are tormented by the know-ledge that they are helpless. Many told us that they had a bad conscience, that they were distressed at not being able to protect their children. People told us: "Our children will ask: 'Why didn't you protect us?'"'⁴ Having witnessed this suffering, the group was heavily critical of Soviet actions following the disaster; they condemned the mass gatherings that went ahead despite the accident, such as the compulsory May rallies in which a particularly large number of children had taken part: 'This was a crime against the children!'⁵ Only 'together' – which was obviously a reference to the former adversaries of the Cold War – Lupandin said, would it be possible to tackle the legacy of these mistakes: 'These children are not just your children – they are also our children, the children of the whole planet!', the group explained.⁶

This chapter examines how the Soviet Chernobyl children became 'children of the whole planet' – how children of the state became children of a civil society that was growing in strength and entering into trans-national networks. The chapter is concerned with the individuals and organizations involved, and with the challenges that they found them-selves facing in the final stages of the Cold War and the social context of the collapsing Soviet Union. What arguments did they use to justify taking responsibility for aspects of people's well-being that had previously been the concern of the state? The chapter connects local perspectives with processes unfolding on regional, Soviet, and international and trans-national levels. The growing interconnectedness also influenced the pro-duction of knowledge, both lay and expert. The findings of the US–Soviet group trip, for instance, not only served to increase public understanding but also fed into academic studies. In political terms, the most important of these were authored by Lupandin and examined sociological aspects of the disaster, focusing on children in the process; the state engaged him to

---

³ HA, Francis U. Macy, Box 4, F. 'Chernobyl Trip Reports '91'.
⁴ Ibid.
⁵ HA, Francis U. Macy, Box 4, F. 'Chernobyl Trip Reports '91'; HA, Francis U. Macy, Box 1, F. '1986 Carl Rogers in USSR'.
⁶ HA, Francis U. Macy, Box 1, F. '1986 Carl Rogers in USSR'.

do this work to help provide the knowledge base for its Chernobyl children programme.[7] Kharash, meanwhile, later published a textbook chapter on dealing with situations of extreme crisis.[8]

4.1 SEEKING HELP FROM ABROAD: ENEMIES NO MORE

As shown by the examples in Chapter 3, initiatives not associated with the state, and individuals acting on their own accord, were already calling for foreign assistance to be accepted in 1989. The BSSR government and Ukrainian leadership, though, did not turn to the world community until February and April 1990, respectively.[9] The Ukrainian and Belarusian declarations of sovereignty in summer 1990 made international cooperation easier, and both republics were able to draw on the several decades of experience that they had gained on the international stage as members of the United Nations.[10] Having a permanent presence in New York, Geneva, and Paris also meant that they were able to draw on existing contacts in their search for help in dealing with the aftermath of Chernobyl.

Shortly before Christmas 1990, during the Forty-Fifth General Assembly of the United Nations in New York, 121 states introduced a Chernobyl resolution that recognized the accident as a disaster of global proportions. The resolution identified safeguarding children as particularly important, and underlined the need for national and international action to support the youngest victims of the disaster. It also called on the United Nations to coordinate efforts to mitigate the effects of the disaster and draw up a programme to that end.[11] The basic problem with all such Chernobyl programmes on a national or international scale presented itself once again here: the ambitious plans were based on unrealistic

---

[7] HA, Francis U. Macy, Box 16, 1991. On the state Chernobyl children programme, see Section 3.3.2.

[8] Adol'f. Kharash, 'Gumanitarnaia ekspertiza v ekstremal'nykh situatsiiakh: Ideologiia, metodologiia, protsedura' in A. I. Dontsov, I. M. Zhukov (eds.), *Vvedenie v prakticheskuiu sotsial'nuiu psikhologiiu: Uchebnoe posobie dlia vysshikh uchebnykh zavedenii* (Moscow: Nauka, 1994), 60–88.

[9] See Sahm, *Transformation im Schatten von Tschernobyl*, 347.

[10] The two republics had been allowed to join the United Nations in an accommodation with the Soviet Union, which was consequently represented by three seats.

[11] UN General Assembly, *International Co-operation to Address and Mitigate the Consequences of the Accident at the Chernobyl Nuclear Power Plant*, 45/190, 21 December 1990, https://digitallibrary.un.org/record/105595/files/A_RES_45_190-EN.pdf (2 May 2024).

funding expectations and therefore destined to fail from the outset. This was not changed by the fact that the General Assembly reaffirmed its support for the programme with a further resolution in 1991.[12]

4.1.1 Off to (Socialist) Paradise

State and non-state organizations started sending Chernobyl children abroad at almost the same time. The Ukrainian Komsomol was one of the first organizations to take the initiative in this regard; it collaborated with the Lenin Children's Fund and several other state institutions, and found a destination that was at once both exotic and ideologically unproblematic: the socialist brother country of Cuba.

In February 1990, Cuba sent three specialists – an oncologist, an endocrinologist, and a haematologist and immunologist – to Ukraine in order to see the children's situation at first hand and choose participants for the first trip. Only a month later, 139 Ukrainian children – as well as about twenty accompanying adults (parents, doctors, functionaries) – found themselves on an aircraft making its way to Havana.[13] Ukraine was unable to fund the trip, so Cuba covered all the costs. The Cubans had lost no time in establishing their own Chernobyl children programme.[14] They stressed that it was not a research programme; its professed aim was solely to provide the children with treatment. Nonetheless, doctors on the island went on to collect incorporated radiation dose data for more than 8,000 children and make it available to the IAEA.[15] Fidel Castro personally welcomed the first cohort at the airport and later visited them at the flagship José Martí Pioneer camp in Tarará, where special preparations had been made for them – an indication of the political significance that the Cuban leadership attached to the operation (Figs 4.1–4.4).[16]

According to their own statistics, the Komsomol and Lenin Children's Fund sent almost 1,500 children from the evacuated zone and other contaminated areas to Cuba between March and December 1990 alone.

---

[12] See Sahm, *Transformation im Schatten von Tschernobyl*, 348.

[13] Petrochenkova, *Ostanovis', mgnovenie!*, 6.

[14] See Feriado-Tour, *Rezul'taty lecheniia detei na Kube*, https://bit.ly/4fqGbvB (2 May 2024).

[15] John M. Kirk, *Healthcare without Borders: Understanding Cuban Medical Internationalism* (Gainesville, FA: University Press of California, 2015), 241–2.

[16] 'Ozdorovlenie "chernobyl'skikh detei": Kubinskii fenomen', *Cuba.com.ua*, www.cuba.com.ua/articulos/cuba-tourist-4.html (2 May 2024).

FIGURE 4.1 Chernobyl children at the beach. Tarará 1990.

This included thirty-eight boys and girls with leukaemia, as well as thirty-one children with other forms of cancer; ninety-five children were treated at specialist hospitals in Havana.[17] A US newspaper

[17] GARF, f. 8009, op. 51, d. 5138, l. 203–4.

FIGURE 4.2 Fidel Castro welcoming Chernobyl children, Havanna 1990.

reported that at least two young patients had died in Cuba, which gives an indication of the generally poor health of the children.[18]

The 1,328 children who did not require hospital treatment recuperated at the José Martí camp, around twenty km from the Cuban capital. It was effectively the Cuban version of Artek, and they spent an average of six weeks there. If needed, they were examined by Cuban doctors, who also provided dental and other prophylactic treatment.[19] The children and their parents later remembered the Cuban doctors as being particularly kind to children – an experience that was very different from the brusque manner in which patients (above all female ones) were generally treated in the Soviet health system.[20]

---

[18] HA, Francis U. Macy, Box 4. Vladimir Lipskii is also reported as mentioning the death of two children; see Carter, Christensen, *Children of Chernobyl*, 18. One contemporary witness, however, only mentions one child that died while being treated in Cuba; Trish Marx, Dorita Beh-Eger, Cindy Karp, *I Heal: The Children of Chernobyl in Cuba* (Minneapolis: Lerner Publications Company 1996), 25.

[19] GARF, f. 8009, op. 51, d. 5138, l. 203–4.

[20] See (for the Cuban doctors' kindness to children), 'Ozdorovlenie "chernobyl'skikh detei"'; Petrochenkova, *Ostanovis', mgnovenie!* and (for the experience awaiting patients from evacuated and/or contaminated areas) Petryna, 'Biological Citizenship'; Evgenija Ivanova, 'Vom Tod zum Leben: Tschernobyl-Politik durch die Gender-Brille' in Arndt (ed.), *Politik und Gesellschaft nach Tschernobyl*, 130–50; Arndt, Steinhausen (eds.), *Opfer der Tschernobylkatastrophe* (with various accounts from contemporaries).

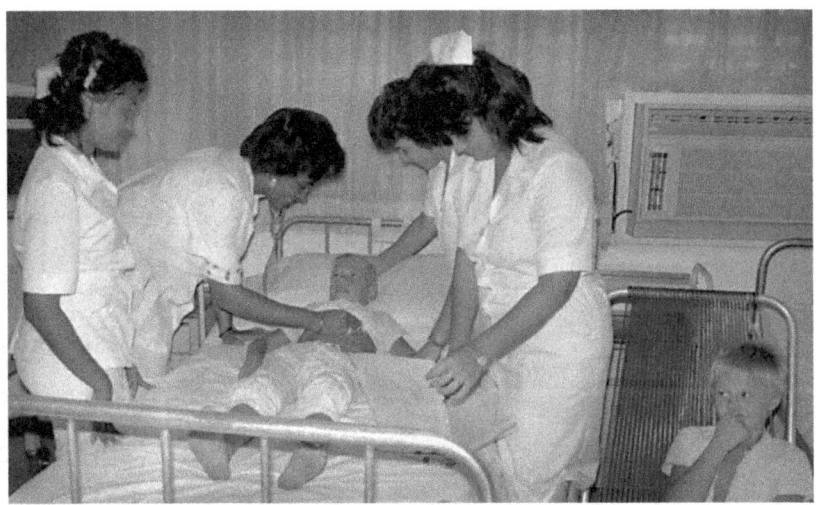

FIGURE 4.3   Cuban doctors and nurses look after a Chernobyl child. Tarará 1990.

Following the visit of the first group, Cuba announced that it was prepared to continue accepting 1,000 children on a regular basis to stay at José Martí. The conversion of the holiday camp into a rehabilitation centre for Chernobyl children was exploited for propaganda purposes: posing as a benevolent father to the nation, Castro made sure the Cuban Pioneers were happy for the camp to be transformed and repurposed, and they duly obliged him. They had previously used the popular complex all year round, but renounced that now out of solidarity with the Soviet Pioneers; more than 500 doctors and medical workers would later look after the visitors from the Soviet Union there.[21] As the Cuban financial contribution was limited to accommodation and medical treatment for the children,[22] the Soviet side had to find money for the flights – a challenge in its own right.

Once it was confident of Castro's backing, the Ukrainian Komsomol planned to send up to 600 Chernobyl children per month, and up to 5,000 per year, to the socialist island. The Komsomol even arranged for the children departing for the Caribbean in autumn and winter 1990 to

[21] Valentina Petrochenkova, 'Nashim detiam v Tarara lushche chem doma', *ZN.ua*, 16 May 1997, https://zn.ua/HEALTH/nashim_detyam_v_tarara__luchshe,_chem_doma.html (2 May 2024).

[22] Nigel Raab notes that the Komsomol 'understood that … it had to offer technical assistance … (cars, buses, computers)' to help the Cubans. Raab, *All Shook Up*, 167–8.

FIGURE 4.4 Fidel Castro thanks Cuban doctors for their commitment to the Chernobyl children. Tarará 1990.

continue their schooling and get Ukrainian food to eat so that it would be easier for them to settle in. The question of how to pay for the journey, however, remained. The Ministry of Civil Aviation wanted the flights paid for in hard currency: 63,000 valuta roubles for each use of a Soviet Iliushin IL-86 widebody.[23] The Komsomol, however, did not hold any hard currency. Requests to the Ukrainian government were unsuccessful; the Central Committee of the Ukrainian Komsomol asked Gorbachev for support instead.[24]

Even though the state authorities could see the appeal of the opportunities offered by rehabilitation in Cuba,[25] they also discussed the potential disadvantages of sending the children there, such as the stress of the long flight and exposure to the tropical sun, which was much stronger than

[23] This was around US$100,000 at the conversion rate up to the end of November 1990 (0.6 roubles for 1 US dollar).

[24] GARF, f. 8009, op. 51, d. 5138, ll. 203–4.

[25] Raab, though, notes that the Chernobyl Union (Soiuz Chernobylia) 'referred to the project as a form of "adventurism" that satisfied the "sensational goals" of individual bureaucrats and took attention away from more meaningful international cooperation'. Raab, *All Shook Up*, 167. Kirk's account paints a different picture. Kirk, *Healthcare without Borders*, 244–5.

what they were used to. Such concerns faded when it became clear that neither Cuban nor Soviet doctors observed any negative effects on the children's health.[26] Even the Ministry of Health in Moscow confirmed that it had not received any reports of negative health impacts resulting from the trips to Cuba.[27]

Once the reservations on the Union level had been dispelled, the Ministry of Health in Moscow made sure the Ministries of Health in the affected republics were on the same page. They supported the recuperation trips to Cuba in principle, as long as two conditions were met: first, only older, 'to all intents and purposes healthy' children between ten and thirteen years of age should go there; second, the warm season should be avoided. The ministries in the republics also told the Moscow ministry that they would be unable to fund, or help fund, the trips. This made planning the second year of trips (1991) considerably more difficult, not least because it was for some time unclear whether the Lenin Children's Fund would lend its support. The future of the state Chernobyl children programme seemed uncertain. It is probably in this context that the Moscow Ministry of Health should be viewed when it said that it would be medically and financially more expedient to establish a network of recuperation facilities at home.[28]

Given the empty coffers, it is no surprise that far fewer than the planned 5,000 children actually made the journey to Cuba every year. A more reasonable estimate would be 1,000 children – the number that Fidel Castro had offered to accept after the arrival of the first group. The website of the Feriado-Tour travel agency, which started organizing trips to Cuba for Ukrainian Chernobyl children in 2012, gives the following breakdown of those who went to Cuba between 1990 and 2011:[29] 17,943 children and 3,776 adults from Ukraine; 2,715 children and 211 adults from Russia; 671 children and fifty-nine adults from Belarus; nine children and two adults from Armenia; and two adults

[26] GARF, f. 8009, op. 51, d. 5138, ll. 203–4.
[27] GARF, f. 8009, op. 51, d. 5138, l. 202.
[28] GARF, f. 8009, op. 51, d. 5138, l. 200.
[29] The numbers vary between sources. One other source reports that 21,000 children spent time recuperating in Cuba in the first thirteen years (to 2003); cf. 'Ukrainskii molodezhnyi Chernobyl'skii Fond: Bozhko Aleksandr Fedorovich', *Who-is-who.ua*, https://who-is-who.ua/main/bookmaket/donetsk/4/96.html (2 May 2024). Another source says that 17,784 children and accompanying adults received medical care in Tarará in the first fifteen years; cf. 'Ukrains'ka hromads'ka blahodiina orhanizatsiia mizhnarodnyi Chornobyl's'kyi fond', *Who-is-who.ua*, https://who-is-who.ua/main/page/kiev/156/116 (2 May 2024).

and two children from Moldova.[30] According to newspaper reports, Cuba had invested around US$350 million in its programme by the time it came to an end in 2011.[31]

The special treatment reserved for the Ukrainians made some of them feel sorry for the Cuban children, whose impoverishment was plain to see. The economic and social hardship on the island meant that everyday life in Tarará could seem paradoxical at times – the visiting children were being given tropical fruit and other food by spotlessly dressed attendants who were unable to provide themselves or their own children with anything like it. In fact, food shortages were affecting parts of the population in the early 1990s – which some visitors from the East took as a sign that the socialist experiment on Cuba had failed.[32]

4.1.1.1 *Cuba's Chernobyl Children in an Independent Ukraine*
The legacy of the Cuban efforts to support the Chernobyl children can still be felt in Ukraine – and not only there (Fig. 4.5) – today. Chernobyl children – now long grown-up – who have been living in Cuba for years and clearly no longer intend to return to Eastern Europe are part of this legacy; one newspaper article came to the conclusion that children who stayed in Cuba for years of treatment ended up hardly being able to remember their Ukrainian homeland at all.[33] Chernobyl children who have returned to Ukraine from Cuba, meanwhile, keep in touch with one another; the Deti Tarara (Tarará Children) association was set up to facilitate this.[34] At their meetings, they look back nostalgically on that

---

[30] No further information about the children's circumstances is given; they are all counted as Chernobyl children. Feriado-Tour, *Kuba – Ukraina: 1990–2011 roki: Vsestoronnaia meditsinskaia programma pomoshchi detiam, postradavshim ot chernobyl'skoi avarii*, https://feriado-tour.com.ua/treatment-in-cuba/childsprograms/chernobil-ruso/ (2 May 2024); cf. also Petrochenkova, 'Nashim detiam'. Similar figures are also to be found in other sources; cf., Author's own records, Belarusian Ministry of Health, 'Dannye po ozdorovleniiu detei za rubezhom v 1990–2009 gg.' 2010; Author's own records, Belarusian Ministry of Education, Department for International Cooperation, 'Svedeniia ob ozdorovlenii detei i podrostkov vsekh regionov respubliki za rubezhom za 1990–2015 gody.' 2016. Kirk's figures are only slightly different. Kirk, *Healthcare without Borders*, 247.

[31] 'Ozdorovlenie "chernobyl'skikh detei"'.

[32] Kirk, *Healthcare without Borders*, 240–1.

[33] Aleksandr Chalenko, 'Nashi na Kube. Domoi ne khotim!', *Segodnia*, 3 September 2009, www.segodnya.ua/lifestyle/fun/nashi-na-kube-domoj-ne-khotim-162124.html (2 May 2024).

[34] Elena Korotkova, 'Kastro potratil $350 mln na lechenie detei iz Chernobylia. Fidel' skazal: 'Pomogat' budem stol'ko, skol'ko budet nuzhno', *Moskovskii komsomolets*, 27 November 2016, https://bit.ly/3Bwf4Bp (2 May 2024).

FIGURE 4.5 Flyer of the exhibition Lost Documents: The Children of Chernobyl in Cuba by photographer Sonia Cunliffe at the Havana National Library in 2017.

'little bit of paradise on earth' and their experiences of recuperation there, the kind-hearted people, and the caring and knowledgeable doctors.[35]

4.1.1.2 *Adding New Destinations*

The visits to the allied island of Cuba got the state's efforts to arrange trips abroad for the Chernobyl children off to a promising start. In June 1991, at the suggestion of the Lenin Children's Fund and the Councils of Ministers of Ukraine, the RSFSR, and the BSSR, the Soviet Council of Ministers decided that recuperation abroad should be extended. The plan was for the state to pay for 10,500 children and accompanying adults to go to other countries, with the assistance of ministries that could arrange transport on the necessary scale. The Council of Ministers attached considerable importance to how the undertaking would be perceived abroad: radio and television were to be used to generate extensive coverage of the humanitarian support the Soviet Union was getting to help its international Chernobyl children recuperate.[36] These hopes were fulfilled: there was considerable public interest, and the arrival of the Chernobyl children caught the attention of the media in all the countries that received them.

4.1.2 Off to the Capitalist Enemy

Half a year after Fidel Castro welcomed Chernobyl children to Cuba in March 1990, the first children set off for the United States. The group that flew to the Soviet Union's erstwhile enemy consisted of eight patients, aged between six and fifteen, from the children's cancer hospital in Kyiv. They were heading for Ashford, Connecticut, and the Hole in the Wall Camp that the actor and philanthropist Paul Newman had created there. Among them was Vova Malofienko. He had been at the Kyiv hospital for several months when it was visited by representatives of the Children of Chornobyl Relief Fund from Short Hills, New Jersey, an aid organization founded by a Ukrainian diaspora couple, Nadia and Zenon Matkiwsky. The group had not actually come with the intention of selecting children for recuperation in the United States – they had intended to see how they might provide material humanitarian help. The hospital and its medical equipment were in a very poor condition, Malofienko reports. Relatives often had to bring food for the children due to problems with the kitchen. Malofienko recalls that this was particularly difficult for his parents because they lived two

[35] Quoted in 'Ozdorovlenie "chernobyl'skikh detei"'.
[36] GARF, f. R8009, op. 51, d. 5138, ll. 240–2.

hours away. They were not allowed to stay with him in the hospital; the six-year-old, who had by then lost his hair, was on his own.[37]

While still at the hospital, the representatives of the Children of Chornobyl Relief Fund decided that special respite in the United States should be arranged for ten children. Vova Malofienko's name was not on the list at first. But, he recalls, his mother implored the doctors so much that they put him on the waiting list. She argued that a lot could change before the trip was due to take place, which was still months away. She was to be proven right: one of the listed children died in the meantime, and the condition of two others deteriorated so much that they were unable to travel. That cloud had a silver lining for Malofienko, who flew to the United States together with a doctor and seven other patients (Fig. 4.6).[38]

The children initially spent a week with the Matkiwskys – Zenon was a surgeon – at their 'massive house with a pool and a tennis court' in Short Hills. Only after that did the children make their way to Connecticut to spend ten days at the Hole in the Wall Camp.[39] The camp, which was designed to resemble scenes from the Old West, had been opened three years earlier to provide respite for seriously ill children – a place where they could be themselves together with others in the same situation, and a chance to have fun without being stared at or kept under constant supervision. Medical treatment was not the primary purpose, but it could be always be provided where necessary.[40]

Newman's food company (which produced salad dressings and sauces, among other things) gave all its profits to good causes; his 'All Profits to Charity' slogan inspired many other celebrities in the United States to become involved in charitable activity as well.[41] The Chernobyl children were among those to benefit from Newman's efforts. He even managed to persuade fellow stars – such as Christopher Reeve, of *Superman* fame, and the Broadway legend Chita Rivera – to join in; Vova Malofienko

---

[37] Malofienko to Arndt, 12 December 2017, 13 December 2017.

[38] Ibid.

[39] Ibid.

[40] See Paul Newman, Aaron E. Hotchner, *Shameless Exploitation in Pursuit of the Common Good* (New York: Nan A. Talese 2003); Carter, Christensen, *Children of Chernobyl*, 18; Hole in the Wall Gang, *Founder & History*, www.holeinthewallgang.org/about/Founder-and-History/ (2 May 2024); Nick Ravo, 'For Victims of Chernobyl, a Respite at Camp', *New York Times*, 10 September 1990, https://bit.ly/4iIStSU (2 May 2024).

[41] Zach Schonbrun, 'Paul Newman Who? Salad Dressing Company Adjusts To Reach Millenials', *New York Times*, 13 November 2016, https://bit.ly/3ZVgApX (2 May 2024); Janet Morrissey, 'Charity That Begins with Spaghetti Sauce', *New York Times*, 2 November 2016, https://bit.ly/3ZYDlt2 (2 May 2024).

FIGURE 4.6 Vova Malofienko with his mother (right) and the doctor Molly Schwenn. Boston *c.*1991.

enthusiastically remembers meeting the two actors.[42] Further Hole in the Wall Camps subsequently appeared in several US states, Europe, and Africa; the one in Ireland was a regular destination for Chernobyl children in the 1990s.[43]

It is possible that the Children of Chornobyl Relief Fund was influenced by the Cuban precedent when it decided to bring Chernobyl children to Newman's camp – the charity had included a detailed Cox News article

[42] Malofienko to Arndt, 12 December 2017, 13 December 2017.
[43] Newman, Hotchner, *Shameless Exploitation.*

about the trips to Cuba in one of its newsletters. The Matkiwskys assumed that this would be a one-off;[44] in fact, Chernobyl children were to go on visiting the United States for many years. Some of them, such as Vova Malofienko, stayed there for good.

The Chernobyl children were integrated into the camp's fund-raising efforts in order to help support the camp and cover their own travel costs. A translator from Ukraine who also worked as a hospital clown, for instance, helped the children of the first cohort prepare a performance of *Little Red Riding Hood*. Two hundred people came to see it; they had paid at least US$500 each for their ticket.[45] The event was not only a financial success but also meant that the Soviet children and those looking after them got to see how initiatives such as the camp were made possible in the United States. In actual fact, some of the activities were very familiar to them, albeit in different contexts: children often danced, sang, and performed plays and poems at schools and other venues in the Soviet Union.

The Ukrainians who helped to look after the children (some came with them, others lived in the United States) were struck by the unique qualities of the camp. 'There are no camps like this in the Soviet Union', enthused the doctor from Lviv who had travelled with the first cohort. The hospital clown observed how important it was that the children were free to enjoy themselves outdoors at the camp: 'The children feel like normal people – boating, fishing, horseback riding. It's not the same in Russia [*sic*].'[46] The distractions clearly worked. Malofienko does remember being tired and suffering from nausea, but he also recalls that 'emotionally it was fun to play outside instead of being locked in a hospital room'. He recalls that the camp team made a real effort to keep the children entertained and take their minds off their illnesses.[47]

Other organizations took in Soviet children suffering from cancer that same summer. The Young Men's Christian Association (YMCA), for instance, invited patients to Camp Catch-a-Rainbow in western Michigan.[48] According to its Russian successor, the Lenin Children's Fund organized trips abroad for around 10,000 Chernobyl children from Belarus, Ukraine, and Russia; roughly 2,000 of them went to Cuba. In 1990 alone, 1,235 Chernobyl children went to seventeen

---

[44] HA, Francis U. Macy, Box 4.
[45] Ibid.
[46] Quoted following ibid.
[47] Malofienko to Arndt, 12 December 2017, 13 December 2017.
[48] Carter, Christensen, *Children of Chernobyl*, 170–2.

countries, some of them as far away as Australia. Another organization, set up by former staff at the Lenin Fund, later took over responsibility for such trips – a further example of how people in state and semi-state institutions could acquire organizational know-how and form networks that made later careers outside the state apparatus possible.[49] In 1992, the – by then independent – Belarusian section of the fund started following the US example: in its 'Raduga Nadezhdy' ('Rainbow of Hope') programme, it arranged for children with cancer to make three-week visits to the elite Zubrenok holiday camp.[50]

The Belorussian section of the Lenin Fund brought representatives from twenty-two countries together in Minsk in summer 1990 to discuss ways of helping Chernobyl children to go abroad. The YMCA, the American Cancer Society, and the Christian Children's Fund (also from the United States) were among the participants. It was hoped that the gathering would foster the development of networks that went beyond simply facilitating trips abroad for children.[51] That is, in fact, exactly what happened: a number of organizations not only invited Chernobyl children to visit their countries but provided humanitarian assistance as well.[52] Looking back, the Belarusian Children's Fund notes that 'this was the moment when – and because of the Belorussian children – the world learned about what the tragedy of Chernobyl really meant for the first time'.[53]

4.1.3 'Charity Is a New Idea to Us'

The Soviet side explored several new methods in its efforts to fund the children's trips, often importing ideas from the West and adapting them to the Soviet context.[54] Tamara Maksimova and Vladimir Maksimov, a prominent husband-and-wife team on the television scene, returned from a visit to the United States with the idea of using a twenty-four-hour show to bring in donations.[55] Soviet state television agreed to the idea and,

[49] Rossiiskii Detskii Fond, *Deti Chernobylia*.
[50] Belorusskii Detskii Fond, *Raduga nadezhdy*, www.bcf.by/ru/programs/rainbow.html (2 May 2024). See also https://bcf.by/portfolio/raduga-nadezhdy/
[51] Carter, Christensen, *Children of Chernobyl*, 68–9.
[52] Ibid., 18.
[53] Belorusskii Detskii Fond, *Istoriia fonda*, www.bcf.by/ru/fund/history/history.html (2 May 2024; webpage inaccessible at time of publication, please also see https://bcf.by/istoriya/).
[54] The title of this section is a quotation from Carter, Christensen, *Children of Chernobyl*, XI.
[55] 'Tol'ko vsem mirom. Televizionnyi marafon', *Argumenty i Fakty*, January 1990, https://annd07.livejournal.com/14506.html (2 May 2024).

in January 1990, Tamara hosted the Soviet Union's first telemarathon (*telemarafon*, as the format was known in order to set it apart from US telethons), with the Lenin Children's Fund as the beneficiary.[56] This first Soviet attempt at televised fund-raising was a runaway success; the fund came away with as much as 100 million roubles.[57]

The telemarathons, which in this form were held only until 1991, took place in close collaboration with state institutions, not least because television itself was still firmly under state control. The Ministry of Communications (Ministerstvo sviazi) supported the technologically ambitious shows, and various state officials took part in the broadcasts. Senior figures in the Orthodox Church, such as the metropolitan of Moscow, lent their support by appearing in the telethons.[58] A total of more than 1,000 people participated in the galas; numerous personalities from the cultural scene were involved, including Soviet pop stars and big names in classical music.[59]

The extent to which the Soviet Union had opened up where foreign influences were concerned can be seen from the Chernobyl Telemarathon, a fund-raising gala that was broadcast on the fourth anniversary of the disaster. Individuals and prominent organizations alike played a leading role in planning the event; they included the Soviet Peace Fund (Sovetskii fond mira; its president was Anatolii Karpov, several times world chess champion), the Soviet trade unions, and the Chernobyl Union (Soiuz Chernobylia). The organizing committee also included the director of the Kurchatov Institute, Evgenii Velikhov; the Belarusian author and Chernobyl activist, Ales Adamovich; and the well-known Russian actress, Liudmila Gurchenko. The project was also international in scope, as can be seen from the involvement of the Californian physician Robert Peter Gale and Yōhei Sasakawa, chairman of the non-profit Nippon Foundation in Japan.[60]

In a widely circulated press release in English, the organizers set out the three main aims of the telemarathon. The first was to increase public

---

[56] Parts of the broadcast can be viewed at Net-film.ru, *Telemarafon Detskogo Fonda (1990)*, https://www.net-film.ru/film-20931/ (2 May 2024). The blog of 'anndo7' contains scans of newspaper reports on the telemarathon: anndo7, 'Telemarafon Detskogo fonda', *LiveJournal*, 7 January 2010, http://anndo7.livejournal.com/14506.html (2 May 2024).

[57] anndo7, 'Telemarafon Detskogo fonda', *LiveJournal*. 7 January 2010, http://anndo7.livejournal.com/14506.html (23 February 2024).

[58] Freedom of religion had been put into law in October 1990.

[59] 'Tol'ko vsem mirom', *Argumenty i Fakty*.

[60] HA, Francis U. Macy, Box 6, F. 'Environmental Trip '90 Program'.

awareness: 'to make the sad lessons of the Chernobyl Nuclear Disaster well known to the peoples all over the world'.[61] Accordingly, the telemarathon included extensive coverage of the scale of the disaster – but it also made reference to events such as the Three Mile Island accident in 1979 and the dropping of the atomic bomb on Hiroshima in 1945.[62] The telemarathon's second aim was to bring to attention the plight of those who lived in contaminated areas and were affected by the latest rules, which either required them to be evacuated or permitted them to relocate if they wished. The fund-raising gala depended above all on footage of 'liquidators' and visibly sick and weak children for its visual impact.[63] The organizers hoped to force a change in the way nuclear power stations were run all over the world, starting with 'public control of Nuclear Power plants by the Citizens of the Planet'.[64] The fact that a call for public and international oversight of Soviet nuclear facilities could be made on state television is an unmistakable reflection of how far glasnost had come.

The third – and, by nature, most important – aim of the telemarathon was to bring in donations. The organizers were seeking solidarity and tangible help. The long list of needs reveals how powerless the Soviet Union was. All contributions were welcome, according to the press release: material as well as financial assistance, moral as well as social support.[65] According to the organizers, there were 'hundreds of thousands of Chernobyl Disaster victims affected by radiation'.[66] This was, in effect, an admission that the Soviet welfare state had failed. The telemarathon appears to have raised more than US$100 million – far more than the first gala of this kind. Some of the takings were earmarked for practically relevant scientific research, 'to raise the operation security of the nuclear power plants all over the world'.[67]

The use of new media formats was characteristic of the late Soviet Union. Regional telemarathons, for instance, also took place in support of various causes, providing an opportunity for state and non-state

[61] Ibid.
[62] Ibid.
[63] Associated Press, 'Protests, Telethon Mark Chernobyl Anniversary', *The Los Angeles Times*, 27 April 1990, www.latimes.com/archives/la-xpm-1990-04-27-mn-183-story .html (2 May 2024).
[64] HA, Francis U. Macy, Box 6, F. 'Environmental Trip '90 Program'.
[65] Ibid.
[66] Ibid.
[67] HA, Francis U. Macy, Box 6, F. 'Environmental Trip '90 Program'.

institutions to explore ways of collaborating.[68] They were, in a sense, the next step on from the 'space bridges' of the 1980s – television talk shows where people in the Soviet Union and the United States were able to interact directly in public, almost in real time, across a satellite link. The awkwardness and stiffness that had marked these broadcasts, above all on the Soviet side, were largely a thing of the past in the telemarathons,[69] which were also genuinely broadcast live (to be on the safe side, Soviet television had broadcast the 'space bridges' with a short time lag so as to be able to cut ideologically problematic content if the need arose).

From 1990, television was actively exploited to help fill the empty coffers of social welfare provision in the collapsing Soviet Union. The new methods were aimed at audiences at home and abroad. On the one hand, the people of the Soviet Union were mobilized; they were shown that glasnost and perestroika had reached Soviet television, and that their fate was now in their own hands. On the other hand, there was an explicit effort to target foreign donors and individuals who would spread the word abroad, as demonstrated not least by the effectiveness of the aforementioned press release, which made it as far as California. A report on the first telemarathon in the weekly *Argumenty i fakty* began by noting that the gala had made it into the *Guinness Book of Records*.[70]

'Charity is a new idea to us and we want to learn from you',[71] Vladimir Lipskii told Michelle Carter, a member of one of the early US groups to visit the Soviet Union with a view to helping the Chernobyl children. Lipskii tried to persuade them 'to write the story of your visits, help us understand why Americans care about the children of Chernobyl, and why you travel so far to help'.[72] Michelle Carter and Michael Christensen took up the suggestion; the result was their book, *Children of Chernobyl*. It was first published in Russian translation in Belarus, but was almost impossible to get hold of there because of the rapidly changing publishing and distribution

---

[68] E.g. the *Chernobyl-Tula* telemarathon of September 1990, which was organized on the initiative of the local Chernobyl association and politicians. GARF, f. 8009, op. 51, d. 4607, l. 85.

[69] See Julia Risch, *Russen und Amis im Gespräch: Die sowjetisch-amerikanische Telebrücke (1982–1989): Ein vergessener Beitrag zur Beendigung des Kalten Krieges* (Berlin: SAXA-Verlag, 2012).

[70] 'Tol'ko vsem mirom', *Argumenty i Fakty*.

[71] Carter, Christensen, *Children of Chernobyl*, XI.

[72] Ibid.

landscape.[73] The English edition, which was published in the United States a year later, reached a much wider readership. Thus, although initially written to inform a Belarusian readership, the book ended up becoming something of a handbook for supporters of the Chernobyl children movement in North America. With its numerous personal stories and the information that accompanied them, the book served US initiatives as a source of both ideas and practical guidance.

While all this was happening, the limits of state efforts to support the Chernobyl children were becoming painfully obvious. In Belarus, the state was only able to arrange 30 per cent of the necessary recuperation places in 1991, even though practically all sanatoriums, holiday accommodation, tourist bases, and Pioneer camps were used (or repurposed) to this end.[74] There simply was not enough money to address the shortfall and establish new recuperation centres for the children in their own country. In the mid-1990s, a total of 945,000 children in Ukraine and 512,000 children in Belarus were officially entitled to an annual recuperation trip. Since 1994, less than half of them had been able to take advantage of this, because the republics were unable to cope with the costs.[75] They found themselves having to look further afield and turn to non-state organizations for help. The general public was increasingly losing faith in state initiatives; for Belarus, David Marples cites an April 1992 survey in which 'only 10 percent of the population had faith in the effectiveness of such aid', whereas '60 percent had confidence in a private charitable trust called Children of Chernobyl'.[76] Indeed, according to the Belarusian government, no less than 82 per cent of the total aid effort for 1993 came from non-governmental organizations.[77]

---

[73] Carter recalls that she first saw a copy of the book in 1995, in the vast Dom Knigi bookstore in Moscow. Michelle Carter, interviewed by Melanie Arndt, 9 April 2016. Belmont, CA.

[74] 'Goskom RB po problemam posledstvii katastrofy na ChAES', 29 October 1991, in Adamushko et al. (eds.), *Chernobyl'*, 415–23, here 422.

[75] Sahm, *Transformation im Schatten von Tschernobyl*, 251.

[76] Marples, in his introduction to Grigorij Medvedev, *No Breathing Room: The Aftermath of Chernobyl* (New York: Basic Books, 1993), 25. Marples does not say which of the many organizations with this name the poll was referring to, but it can be assumed that it was the 'For the Children of Chernobyl' fund run by Gennadii Grushevoi (with whom Marples was in frequent contact), not least because the fund published the results of a similar survey in its *Demos* journal.

[77] Sahm, *Transformation im Schatten von Tschernobyl*, 358.

4.2 NON-STATE ORGANIZATIONS STEP IN AT HOME AND ABROAD

4.2.1 'Detiam Chernobylia' ('For the Children of Chernobyl')

'I give my child this milk, and when it drinks, I turn away in tears', said a young mother, Marina Bortsova, in a recording that was played at a demonstration in Minsk on 25 July 1989.[78] The protest was organized by the Belarusian Popular Front 'Revival' (Belaruski Narodny Front 'Adradzhenne', BNF) – the first demonstration for which official approval had been given – and would enter the Belarussian history as the first environmental demonstration in the capital. Not all that far from the centre of Minsk, around 10,000 Belarusians were packed together in front of the Planeta hotel. The atmosphere was tense. A substantial police presence, including dogs, was poorly concealed nearby. Bortsova could be heard explaining that she had no choice but to give her child milk from her cows, in full knowledge of the fact that they were grazing on contaminated pastures – it was simply not possible to get hold of 'clean' milk in her village.[79]

Bortsova had been recorded by a small group of men from the BNF who investigated the effects of the Chernobyl explosion on the contaminated areas of the BSSR in summer 1989. One of them was Gennadii Grushevoi (Bel. Henadz' Hrushavy), a professor of philosophy in Minsk who chaired the BNF's Chernobyl children committee and later set up the Belarusian Charitable Fund 'For the Children of Chernobyl' (Ru. Belorusskii Blagotvoritelnyi Fond 'Detiam Chernobylia'; Bel. Belaruski Dabrachynny Fond 'Dzetsiam Charnobylia') – the first civil society initiative that actively made helping the Chernobyl children its primary objective.

The fund was one of the first non-state organizations of any kind in the BSSR. The organization spent the first year and a half of its existence operating illegally. It was officially registered under Grushevoi's leadership on 20 November 1990, six weeks after the formation of non-state organizations became possible under the 'public organizations' law that

[78] Irina Grushevaia gives this date for the demonstration, recalling that it took place a day after Gennady Grushevoi's birthday on 24 July. Irina Gruschewaja, Alexander Tamkowitsch, *Der Tschernobyl-Weg: Von der Katastrophe zum Garten der Hoffnung* (Berlin: RMF, 2017), 42; Tamkovich, *Filasofiia dabryni*, 33. Zaprudnik, Sahm, and Dalhouski date the demonstration to 26 July. Cf. Jan Zaprudnik, *Belarus: At a Crossroads in History* (Boulder: Westview Press, 1993), 242; Sahm, *Transformation im Schatten von Tschernobyl*, 223; Dalhouski, *Tschernobyl*, 175.

[79] Quoted in Tamkovich, *Filasofiia dabryni*, 33.

came into force on 9 October. The founding members included bodies such as the Union of Cinematographers (Soiuz kinematografistov) and twelve citizens; Grushevoi's wife, Irina, was among them. The diversity of this list is typical of the alliances involved in the formation of non-state initiatives in late socialism.[80] In the next twenty-two years of its existence, the foundation's programme handled around 600,000 trips abroad for Belarusian Chernobyl children – roughly two-thirds of all the trips made by young Belarusians.[81]

In what follows, the 'For the Children of Chernobyl' fund is used as an example with which to illustrate how the late phase of perestroika saw non-state initiatives beginning to draw attention to the catastrophic environmental situation in the Soviet Union and take over various aspects of the state's role in looking after its citizens – most of all when it came to sending Chernobyl children abroad for treatment or recuperation. This change took place in a context of political and social upheaval where the trail of environmental destruction left behind by the Soviet system was coming to light. Those involved sought to draw international attention to their situation. For large parts of Western European – let alone North American – society, Belarus was a 'blank space on the map of Europe'.[82] Initiatives such as the 'For the Children of Chernobyl' fund played a key role in filling in some of the missing contours.

4.2.1.1 *Beginnings: The Belarusian Popular Front*

The easing of Chernobyl-related censorship made it possible for figures in the Popular Front to turn their attention to the legacy of the accident. The new movement, which was still illegal to begin with, attracted a variety of opponents of the regime in Minsk and much further afield; regional groups facilitated the transfer of information between centre and periphery and thus came to play an important role in the Chernobyl discourse. It was members of the BNF from Khoiniki and Narouvlia who informed Gennadii Grushevoi (he had, by his own account, been involved in the BNF since 1988) about the scale of the disaster and the situation in the contaminated southern part of the republic.[83] The BNF was formally established in

---

[80] Author's own records, 'BBF', extract from document lodged with the Supreme Soviet of the Republic of Belarus (1993).

[81] Aleksandr Tomkovich, 'Negromkaia data. Oni spasli nashikh detei … ', *Svobodnye Novosti Plius*, 1 June 2015, www.sn-plus.com/ru/page/society/5825 (2 May 2024).

[82] Cf. Thomas M. Bohn, Victor Šadurskij, *Ein weißer Fleck in Europa: Die Imagination der Belarus als Kontaktzone zwischen Ost und West* (Berlin: De Gruyter, 2011).

[83] Tamkovich, *Filasofiia dabryni*, 29.

Vilnius in June 1989, but the Chernobyl children committee – the predecessor of the 'For the Children of Chernobyl' fund – had existed under its auspices since April.[84] Overall, 1989 was to be 'a veritable Chernobyl year' for the BSSR.[85]

The committee's primary aim was to obtain information. It was to this end that Grushevoi and other BNF members set off for the southern BSSR oblast of Mahileu in July 1989, hoping to carry out interviews and gather information. Despite the constant presence of the KGB, which sometimes attempted to interfere with their activities, they interviewed individuals such as the young mother whose words were to shock protestors at the demonstration at the end of July. Gennadii Grushevoi later described what he had witnessed on this trip as the formative experience behind his activism on behalf of the Chernobyl children.[86] Irina explained that the committee focused specifically on the children because Chernobyl had made clear that the slogans of 'happy Soviet childhood' were just 'empty platitudes' – the 'true picture' had become clear after Chernobyl.[87] Grushevoi was never able to forget what he had seen in the contaminated areas: 'And I remember – I keep remembering! – those children who were running around in Chudiany . . . What happened to them? Where are they now?'[88]

The demonstration outside the Planeta hotel at the end of July 1989, at which the BNF representatives presented what they had found on their trip, was followed by the committee's first practical intervention: it managed to get a children's home in the town of Slauharad in the Mahileu oblast temporarily relocated. According to the 'For the Children of Chernobyl' fund, the building was located in an area that was contaminated with 21 Ci/m² and thus in a zone where there was a right to resettlement.[89] Grushevoi was not prepared to accept this state of affairs. At first, he asked the Belorussian section of the Lenin Children's Fund to arrange for the children to be taken somewhere else. They expressed regret at the orphans' predicament – they even tried to get the BSSR Ministry of

---

[84] Author's own records, 'BBF' (1994).

[85] Tatjana Kasperski, 'Nation versus Gedächtnis: Die Nationalisierung kollektiver Vorstellungen über Tschernobyl als Faktor zum Vergessen der Katastrophe' in Arndt (ed.), *Politik und Gesellschaft nach Tschernobyl*, 152–81, here 158.

[86] Author's own records, Presentation of the Fund.

[87] Tamkovich, *Filasofiia dabryni*, 41; on 'happy Soviet childhood', see Chapter 1.

[88] Quoted in T. Nichiporuk, '"Ia otvetstven za to: chtoby pobezhdalo Dobro"', *Belaruskaia Maladzezhnaia*, [1996], 3. Chudiany, a village in the Khoiniki raion, was later evacuated and buried due to the level of contamination there.

[89] See Sahm, *Transformation im Schatten von Tschernobyl*, 224.

Education to help following Grushevoi's request – but they explained that there was nothing they could do themselves due to a shortage of available children's homes in 'cleaner areas'.[90] Disappointed by this lack of urgency, Grushevoi then turned to the BSSR Council of Ministers – with success: in September 1989, the children from Slauharad and their carers were moved to a holiday facility in the village of Aksakaushchyna (Ru. Aksakovshchina) near Minsk, where the BSSR Ministry of Health had opened a research institute on the medical effects of radiation exposure the previous year. It was in the context of funding this initiative that Grushevoi made his first foray into soliciting private donations. The author Ales Adamovich donated his speaker's fee from an appearance in Japan, the actress Margarita Terekhova donated part of a performance fee, and the Union of Cinematographers also provided financial support.[91]

In contrast to the original plan, however, the children were not able to stay in Aksakaushchyna until suitable permanent accommodation could be found for them in an uncontaminated location. Instead, they had to go back to their old facility in Slauharad after only two months.[92] These were not the only children forced to remain in homes in contaminated locations; by 1996, according to Iouri Pankrats of the 'For the Children of Chernobyl' fund, those in such a situation included 5,026 children with disabilities and their carers, spread across 26 homes in places where the right to resettlement applied.[93]

Despite the ultimately disappointing outcome, the fund saw this first intervention as an overall success and took encouragement from it. Grushevoi recalled that it had shown him what was needed: a structure independent of the state that would help those affected and enable them to live better lives.[94] In parallel to her husband's activities, Irina Grushevaia (a lecturer in German) went about soliciting support for the Chernobyl children in the West – successfully, as soon became apparent. Her first visit to the West took place in the second half of 1989, when she went to Düsseldorf with a party of archivists to support a documentary film

[90] Tamkovich, *Filasofiia dabryni*, 30.
[91] Ibid., 35; Gruschewaja, Tamkowitsch, *Der Tschernobyl-Weg*, 39.
[92] Tomkovich, 'Negromkaia data'.
[93] Pankrats presented a document with these figures to the Chernobyl session of the Permanent Peoples' Tribunal in Vienna in 1996. Belorusskii blagotvoritel'nyi fond 'Detiam Chernobylia', *Chernobyl'*, *Posledstviia dlia okruzhaiushchei sredy, zdorov'ia i prav cheloveka*, 1996, 35.
[94] Tamkovich, *Filasofiia dabryni*, 31.

project on the German invasion of the Soviet Union. She returned the following year to act as an interpreter for interviews with contemporary witnesses.[95] She used her fee and her days off to seek support for the Chernobyl children in the surrounding area.[96] Many other trips to the West were to follow, enabling Grushevaia to make new contacts and develop an international network of supporters; from an early date, this included the Ecumenical Forum of Christian Women, an international network of women from various Christian churches.[97]

The Orthodox priest Igor Korostelev was an important Christian partner closer to home. As a priest in the Orthodox community of the 'Joy of All Who Suffer' Mother of God Icon (Khram ikony Bozhai Matsi 'Usikh tuzhlivykh Radasts'), Korostelev supported the 'For the Children of Chernobyl' fund from its earliest days. He was based in Frunzenskii, a Minsk district typified by Soviet-era apartment blocks, where the city had housed a particularly large number of families from contaminated areas. Construction of the church began in the early 1990s; Korostelev used an old army tent for several years before work had progressed sufficiently to start using it. He became a point of contact for people who had gone through resettlement, as well as for numerous initiatives at home and abroad.[98] The 'For the Children of Chernobyl' fund facilitated some of his collaborations with foreign partners, but he also remained open to other aid initiatives that worked with the fund on a temporary basis, if at all.

Although the 'For the Children of Chernobyl' fund collaborated with numerous Christian organizations and Gennadii Grushevoi was chairman of the Belarusian Christian Democratic Party (Belorusskaia khristiansko-demokraticheskaia partiia) from 1996 to 1999, religion did not really play a significant role for the Grushevois. Their own remarks, even at Christian events, are indicative of this; they made their case in humanitarian terms, and rarely along specifically religious lines.[99]

4.2.1.2 *From Philosopher to Activist, from Lecturer in German to Networker*

Gennadii Grushevoi framed the path from philosopher to activist as one he found himself compelled to take.[100] On the day of the accident, he

---

[95] Irina Grushevaia to Melanie Arndt. Email, 29 October 2019.
[96] Tamkovich, *Filasofiia dabryni*, 145.
[97] Gruschewaja, Tamkowitsch, *Der Tschernobyl-Weg*, 312.
[98] See 'Liudi stroiat khramy', *Sobor.by*, http://sobor.by/kamen.php (2 May 2024).
[99] See Arndt, *Tschernobylkinder*, 233.
[100] Nichiporuk, "'Ia otvetstven za to: chtoby pobezhdalo Dobro'".

recalled, he was – like many of his compatriots – making the most of the
unusually good weather, out in the fresh air with his thirteen-year-old
daughter and eight-year-old son. It was not until near the end of 1988, he
wrote, that he understood what 'Chernobyl' really meant and felt anything akin to fear. Prior to that, he explained, he had always distinguished
the malign 'military' from the supposedly 'peaceful' atom: whereas the
'military atom' brought death and destruction, as at Hiroshima and
Nagasaki, in his mind the 'peaceful' use of nuclear energy had been akin
to 'a harmless lightbulb'.[101] He had soon learnt about the accident from
university colleagues and acquaintances from the Homel oblast, he said,
but had 'listened to and treated' the news 'like any ordinary person who
thinks they can't control or change anything'; 'OK', he had thought, 'an
accident. Well then, we'll deal with it.'[102] Many others have described
their response at the time in similar terms.

Grushevoi also recalled how his own indifference was later turned
against him when he criticized the failure of those making the decisions
in 1986 to live up to their responsibilities. He rebutted this with the
argument that, unlike him, the functionaries at the time must already
have known what had to be done and thus had no excuse for their failures
when it came to helping people and keeping them safe. Instead of trying to
avoid facing the facts, he said, in 1989 he had come to 'see things through
the eyes of those' who had been left to suffer because of these failings.
According to Grushevoi, there was only one difference between opponents and advocates of nuclear power: the former had experienced what it
meant to become 'victims of the atom'.[103]

Both the Grushevois emphasized that their activism was guided by an
inner sense of responsibility. At the same time, their actions provided them
with a source of orientation and stability during the transformations that
this 'showcase' Soviet republic was experiencing.[104] For Gennadii, perestroika opened up new spaces and an alternative career path outside state

---

[101] Gennadi Gruschewoi, 'Monolog über kartesianische Philosophie' in Swetlana
Alexijewitsch, 'Stimmen aus Tschernobyl', *Aus Politik und Zeitgeschichte* 13 (2006),
3–11, here 7. See also Melanie Arndt, 'Grün nach der Katastrophe? Die Entwicklung der
Umweltbewegungen in Litauen und Belarus nach Tschernobyl' in Martin Sabrow (ed.),
ZeitRäume 2009 (Göttingen: Wallstein, 2010), 8–21, here 10.
[102] T. Nichiporuk, '"Ia otvetstven za to: chtoby pobezhdalo Dobro"'.
[103] Quoted in ibid. Grushevoi, however, was wrong in assuming that all those adversely
affected by nuclear accidents are opposed to nuclear power. As my interviews with
Chernobyl children demonstrate, this is far from the case. See Section 5.6.
[104] For the BSSR as a 'showcase' republic, see Kathleen Mihalisko, 'Belarus: Retreat to
Authoritarianism' in Karen Dawisha, Bruce Parrott (eds.), *Democratic Changes and*

structures. According to his wife, he had refused to join the Communist Party after becoming a philosophy lecturer, and suffered from episodes of depression when his academic career consequently ground to a halt. He was no longer afflicted by them now.[105] Irina described her activism as a way of coping with the upheaval of *katastroika*: 'We were able to use it [this period] to grow as individuals and to change things. In the country and in people's minds. We were able to help many to realize that nobody is going to change our lives for us – we are the only ones who can do that.'[106]

The Grushevois played charismatic leading roles in the 'For the Children of Chernobyl' fund and its offshoot, the International Association for Humanitarian Cooperation (Fig. 4.7). This is apparent from their own remarks, as well as in how they were seen by those who joined them on their journey. Gennadii has been described as a 'voice for Chernobyl', 'a man with society in the palm of his hand', a 'prophet', and a 'legend of a man' (*chelovek-legenda*).[107] Typical saviour narratives have been used to frame his achievements, but the legacy of Soviet practices of heroization can also be felt here as well. Drawing on critical leadership studies,[108] the Grushevois' status as leaders can be understood as a social phenomenon embedded in a multidimensional collective context.[109] They engaged in a 'management of meaning'[110] by projecting their actions as significant and worthy of support inside and outside their organization, at home and abroad. At the same time, this management of meaning was also shaped by the way the individuals and institutions around them saw things, and by the expectations they had and/or were assumed to have.

Authoritarian Reactions in Russia, Ukraine, Belarus, and Moldova (New York: Cambridge University Press, 1997), 223–81, here 237.

[105] Tamkovich, *Filasofiia dabryni*, 29.

[106] Ibid., 98.

[107] Ibid. 164, 303; Valentina Smolnikowa, *Nachruf*, 10 February 2014, https://bit.ly/3P8z EL3 (6 November 2019); Inga Ostrovtsova, 'Sad nadezhdy: Fil'm o beloruse, kotoryi spas 600 tys. detei', *Belsat*, 22 April 2016, https://bit.ly/3P2qzn3 (2 May 2024).

[108] See, e.g., Joyce K. Fletcher, 'The Paradox of Postheroic Leadership: An Essay on Gender, Power, and Transformational Change', *The Leadership Quarterly* 15, no. 5 (2004), 647–61; David Collinson, 'Critical Leadership Studies' in Alan Bryman (ed.), *The SAGE Handbook of Leadership* (Los Angeles: Sage, 2014), 179–92; Neil Sutherland, Christopher Land, Steffen Bohm, 'Anti-leaders(hip) in Social Movement Organizations: The Case of Autonomous Grassroots Groups', *Organization* 21, no. 6 (2014), 759–81; Mats Alvesson, André Spicer, 'Critical Leadership Studies: The Case for Critical Performativity', *Human Relations* 65, no. 3 (2012), 367–90.

[109] Sutherland, Land, Bohm, 'Anti-leaders(hip) in Social Movement Organizations, 763.

[110] See Linda Smircich, Gareth Morgan, 'Leadership: The Management of Meaning', *Journal of Applied Behavioural Studies* 18, no. 3 (1982), 257–73.

FIGURE 4.7 Gennadii Grushevoi and Irina Grushevaia in front of the foundation's bus, 1990s.

At home, and even more so abroad, the Grushevois became sought-after contacts for the media, politicians, and civil society organizations alike. They were asked for their views not just on Chernobyl and sending children away to recuperate, but also on other aspects of Belarusian politics and society. They became accomplished at setting the agenda: issues they raised would go on to get attention in their international networks – and thus also in the public sphere in the West.

4.2.1.3 *The First Trips Abroad for Children and the First Independent Humanitarian Aid Efforts*

The first Chernobyl children ever to go abroad did so over Christmas 1989, with the help of the BNF's Chernobyl children committee. On 20 December, a group of children from the village of Stralichava (Ru. Strelichevo) in the Khoiniki raion (Homel oblast) set off to spend twenty days in India.[111] They had been invited by Yogesh K. Gandhi's Gandhi

---

[111] Author's own records, Presentation of the Fund; Tomkovich, 'Negromkaia data'.

Memorial International Foundation, which had previously organized exchanges between schoolchildren from Moscow and the United States.[112] This year, the Moscow children were flying to India instead – and they were joined on the aircraft by children from a radiation-afflicted village in the BSSR.[113] Party circles in the BSSR were dumbfounded, Irina Grushevaia recalls: 'It was as if a bomb had gone off. How had this been approved?'[114] Grushevoi had bypassed the Belorussian authorities and handled the travel formalities with Moscow directly.[115] When the children returned from India, the BNF's Chernobyl children committee concluded from the improvement in their condition that trips abroad were 'the simplest and swiftest form of help', Grushevaia recalls.[116]

Grushevaia's lobbying in West Berlin began to show results at around the same time as the trip to India. Only a month after she visited the Protestant Patmos Church in Berlin-Steglitz in spring 1990, a first group of children from Vetka, Slauharad, and Narouvlia, as well as a group of children with leukaemia from various locations in the BSSR, were on their way to West Berlin at the invitation of the church. The BNF committee had collaborated with the BSSR's Friendship Society (Obshchestvo druzhby s zarubezhnymi stranami) in making preparations for the trip: the Friendship Society had taken care of travel formalities for the 150 children and accompanying adults (the fund was unable to do this itself because it was not officially registered at this point in time).[117]

According to the BNF Chernobyl children committee's own figures, it had already sent 6,000 children to 5 different countries before becoming an officially registered association in November 1990.[118] The numbers were to increase rapidly in the years that followed. The exact figures vary, however, and comparison is complicated by the different counting

[112] Yogesh K. Gandhi's claim to be related to Mahatma Gandhi is dubious. His foundation also came under increasing scrutiny from the mid-1990s. In 1999, a San Francisco court found him guilty of tax evasion, mail fraud, and breaking campaign donation law. The foundation ceased operations that year. See Alan C. Miller, 'Donor to Democrats Sentenced', *The Los Angeles Times*, 21 December 1999, www.latimes.com/archives/la-xpm-1999-dec-21-mn-46169-story.html (2 May 2024).
[113] Tomkovich, 'Negromkaia data'.
[114] Quoted in ibid.
[115] Gruschewaja, Tamkowitsch, *Der Tschernobyl-Weg*, 49.
[116] Tamkovich, *Filasofiia dabryni*, 37; Author's own records, 'Die Hoffnung im Land der Hoffnungslosigkeit: Vortrag auf CEC/CCEE Beratung "Environment and Development. A Challenge to Our Lifestyles"', Orthodox Academy Crete.
[117] Tamkovich, *Filasofiia dabryni*, 236; Gruschewaja, Tamkowitsch, *Der Tschernobyl-Weg*, 355.
[118] HA, Center for Civil Society International, Box 48–4, [1993].

systems that were employed. The 'Ten Years after Chernobyl' report published by the 'For the Children of Chernobyl' fund cites a total of 150,000 Belarusian Chernobyl children who had gone to recuperate in 19 countries by the time of its release in March 1996. According to the report, more than half of them – over 75,000 children – had gone abroad with the assistance of the fund.[119] Most of that number had been to Germany (*c.*56,000), followed by Italy (*c.*8,500), Austria (*c.*4,000), and Belgium and Poland (*c.*3,000 each).[120]

In addition to the first children's trips abroad, the BNF's Chernobyl children committee also organized an early, large-scale humanitarian effort to help Chernobyl children at home. After being approached by Grushevoi, the Moldovan SSR donated 17 metric tonnes of juice and fruit purée. The committee exploited the BNF's regional networks in order to get these supplies to where they were needed: committee representatives distributed them in the towns of Vetka and Slauharad (in the Homel oblast) and Khoiniki (in the Mahileu oblast).[121]

The 'For the Children of Chernobyl' fund should have received its first humanitarian aid from outside the collapsing Soviet Union from the Patmos Church in West Berlin in 1990 – but the consignment of medicine never arrived. It was delivered to the Belorussian section of the Soviet Peace Fund (Sovetskyi fond mira) instead;[122] the Grushevois suspected that this was because humanitarian aid was meant to be distributed through state channels.[123] Some individuals involved in the Peace Fund were also close to the 'For the Children of Chernobyl' fund, but successful collaboration between the two organizations proved impossible. The Grushevois felt that the Peace Fund's methods were corrupt, greedy, and opportunistic.[124]

4.2.1.4 A 'Political Chernobyl' and a New Old Enemy: the State

Incidents such as the non-appearance of the medicine donated from West Berlin became increasingly common as the years went by. This only

---

[119] '10 let posle Chernobylia. Situaciia, problemy, deistviia' in Belorusskii blagotvoritel'nyi fond 'Detiam Chernobylia' (ed.), *III Mizhnarodny Kangres 'Svet paslia Charnobylia': Osnovnye nauchnye doklady* (Minsk, 1996), 7–42, here 38.

[120] For details, see Arndt, *Tschernobylkinder*, 239, 442–3.

[121] Author's own records, Presentation of the Fund; Leanid Mindlin, *Sad nadzei*. Belarus, 2016: Belsat.

[122] On the Peace Fund (since 1992, the International Association of Peace Funds – Mezhdunarodnaia assosiatsiia fondov mira), see Chapter 3.

[123] Author's own records, Presentation of the Fund

[124] Tomkovich, 'Negromkaia data'.

exacerbated the 'For the Children of Chernobyl' fund's distrust of state and quasi-state institutions. Such tensions were a constant presence throughout the existence of the fund, which would ultimately refuse to countenance any form of collaboration with state institutions. In particular, the situation deteriorated when Aliaksandr Lukashenka took office as president in 1994. Lukashenka used manipulated referendums to introduce elements of an increasingly authoritarian rule, putting democratic institutions that had barely established themselves to the test.[125] Shortly after elections to the republic's Supreme Soviet, he replaced it with a bicameral parliament from which almost half of the deputies who had just been elected were excluded; Gennadii Grushevoi was one of them.[126] As he extended his autocratic powers, Lukashenka also began monitoring civil society initiatives and restricting their activities. His opponents were soon referring to a 'political Chernobyl' because of this,[127] but the president was not to be deterred. He used bureaucracy, propaganda, and the methods of a police state to cement his position.

The constantly changing legal situation made it harder for Belarusian Chernobyl initiatives and their foreign partners to operate. It caused considerable unpredictability and uncertainty, requiring frequent rethinking and adjustments so that programmes could run as smoothly as possible. The legal changes also resulted in an increasing dependence on goodwill on the part of state authorities. In addition to dealing with widespread arbitrariness in the behaviour of state actors, members of non-state bodies found themselves constantly having to turn to the authorities for guidance and information; in many cases, they were dependent on the benevolence of individual officials. The officials, meanwhile, were often unsure of their own position and found stonewalling preferable to the risk of having to face difficult questions from their superiors. It is likely that resentment regarding the activists' travels and contacts in the West – of which fashionable clothes and an apparent sense of entitlement could serve as a reminder – was also a factor here. The authorities were

---

[125] Astrid Lorenz, 'Politischer Wandel in Belarus: Tendenzen, Probleme, Perspektiven', *WeltTrends* 29 (2001/2002), 59–77, here 67.

[126] See Melanie Arndt, *Opposition gegen das Regime Lukashénkas in Belarus (1994–2001)*, unpub. Master's thesis, University of Potsdam, 2003, 99; Natchyk, 'Referendum' in Vital' Silitski et al. (eds.), *Nainoushaia historyia belaruskaha parliamentaryzmu* (Minsk: Analitychny hrudok, 2005); Uladzimir Rouda, *Palitychnaia sistema respubliki Belarus'* (Vilnius: EHU, 2011), 144–52.

[127] See Melanie Arndt, 'Einleitung. Ökologie und Zivilgesellschaft' in Arndt (ed.), *Politik und Gesellschaft nach Tschernobyl*, 10–24; Tatjana Kasperski, 'Nation versus Gedächtnis', 178.

constantly making life difficult for Chernobyl organizations and threatening to revoke their operating permits on trivial grounds. Transitional
periods in which to become familiar with new regulations were not
generally provided.[128] The ever-changing rules and procedures placed
Belarusian NGOs in the unenviable predicament of constantly having to
explain themselves on three sides so as not to fall out of favour – not just to
the authorities, but also to foreign partner organizations and to their own
workers and volunteers. Foreign partner organizations, which had never
experienced life under socialism, were frequently completely at a loss to
understand how things got done. The situation could become particularly
fraught if they – knowingly or not – infringed on new regulations by acting
over-hastily and on their own initiative, thereby jeopardizing the position
of their local partners. The Belarusian organizations, on the other hand,
had to keep their own workers and volunteers motivated so that they
remained onboard in such difficult circumstances, rather than simply
withdrawing from civil society activity.

The state also used the media to undermine the legitimacy of NGOs.
The information boom around 1990 was followed by a reversion to
extensive restrictions on freedom of speech and press freedom under
Lukashenka. State and non-state media – the latter were generally
opposed to Lukashenka's regime and provided a voice for NGOs – were
still getting involved in highly politicized matters in the mid-1990s. Even
so, independent media outlets were increasingly being banned or finding it
harder to reach their audiences. It was impossible to get hold of independent newspapers outside the main urban centres, and independent journalists were disadvantaged when it came to accessing information, as well as
being harassed and obstructed when going about their work.

The trend towards centralization and primacy of the state where
Chernobyl-related aid was concerned came to a head in June 1997,
when Lukashenka established the Department for Humanitarian
Assistance under the President of the Republic of Belarus (Departament
po gumanitarnoi pomoshchi pri Prezidente Respubliki Belarus). This
move provided the 'For the Children of Chernobyl' fund with a further
impetus in its drive to set itself apart from the state. Like other non-state
organizations, the fund interpreted this new entity as a direct attack on its
activities – the department was, after all, defined as the 'state organ
responsible for the implementation of humanitarian programmes,

---

[128] Author's own records, 'Gennadij Gruschevoj, "Die Kinder von Tschernobyl" heute',
25 November 2006.

including children going abroad to recover', and had been given far-reaching powers.[129] In addition to the fund, there were around eighty other NGOs in Belarus in 1997 that were involved in organizing humanitarian aid for people harmed by Chernobyl. Around fifty of them had charitable status.[130] It seemed inconceivable to both sides – civil society and the state – that state and non-state efforts could coexist successfully.

The establishment of the department did have an adverse impact on the work of NGOs. Hardly anything could be done now without its seal of approval – and, in many cases, the signature of the president himself.[131] Donations could no longer be given to recipients in person as a matter of course – and precisely this personal connection was important to the 'For the Children of Chernobyl' fund and its donors because of the sense of trustworthiness and authenticity that it created.[132] Humanitarian aid was now often stuck in the hands of the customs authorities indefinitely. If it was not released soon enough, people who had travelled from afar to hand over the goods in person had to return home with nothing to show for their efforts.[133] A number of the fund's foreign partners turned away from Belarus and focused their efforts on Ukraine and Russia instead because the rules there were less rigid – or, at least, relatively less opaque.

According to Grushevoi, the volume of humanitarian aid dropped to a third of what it had been.[134] It is, however, worth noting that aid shipments to contaminated areas had already started to decline before the new department entered the picture. This is evident from expressions of discontent on the ground, such as a letter drawn up by sixty members of the soviet of the Homel oblast and the chairman of the oblast executive committee.[135] There were clearly other factors involved than Lukashenka's actions alone. The 'For the Children of Chernobyl' fund ended its relationship with several partners because they worked with state institutions; in its eyes, doing so amounted to endorsing a morally bankrupt regime. The tone was not always pleasant. It was not all that

[129] Prezident Respubliki Belarus, *Polozhenie o Departamente po gumanitarnoy pomoshchi pri Prezidente Respubliki Belarus n. 404*, 24 July 1997.
[130] Aleksandr Tomkovich, 'Naezd' na dobrotu, *Svobodnye Novosti*, 25 April 1997, 1–2, here 2.
[131] Tamkovich, *Filasofiia dabryni*, 146.
[132] See the discussion of prosocial behaviour and self-interest in Lingelbach, *Spenden*, 20.
[133] Tomkovich, 'Naezd' na dobrotu', 2; Natal'ia Radina, 'Izderzhki total'nogo kontrolia. Blagotvoritel'nuiu pomoshch liudi okazyvaiut liudiam, no ne totalitarnomu gosudarstvu', *Nasha Svaboda*, 18 May 2000, https://bit.ly/3VGkDnK (2 May 2024).
[134] Ibid.
[135] Sahm, *Transformation im Schatten von Tschernobyl*, 361.

long since the Grushevois themselves had become politically active, they were constantly concerned that their country's new-found freedoms were under threat, and they were highly motivated individuals. In this context, the fact that many aid initiatives did not wish to take sides in the same way was simply not good enough for them. A disappointed Grushevaia ended up alienating Western organizations that were not overtly political and felt they were being unfairly rebuked. Anyone who was not with the fund was against it – and against democratic values.[136] Even partners that demonstrated their allegiance by choosing to continue working with the fund found its approach problematic.[137] Many of the fund's own people were put off or felt pressurized by its inflexible stance and left. Some of them set up initiatives of their own and burnt their bridges with the fund – not least those who used its databases of contact information to further their own ends.[138]

Iouri Pankrats, the coordinator of the fund's North America programme, later said that he was not comfortable with the uncompromising stance the fund adopted at times, but he also displayed an appreciation of the reasoning behind it. Competition over limited resources, he explained, had fed into the decision to keep the state at one remove, as had the desire to ward off corruption – representatives of the state, he said, had attempted often enough to influence the composition of groups and tried to get children of the *nomenklatura* included or given preferential treatment.[139] The pressure this created extended to many of the fund's personnel; the adoption of an unambiguous position was intended to shield them.[140] Pankrats is convinced that Grushevoi was always committed to equality, even if it meant sending the children of state functionaries abroad if they lived in or came from contaminated areas.[141] It appears that the fund actually helped the sons of Lukashenka and Ivan Titenkov (head

[136] Karola Klatt, '"Dort arbeiten, wo der Staat versagt": Wie sich zivilgesellschaftliche Gruppen in einer Diktatur behaupten: Ein Gespräch mit der belarussischen Bürgerrechtlerin Irina Gruschewaja', *Institut für Auslandsbeziehungen*.

[137] UWLSC, HLOR, acc5911-001, Box 36, F. 'Cancer Materials/Etiology USA, WHO IPHECA Pilot Project'; Nadya Neal, *Snapshots of the Clouds: An American Story Fifty Years after Hiroshima – Ten Years after Chernobyl*, unpub. manuscript, 1996, 106. The relevant passage was not included in one of the later revisions of the book (Nadya Neal, *On Silent Clouds of Butterfly Wings: Growing Up in the Nuclear Age*, 3rd ed. (Chattaroy, WA: Booksurge Publishing, 2009)).

[138] Irina Grushevaia, interviewed by Melanie Arndt, 10 May 2019. Berlin.

[139] Iouri Pankrats, interviewed by Melanie Arndt, 26 March 2016. Vancouver.

[140] Grushevaia to Arndt, 29 October 2019.

[141] Pankrats, interviewed by Arndt.

of the presidential business administration from 1994 to 1999) go abroad;[142] conversely, it seems that Vladimir Konoplev (later an associate of Lukashenka and speaker of the House of Representatives) at one point led a local group of the fund in the Shklou (Ru. Shklov) raion, where Lukashenka's political career began.[143]

Rather than being specifically Belarusian or Eastern European, efforts on the part of the state to regulate humanitarian aid from abroad were also to be found further afield, particularly in situations of political or social upheaval.[144] Almost at the same time as Belarus, Ukraine introduced a registration requirement for Chernobyl NGOs. This was intended to prevent favourable provisions for humanitarian aid, such as exemption from import duty, from being exploited for commercial purposes, and to end the practice of getting rid of unwanted goods by sending them to the country as humanitarian aid (e.g. food or medicine that was beyond its expiry date or had insufficient labelling).[145] The extent of the Belarusian state's efforts to impose its authority, however, often went beyond what happened in states that were either democratic or moving toward democracy, including Ukraine. The aim in Belarus, as was readily apparent from the obfuscation and lack of transparency, was to control the structures of civil society by controlling humanitarian aid.

4.2.1.5 All in It Together: The International Network of the 'For the Children of Chernobyl' Fund

Less than five years after its foundation, the 'For the Children of Chernobyl' fund had already made connections with thirty partner organizations in twenty-three countries.[146] Its network included not only well-established organizations but also smaller initiatives in East Central and Western Europe and the United States. These smaller groups were often set up spontaneously after people heard about the plight of the Chernobyl children through media reports or talks. In Belarus, they generally worked with the 'For the Children of Chernobyl' fund alone, while more

---

[142] Tamkovich, *Filasofiia dabryni*, 148.

[143] Lukashenka was elected to represent the Shklov constituency in the twelfth BSSR Supreme Soviet in March 1990. Valerii Karbalevich, 'Put' Lukashenko k vlasti' in Dmitrii E. Furman (ed.), *Belorussiia i Rossiia: Obshchestva i gosudarstva* (Moscow: Prava Cheloveka, 1998), 226–57, here 232.

[144] See Robert Jacobi, *Die Goodwill-Gesellschaft: Die unsichtbare Welt der Stifter, Spender und Mäzene* (Hamburg: Murmann, 2009), 128.

[145] On this, see also Sahm, *Transformation im Schatten von Tschernobyl*, 358.

[146] Thus the fund's own account in Private Archive Pankrats, 'Pankrats (BBF), The Belarusian Charitable Fund "For The Children of Chernobyl"', 15 June 1994.

established organizations collaborated with numerous partner organizations in the affected areas. Most of the Western organizations focused on recuperation for children and, beyond that, supplying food and other goods that were needed (or thought to be needed) in Belarus. They then found (sometimes creative) ways of bringing them into the country – for example, in hand luggage in the early years.

A small number of foreign organizations, such as Action Reconciliation Service for Peace (Aktion Sühnezeichen Friedensdienste) in Germany, and religious groups, such as the Mormons, sent volunteers to Belarus to work for extended periods at orphanages and similar institutions, or for the 'For the Children of Chernobyl' fund. Many volunteers worked with Chernobyl children in person during their visits to Belarus, especially ones who had disabilities or were seriously ill with cancer. These children did not travel, and it is impossible to reconstruct with certainty how the collective term 'Chernobyl children' came to be applied to them as well. Various possibilities present themselves. The institutions involved may have introduced the term simply because they saw the children as victims of the accident, or in the hope that it would help to attract attention and garner support abroad. It is also conceivable that foreign aid organizations were behind the introduction of the term. What is clear is that it was, in very many cases, not possible to identify a direct link between the disaster and these children's illnesses or disabilities.

The 'For the Children of Chernobyl' fund made much of being not just the oldest NGO but also an efficient and readily approachable organization with which collaboration was straightforward. This spoke to a fundamental need for security on the part of many foreign parties in their efforts to navigate the post-Soviet, post-disaster space. They were not familiar with it, and almost everything they had heard about it had been negative. The fact that one Chernobyl children organization sounded much like the other only added to the disorientation: For the Children of Chernobyl, The Children of Chernobyl, Chernobyl Children, For the Chernobyl Children, and so on. The accusations of corruption that were constantly being levelled against Chernobyl children organizations were present at the back of outsiders' minds as well, adding to the desire to work with organizations that were as reliable and reputable as possible. The meteoric rise to prominence of the 'For the Children of Chernobyl' fund also brought it recognition from representatives of non-state institutions, writers and other cultural figures, and politicians from various countries. Grushevoi himself gave addresses in several parliaments, including Switzerland, Germany, Norway, Belgium, and Poland.[147] Despite the international recognition, the fund remained, as it had been from the beginning, wedded to a discourse according to which it was under constant threat. This increasingly began to define how it perceived others as well. Behind every criticism, behind every bureaucratic stumbling block, the fund saw schemes of the 'old *nomenklatura*' and 'various foreign "advisors" in their thrall'.[148] The sources do not permit conclusions to be drawn with any certainty regarding the extent to which these fears corresponded to reality. One way or the other, unease and distrust are representative of the fragile constitution of post-Soviet society, as evidenced not least by the public interest in religion and spirituality during these years of upheaval.

What cannot be denied is that there were a number of acts of repression directed against the fund and other non-state organizations. Even before the fall of the Soviet Union, Grushevoi was fined for having been involved in organizing the first 'Chernobyl March' in September 1989. In the mid-1990s, he was the subject of serious accusations relating to the embezzlement of donations. Seven years later, in spring 1997, the KGB audited the 'For the Children of Chernobyl' fund (see section on 'The Crisis of 1997–8'). The fund inevitably saw the fact that the KGB, rather than another agency, undertook the audit as a calculated affront and attempt at intimidation. In 1999, all NGOs, political parties, and trade unions were required to re-register. The 'For the Children of Chernobyl' fund succeeded in being approved again, but the overall number of non-state organizations was almost halved by this measure,[149] before returning to the level of 1999 again a few years later.

4.2.1.6 From Homo Sovieticus to Puer Post-Sovieticus

The 'For the Children of Chernobyl' fund channelled most of its resources into rehabilitation and recovery trips for children.[150] From the early 1990s, though, it was also involved in other areas: in its heyday, it employed up to 32 people in Minsk, handling more than 100 projects.[151] Local and

---

[147] HA, Center for Civil Society International, Box 48–4, [1994].

[148] Author's own records: Presentation of the fund.

[149] Nelly Bekus, *Struggle over Identity: The Official and the Alternative 'Belarusianness'* (Budapest: Central European University Press, 2010), 115.

[150] Author's own records, 'BBF, Verwendung der Geldmittel der Belarussischen gemeinnützigen Stiftung "Den Kindern von Tschernobyl" im Jahre 1992, 1993'. (1993/94); Author's own records, 'BBF, Ispol'zovanie denezhnykh sredstv BBF za 1 polugodie 1994g.' (1994).

[151] Tamkovich, *Filosofiia dabryni*, 148; Gruschewaja, Tamkowitsch, *Der Tschernobyl-Weg*, 179.

regional groups with around 8,000 members in total were active in 71 of the republic's 110 raions.[152] Activities that directly involved helping children included building the first Belarusian SOS Children's Village, near Minsk; helping children with diabetes learn how to manage the illness; the first kindergarten groups for children with cerebral palsy; and mother-and-child recuperation trips for children who were still too young to travel abroad on their own.[153]

In many cases, the fund was attempting things that had never been done before, particularly when it came to supporting orphans and integrating children with disabilities into society. Even in Minsk, children's homes were often in a shocking condition in the early 1990s, not least compared to facilities in Western Europe or the United States. The fund's personnel got numerous ideas from their foreign partners, many of whom brought experience of working in such contexts with them to Belarus.[154]

The 'For the Children of Chernobyl' fund was also active outside child welfare in the narrow sense. At the beginning of the 1990s, it opened a facility producing 'clean' (i.e. uncontaminated) baby food at the Lomonossov kolkhoz in the Brest oblast.[155] Production got off to a shaky start and presented the fund with considerable challenges, but Grushevaia recalls an almost boundless optimism and a conviction that 'we'd manage everything'.[156] It was in this spirit that the fund organized material assistance for families who were (or had been) resettled, Chernobyl 'invalids', and 'liquidators', as well as obtaining equipment and medication for hospitals. In addition to trips for Chernobyl children, the fund also arranged training visits to foreign medical facilities for doctors and nurses; 146 people took advantage of this in the first 3 years alone.[157]

The fund was financed by Western donations and adopted what it called the Western model. This went hand in hand with a clear sense of belonging to (Western) Europe.[158] Time after the time, the fund took up arms against the *homo sovieticus*. It wanted to overcome the Soviet

[152] Tamkovich, *Filasofiia dabryni*, 150.

[153] Author's own records, I. A. Setschko, 'Das Programm "Kindergarten"' (1994); Author's own records, BBF, 'Im "Mutter-Kind"-Programm' (1994); Author's own records, Kinder Tschernobyl Solingen e.V., Übersicht. Diabetes-Schule (1994).

[154] Personal observations of the author.

[155] On this, see also Tamkovich, *Filasofiia dabryni*, 294.

[156] Ibid.

[157] Author's own records, BBF 'Report of Activities in the First Three Years (in German)' (1993).

[158] HA, Enid Schreibman, Box 3, F. 'Lessons of Chernoble [sic]' 1996.

mentality through self-help.[159] In the eyes of the writer Sviatlana Aleksievich, the fund succeeded in this; in her words, it had 'saved' the children twice over: 'It not only healed them, but opened up the world for them.' Similarly to the fund itself, Aleksievich combined her endorsement of its work with criticism of her own society: 'People have learnt to appreciate that their lives matter. Their culture didn't teach them this; it was the world that showed them how.'[160]

The Chernobyl children who went to other countries had a special role to play here: one that harked back conceptually to Soviet times. The idea was that these *pueri post-sovietici* would act as models and educators, just like their Soviet predecessors were supposed to: only now, it was their experiences abroad that they were meant to pass on to their families. In this way – so it was hoped – the families in turn would be able to put their existence as *homines sovietici* behind them and democratize society. At least in the period considered by the present book, however, that is not what came to pass. Re-educating political figures – or most of the general public for that matter – was not quite as simple as it seemed. In 2016, Grushevaia resigned herself to having lost the battle, concluding that the 'For the Children of Chernobyl' fund had ultimately been unable to overcome the persistence of Soviet mentalities and ways of doing things.[161] This bleak assessment does not, however, apply everywhere and may need to be reconsidered in light of events in 2020. Belarusian society and politics were, and still are, influenced by what the Chernobyl children and everyone else involved in the Chernobyl movement experienced abroad. This remains the case even if the democratization of society as a whole has not transpired in the way that was hoped, and is instead being prevented by force.

4.2.1.7 *Caught in the Crossfire*

In 1996 and 1997, the 'For the Children of Chernobyl' fund became embroiled in a scandal that had been brewing for many years and threatened to cast the entire Chernobyl movement into disrepute. Various German initiatives accused the fund of having misappropriated donations and engaged in deliberate deception. A number of initiatives severed their links with the fund as a result. Supporters and detractors of the fund were at loggerheads with one another; both sides turned to the courts on several

---

[159] Author's own records, Presentation of the fund.
[160] Tamkovich, *Filasofiia dabryni*, 165.
[161] Ibid., 230.

occasions to defend themselves against the accusations and allegations that were being thrown back and forth. Was the KGB behind it all?

The climax of public interest in the affair was a programme broadcast on German television on the tenth anniversary of the disaster in 1996. In it, a journalist claimed that the Grushevois had been taking the opportunity to line their own pockets.[162] The fund's German network was outraged and responded that it was all a misunderstanding related to a technicality of currency exchange rates; the rebuttal also claimed that evidence had been forged, that 'the entire response to Chernobyl was being defamed and slandered', and that it was ultimately the Chernobyl children who were suffering as a result.[163] In the eyes of the fund's supporters, the programme was part of a 'slanderous smear campaign' that had begun in 1991, instigated by the fund's opponents.[164] The fund believed that the strings were being pulled by the Deutscher Verband für Tschernobyl-Hilfe (German Chernobyl Aid Union) and its president, Edmund Lengfelder, and the Belarusian KGB, which the fund thought was using the German media to push its agenda. The Grushevois had collaborated with Lengfelder, a radiation biologist, when they first became active in civil society. According to Grushevaia, they had parted company with him because he thought research was more important than direct assistance. They now suspected their former partner of colluding with the KGB to destroy the fund.[165]

It is impossible to say with any certainty who was really at fault in the whole affair. It is more productive to see it as a reflection of the upheaval during the 1990s. This was a time when various different ideas about what the future should look like were jostling for position on every level – between East and West, but also within the East and within the West. The shifts and ruptures involved were laid bare by the controversy over the alleged embezzlement of donations, which resembles a maze of suspicions, misunderstandings, insinuations, and evasion on the part of all concerned. Finding a way out of these self-perpetuating labyrinthine arguments was particularly difficult when they were based on assumptions that had not been verified – or could not be. The conflict was shaped by the fact that those involved had ways

---

[162] Alois Theisen, *Die Kinder von Tschernobyl*, ZDF-Frontal, 30 April 1996.

[163] Author's own records, Statement on the ZDF program 'Frontal', 30 April 1996.

[164] Ibid.

[165] Grushevaia, interviewed by Arndt; Author's own records, Statement on the ZDF program 'Frontal', 30 April 1996. The theory that the charge of misappropriating donations was invented by the KGB is also endorsed by Marples, *Belarus*, 101.

of communicating[166] and experiences of state institutions that were sometimes very different, which meant that there were a number of vulnerabilities to contend with. Looking back, Irina Grushevaia captured the problem in a nutshell: 'They couldn't understand why we have to be so opposed to the state, why we are so critical ... And I couldn't understand what they don't understand.'[167]

4.2.1.8 The Crisis of 1997–8

In the words of Irina Grushevaia, 19 March 1997 was the 'darkest day' in the history of the 'For the Children of Chernobyl' fund.[168] It marked the start of an investigation by the Belarusian KGB that was to continue for four months. The fund was accused of financial irregularities, misappropriation of donations, tax evasion, and funding the opposition (it had given financial support to workers who lost their jobs as a result of the Minsk metro strike in 1995).[169] Belarusian state television also broadcast a programme attacking the fund. The Grushevois were both in Germany at this point.[170] The effort that the fund had put into cultivating networks since the late 1980s now started to pay off in the form of international condemnation of the KGB's appearance at its offices. The reach of the fund's networks became clear as well. Non-state organizations in various countries were unhappy about the Belarusian security service's 'assault' on the fund and mobilized politicians to intervene on its behalf. The investigation was raised in parliament in several countries, and there were threats of cutting financial aid for Belarus. Numerous foreign media outlets reported on the affair, even if the tone was often overly dramatic.[171] In June 1997, the European Parliament passed a resolution that called for an end to interference with the fund's activities.[172]

In summer 1997, while the KGB's investigations were ongoing, Grushevoi found it necessary to go into 'forced political exile',[173] as his

---

[166] On this, see also Svetlana Boym's concept of 'communication with half-words' in the Soviet Union. Boym, *Common Places: Mythologies of Everyday Life in Russia* (Cambridge, MA: Harvard University Press, 1994); Arndt, *Tschernobylkinder*, 259–60.
[167] Grushevaia interviewed by Arndt.
[168] Tamkovich, *Filasofiia dabryni*, 150.
[169] See Sahm, *Transformation im Schatten von Tschernobyl*, 360.
[170] Marples, *Belarus*, 101.
[171] See: 'Kinder von Tschernobyl', Geheimdienst besetzt Stiftungsbüro in Minsk', *Frankfurter Rundschau*, 22 March 1997, 5.
[172] Sahm, *Transformation im Schatten von Tschernobyl*, 360.
[173] Tamkovich, *Filasofiia dabryni*, 225; Gruschewaja, Tamkowitsch, *Der Tschernobyl-Weg*, 336.

wife described the time he spent in Germany. He wanted to avoid the risk of being arrested. Only around a year later, when the accusations raised against the fund had been dropped as unfounded, did he return to Belarus for good. By her own account, Grushevaia has been living in Germany for political reasons since 2008.[174]

The fund realigned the scope of its activities after Grushevoi's return. There was, for the most part, a clear division of tasks when it came to recuperation trips for children. Much of the work involved in organizing their visits abroad was handed over to Grushevaia's International Association for Humanitarian Cooperation (IAHC). This gave the fund the opportunity to turn its own attention to new fields, such as establishing youth centres ('workshops for the future' – *masterskie budushchego*) and places to bring generations together ('bridges between generations' – *mosty pokolenii*). The fund opened its first 'workshop for the future' in the Malinovka area of Minsk, next to Frunzenskii, which many people from contaminated areas had made their home.[175] The same is true of Malinovka, where the fund and Grushevoi were very popular; its voters elected him to the Supreme Soviet in the first round of the 1995 elections, something that only three candidates managed.[176]

The forty regional sections of the fund that were left at this point set up youth clubs modelled on the 'workshop for the future' in Malinovka, with the aim of encouraging young people to develop a sense of independence and initiative.[177] In this respect, then, there was a contrast with the children's trips, which were funded from abroad and did not give the young travellers much scope to get involved on their own account. On the other hand, the 'workshops for the future' and 'bridges between generations' were often unable to find suitable venues for their gatherings and initiatives because, according to the fund, state institutions refused to support them.[178]

Eventually, the fund had to let almost all the paid staff at its head office in Minsk go. The fund's official explanation was that, as a charitable association, it was only permitted to use its money for humanitarian purposes, and that wages, office rent, and telephone bills did not count

---

[174] See her profile on the BEST-Sabel Berufsakademie website: http://bsb-hochschule.de/Ue ber_Uns/Unser_Team/Irina_Grouchevaia.php (8 April 2017).

[175] Private Archive Grushevaia, BBF, IAHC, 'The Youth Program "Workshop of the Future"' (2008).

[176] Natchyk, 'Referendum', 89.

[177] Tamkovich, *Filasofiia dabryni*, 226.

[178] Ibid., 229.

as such.[179] The IAHC now took over responsibility for the practical side of children's trips abroad. This parallel organization, which was run by Grushevaia and did not have charitable status, had been in existence since April 1992. Prior to 1998, it had, among other things, organized trips for children making repeat visits to the same host family and arranged visits to Belarus for host parents. Changes to the law meant that both these activities were no longer classed as charitable in nature; they were treated as tourism instead.[180] Members of the fund protested against this classification: the Chernobyl children were not tourists in their eyes.[181]

In practice, organization of trips was split between the 'For the Children of Chernobyl' fund and the IAHC; the processes involved were not always easy for outsiders to follow. The fund still obtained official invitations for the children from host organizations, decided on the composition of groups, found interpreters and/or accompanying adults, and collated the necessary documents. Once a board meeting had confirmed that preparations were complete, the IAHC took over. The new structures caused confusion and mistrust among foreign partners, not least because they were now faced with higher costs. As the IAHC did not have charitable status, it had to pay taxes as a business. In order to cover the additional outgoings, the 'For the Children of Chernobyl' fund asked its foreign partners to contribute more for every child that travelled.[182]

By openly affirming its Western orientation, the fund provoked an increasingly authoritarian state that was increasingly shutting itself off from the West. In turn, the state's uncompromising approach to non-state institutions caused the fund to become more and more radical. It was now clearly distancing itself from initiatives that collaborated with state institutions, and demanded unconditional loyalty from its partners. The fund even refused to accept vitamins and medicine that entered Belarus with the help of state institutions on the grounds that this

---

[179] Cf. Author's own records, G. W. Gruschewoj, Letter to partner organisations. 11 April 1998. In fact, registered associations were permitted by law to spend their funds – including donations from abroad – on such expenses. It is not clear whether this was treated differently in practice. It is also possible that other factors may have been at play: donors might have wanted their donations be used for a specific purpose, for instance, or certain stipulations might have been attached to international agreements and programmes. Cf. Prezident Respubliki Belarus, *Zakon Respubliki Belarus' ob obshchestvennykh ob"edineniiakh. No. 3254-XII*, 4 October 1994, http://pravo.by/document/?guid=3871&po=v19403254 (2 May 2024).

[180] Gruschewaja, Tamkowitsch, *Der Tschernobyl-Weg*, 333–6.

[181] Ibid., 334.

[182] Ibid.

furthered the interests of a political regime that was not doing enough for the Chernobyl children itself.[183]

The situation in Belarus took a dramatic turn for the worse at the end of the 1990s. Four opponents of Lukashenka disappeared in early summer and autumn 1999.[184] Hans-Georg Wieck, head of the OSCE mission in Belarus before he was forced to leave the country in 2001, did not rule out politically motivated murder.[185] He agreed with the conclusions reached by the journalist Dzmitryi Zavadski in his investigations; Zavadski himself had disappeared in summer 2000.[186] Meanwhile, the state and the 'For the Children of Chernobyl' fund remained at loggerheads; 2002 brought a formal warning to rectify petty matters such as the failure to include the fund's legal status on a doorplate and the use of a monochrome variant of the fund's logo, which was technically supposed to be in colour.[187] Grushevoi appealed – as he did in all such conflict situations – to the international community to put pressure on the Belarusian government in the name of the fund.[188]

In 2004, Lukashenka once again held a referendum to gain approval for a further constitutional change: there was now no limit on the number of terms the president could hold office for.[189] In his official speech to mark the signing of the outcome into law in November 2004, he engaged at length with 'protecting the rights of children and young people'. Initially, the speech sounded much like the *Pravda* article that preceded the establishment of the Lenin Children's Fund in 1987.[190] Lukashenka denounced parents who neglected their children, and called for greater support for children in institutional care. However, he then turned his attention to provisions for children to recuperate abroad. The government, he said, was going to put an end to the 'unchecked nature' (*beskontrol'nost*) of these arrangements 'once and for all'.[191]

---

[183] Author's own records, G. W. Gruschewoj, Letter to BAG (1996/1997).

[184] See Arndt, *Tschernobylkinder*, 270; Andrew Wilson, *Belarus: The Last Dictatorship in Europe* (New Haven: Yale University Press, 2011), 190–2.

[185] See Arndt, *Opposition*, 110.

[186] Wilson, *Belarus*, 190–2.

[187] See Author's own records, Letter to BAG; 'Fond "Detiam Chernobyl'ia" vstal vlastiam peperek gorla', *Belorusskaia Delovaia Gazeta*, 20 May 2002.

[188] Ibid.

[189] For the 2004 referendum in detail, see David R. Marples, *The Lukashenka Phenomenon: Elections, Propaganda, and the Foundations of Political Authority in Belarus* (Trondheim: Norwegian University of Science and Technology, 2007).

[190] See Section 3.1.

[191] Aleksandr Lukashenko, *Vystuplenie na tseremonii podpisaniia resheniia, priniatogo respublikanskim referendumom*, 17 November 2004, https://bit.ly/3DnScVh (2 May 2024).

Lukashenka explained his decision by referring to the fact that, of the 130 organizations licensed to facilitate recuperation for children, 115 were non-state organizations and 9 were religious ones – an imbalance that concerned him. He did not, he said, wish to suggest indiscriminately that these organizations were behaving 'unscrupulously and exploiting the suffering that Chernobyl has caused us'. However, he said, there were 'facts' to show that these organizations were squeezing money out of parents in return for additional services. In good Soviet style, he criticized local bodies for not doing their job properly and underlined his dissatisfaction with the fact that only four state organizations were engaged in sending children abroad. Such trips should, Lukashenka made clear, be the 'prerogative of the state'.[192]

The president was essentially of the view that going abroad was unnecessary: 'We should let our people, our children recuperate inside the country – we have plenty of scope to facilitate this.' Children, he argued, should go abroad only in exceptional circumstances, such as if recuperation or specialist operations were not possible in their own country: 'to the United States, Germany, Israel, wherever you like . . . but under state control'. Lukashenka announced that the number of such cases would be radically reduced and careful tabs kept on children leaving the country. Two months prior to the president's speech, the minister of education had already started signing every travel application personally. 'This process needs to be reduced to a minimum – practically to nothing', Lukashenka declared, adding that 'we should raise and educate our children ourselves'. He framed this reduction as a service to the nation: 'We're fighting to become more Belorussians, and at the same time we're sending thousands of little children [*detishki*] abroad. This is nothing short of shameful for the state. And it is time to put an end to it once and for all.' Why? Because the children were being corrupted in the West, which he set clearly apart from 'people of our nature, our mentality':

Surely you can see how the children come back from there, what this lifestyle is bringing us? It's not as if this consumerist lifestyle hasn't arrived here already anyway – it has swamped the country and all our young people, just like they said it would in Soviet times. And these little children [*detishki*], these tots [*malyshi*], come back from there as consumers on steroids. We have no need for this sort of upbringing.[193]

[192] Ibid.
[193] Ibid.

Lukashenka's conclusion was that those who wanted to help would do better to send money and monitor its use. 'Clothes, things, *sukhari* [sweet, dry rusks], and such like' were not required in Belarus, he said; the country had plenty 'of this junk [*barakhlo*]' as it was. He explained that medical equipment was what was most urgently needed, but made clear that there was no shortage of well-trained doctors who could use it and perform 'any operation' with it. He made no mention of the fact that the country had actually been receiving humanitarian aid of this kind for years. The speech set alarm bells ringing among the non-state Chernobyl aid organizations.

Faced with the new developments, Grushevoi noted in 2006 a 'progressive, deliberately planned destruction of the structures of non-state organizations'.[194] Lukashenka, he said, was having sports facilities built to demonstrate that everything was fine and the state was taking good care of its citizens – but, in Grushevoi's eyes, the country needed rehabilitation centres rather than sports facilities, which were only of use to the ever-decreasing number of healthy children anyway. Besides, he added, professional sportspeople and most of Belarus's Olympic champions had long since been training in other countries. He expressed his frustration at the fact that it was practically impossible to influence the general public in a situation where the mass media were inaccessible to non-state organizations.[195] As a result, he said, the public were less and less prepared to support the fund's voluntary work. The authoritarian regime was grounded more in fear than anything else: 'Some have accepted this power out of fear; others are afraid of changes – and they are right to be afraid that something is changing.'[196]

Grushevoi believed that the approach being taken by various international organizations was part of the problem. He was, for instance, of the view that the Cooperation for Rehabilitation programme (CORE), initiated by the United Nations Development Programme in 2003, had had a negative impact on Belarusian society. He was not alone in his criticism of CORE's methods. Representatives of the anti-nuclear movement in Belarus and further afield criticized the programme for putting resettlement over people's well-being and making health a purely individual responsibility.[197] This criticism followed from the fact that the

---

[194] Author's own records, Grushevoi, '"Die Kinder von Tschernobyl" heute', Speech at BAG meeting, 25 November 2006.

[195] Ibid.

[196] Ibid.

[197] See Kuchinskaya, *Politics of Invisibility*, 33; Piotukh, *Biopolitics*, 137–8; Karena Kalmbach, 'Frankreich nach Tschernobyl: Eine Rezeptionsgeschichte zwischen

programme's main goal was to make life in contaminated areas safer. Behavioural guidelines for minimizing risk (e.g. by preparing contaminated food in certain ways) were treated as a cure-all, irrespective of how effective they really were. The programme focused on motivating people to stay in contaminated regions and persuading those who had left to go back and make them agriculturally productive again.[198]

A similar focus on 'reviving' contaminated areas was adopted by the Belarusian section – established by presidential decree in 2007 – of the Russo-Belarusian Information Centre for Problems Associated with the Consequences of the Disaster at the Chernobyl Nuclear Power Plant (Belorusskoe otdelenie Rossiisko-belorusskogo informatsionnogo tsentra po problemam posledstvii katastrofy na ChAES, BORBIC). The centre published a magazine with the programmatic title *We're Revitalizing our Native Soil* ('Vozrozhdaem rodnuiu zemliu'); its covers are replete with pictures of happy and healthy children, unspoilt nature, and bountiful agricultural production – all in contaminated parts of the republic. Like CORE, BORBIC said that such produce could be enjoyed safely as long as a few simple rules were followed. For instance: 'Pickling, marinading, and salting will reduce the amount of radionuclides by 15–20 per cent ... The pickling liquor, brine, or marinade must not be consumed.'[199] There are plenty of examples to show that the general public largely ignored such rules in their everyday lives. The best way to avoid being exposed to radiation, of course, was to not live in a contaminated area, but this was not pointed out.[200] The similarities with the CORE programme were not a coincidence: BORBIC was headed by the former CORE coordinator Zoia Trafimchik. She later went on to become deputy head of Gosatomnadzor, the Belarusian nuclear regulator, which was established in late 2007 as part of the presidential decree authorizing the construction of Belarus's first nuclear power plant.[201]

"Nichtereignis" und "Apokalypse"' in Arndt (ed.), *Politik und Gesellschaft nach Tschernobyl*, 237–55, here 245.

[198] Kuchinskaya, *Politics of Invisibility*, 33.

[199] Author's own records, Filial Belorusskoe otdelenie Rossiisko-belorusskogo informatsionnogo tsentra po problemam posledstvii katastrofy na Chernobyl'skoi AES, Informatsionnye plakaty. 2009.

[200] See Chapter 2.

[201] Gosatomnadzor, *Rukovodstvo i struktura Gosatomnadzora*, https://bit.ly/3VK4ayW (2 May 2024); Gosatomnadzor, *Istoriia sozdaniia i perspektivy razvitiia*, https://bit.ly/4gG5cUn (2 May 2024).

While state agencies in Belarus were bringing more and more aspects of Chernobyl aid under their control, support for the 'For the Children of Chernobyl' fund abroad began to wane from the early 2000s. Several organizations weighed the cause of the Chernobyl children against addressing new humanitarian crises and decided to change their priorities. There were widespread difficulties finding people to assist in the draining work associated with the children's trips. The rental agreement for the fund's offices was terminated in 2005, and the chaos that followed only accelerated the withdrawal of foreign partners. Finding a new site was difficult because the cost of most leases was beyond what the fund could afford; in the end, it ended up in a private flat outside the centre of Minsk. It was technically operating illegally there, but still managed to continue its work.[202]

The bureaucratic difficulties only got worse. Since autumn 2005, the fund had no longer been allowed to accept money to pay for coaches to take children abroad. Foreign initiatives now had to settle these bills directly, even if the coaches were arranged by the fund – a cumbersome procedure that placed an additional burden on smaller initiatives, in particular. In addition, constant technical difficulties in the new office made reliable communication problematic; for a time, none of the telephones were working.[203]

Quite apart from the general political situation in Belarus, incidents that had nothing to do with politics also made it harder for the fund to do its work. On 1 July 1997, a double-decker coach crashed because its Belgian driver had been going too fast on a road that was in a poor state of repair, costing the lives of four children; sixty-four other children and one accompanying adult were injured.[204] Despite the fact that this recuperation trip had been organized by a different Belarusian association – World. Contact. Chernobyl (Susvet. Kantakt. Charnobyl) – Grushevoi lost no time in making himself heard in the independent media. He drew up an appeal to all NGOs involved in humanitarian aid work; its message will not have been lost on the Belarusian government either. He was adamant that the accident and the associated suffering must not lead to a loss of heart or resolution: the vital humanitarian work for the victims of the Chernobyl disaster had to continue. Grushevoi was worried that the

---

[202] Author's own records, '"Die Kinder von Tschernobyl" heute', Kassel 2006.
[203] Gruschewaja, Tamkowitsch, *Der Tschernobyl-Weg*, 390–1.
[204] Quoted following Petr Chaly, 'Vinovnikom gibeli chernobyl'skich detei budet priznan bel'giiskii shofer'', *Narodnaia Volia*, 9 July 1997, 1.

state would use the accident as a pretext for imposing further restrictions on non-state efforts to support the Chernobyl children;[205] the heated atmosphere of 1997 was such that he felt any excuse would be used to further discredit the work of Chernobyl NGOs. As it turned out, the accident did not directly lead to any new curbs on their activity.

4.2.1.9 Mission Complete? The End of the 'For the Children of Chernobyl' Fund

After ten increasingly difficult years, Grushevoi took stock of the fund's activities at a meeting with German partner initiatives in winter 2006. He called for 'joint reflection' on how to take their work forward. He was also trying to ascertain how the partner initiatives saw themselves – clearly in order to assess the extent to which it was still in line with his own ideas. The most important premise for him was that state intervention was to be avoided at all costs.[206]

The problematic office situation and constantly changing restrictions on organizing trips abroad for Chernobyl children led more and more foreign partners to pull back more and more emphatically.[207] Nonetheless, the fund managed to survive for six more years, during which it continued to arrange trips abroad for children. Then, in 2012, Grushevoi announced at a press conference in Minsk that the fund had done what it set out to do and would consequently be winding up its activities: 'The Chernobyl children are grown-ups now, they have children of their own, and there is little that we can do for them now. Their destiny is in their own hands now.'[208] The fund was not, he said, a 'travel agency' that existed to 'keep sending this or that group abroad'.[209]

Grushevoi himself, who was at this point suffering from the effects of leukaemia, now focused his energy on Platform Innovation, an organization concerned with the welfare of inmates in Belarusian prisons.[210] The 'For the Children of Chernobyl' fund officially remained in existence until

---

[205] Gennadii Grushevoi, 'Eto gore ne dolzhno lishit' Vas muzhestva i reshimosti prodolzhat' dal'she nelegkoe, no takoe nuzhnoe delo gummanoi pomoshchi liudiam', *Narodnaia Volia*, 9 July 1997, 1.

[206] Author's own records, '"Die Kinder von Tschernobyl" heute', Kassel 2006.

[207] Tamkovich, *Filasofiia dabryni*, 230.

[208] Quoted in 'Blagotvoritel'nyi fond "Detiam Chernobylia" prekrashchaet deiatel'nost'', *Naviny.by*, 21 December 2012, https://naviny.by/rubrics/society/2012/12/21/ic_new s_116_407752 (2 May 2024).

[209] Ibid.

[210] Private Archive Grushevaia, 'BBF, IAHZ, BAG, Projekt "Malinowka – Beratungsstelle für Mädchen und Frauen"' (2008).

2015, when it was legally dissolved a year after Grushevoi's death; its activities had already ceased in March 2014, two months after he died. Olga Dashkevich and Irina Pobiagina, who had been part of the fund's paid staff since 1996, set up an association called Joy for the Children (Radasc dzetsiam).[211]

Irina Grushevaia was in no doubt about the fund's greatest achievement: 'In short, a civil society appeared.' The fund had also, she said, initiated a twofold 'silent revolution':[212] in the society to which it belonged and in the minds of those who contributed to its work. Nobody, Grushevaia rightly observed, forced people abroad to help those harmed by the Chernobyl disaster: 'They came to Belarus of their own free will, to see everything for themselves and make up their own minds about it. There was a "silent revolution" in their minds too.'[213] She also felt that the movement to support the Chernobyl children had led to a shift in how 'Soviet people' were perceived abroad. The general tenor of this assessment is certainly true: the movement had brought former enemies who had been on opposite sides in the Second World War and the Cold War closer together. The foundations for the emergence of a civil society had already been laid earlier, however, and the societies of East and West were still far from 'revolution', as she herself observed with regret.

4.3 'FRIENDS ACROSS THE SEA': CHERNOBYL CHILDREN IN NORTH AMERICA

4.3.1 'But Still Remain | Some Gulfs to Span'

> But still remain
> Some gulfs to span;
> They're wide – but can be bridged by Man.
> A noble bridge, then, let us start
> From land to land
> And heart to heart.

The 'For the Children of Chernobyl' fund used these lines by the poet Mikhail Dudin in a flyer aimed at the North American public in the early

---

[211] Tamkovich, *Filasofiia dabryni*, 256.

[212] Ibid., 102, 108. The Chernobyl children had already been framed as ushering in a 'silent revolution' in 1996: Schuchardt, Kopelev, *Stimmen der Kinder*.

[213] Tamkovich, *Filasofiia dabryni*, 103.

1990s.[214] In so doing, the fund was deploying a long-established trope in how the East imagined itself and was imagined by others: Russian poetry, the supposed epitome of the 'Russian soul'. The flyer introduced the fund and its activities, drawing on a survey about its work that had been published in the magazine *Demos* (which the fund published jointly with the IAHC).[215] According to the flyer, 80.7 per cent of those questioned in this 1994 survey thought the fund was the most effective source of help for people harmed by Chernobyl. This was an increase of 20 percentage points in the space of two years: in a similar survey from 1992, only 60 per cent of those surveyed were of this view.[216] In the intervening period, the fund had extended its international network and facilitated recuperation trips abroad and other forms of assistance for many more Chernobyl children. This added to the organization's appeal in Belarusian society. The fund had already started collaborating with initiatives in Canada and the United States; to begin with, these were primarily Christian organizations, some of which already had a strong presence in the humanitarian sector (e.g. World Vision International and World Aid).[217]

The fund's North America programme, 'Friends across the Sea' ('Druzia cherez okean'), was coordinated by a linguist called Iouri Pankrats. He was teaching English at the Linguistic University in Minsk when he met Irina Grushevaia, who persuaded him to get involved with the fund. While Grushevaia was busy publicizing the plight of the Chernobyl children and the work of the fund in Germany and elsewhere in Western Europe in the early 1990s, Pankrats was doing something much the same thing in North America. 'I could hardly stop talking', he said of all the presentations and interviews he gave at the time.[218] He too

---

[214] Quoted in Private Archive Pankrats, 'The Belarusian Charitable Fund "For The Children of Chernobyl"'. The translation is by Pankrats. For the original text, cf. Mikhail Aleksandrovich Dudin, *Derevo dlia aista: Stichotvoreniia 1968–1978* (Moscow: Molodaia Gvardiia, 1980), 238.

[215] See Section 4.2.1.

[216] Private Archive Pankrats, 'The Belarusian Charitable Fund "For The Children of Chernobyl"'; for the earlier survey, see the end of Section 4.1.

[217] HA, Center for Civil Society International, Box 48–4; Private Archive Pankrats, 'The Belarusian Charitable Fund "For The Children of Chernobyl"'. On the role of these organizations in global humanitarian aid, see Rachel M. McCleary, *Global Compassion: Private Voluntary Organizations and US Foreign Policy since 1939* (Oxford: Oxford University Press, 2009).

[218] Pankrats, interviewed by Arndt.

was successful. As a Canadian journalist put it, 'Pankrats has been instrumental in bringing to the world the plight of the children of Chernobyl.'[219]

Pankrats, who emigrated to Canada with his family in 1996, described how he tried to collaborate with state institutions in the early days of the North America programme: in 1990, he asked Anatolii Karpov, president of the Soviet Peace Fund, to help the 'For the Children of Chernobyl' fund with some Chernobyl children who had been invited to North America and needed a way of getting to the airport from the train station in Moscow. The fund did not have the means to arrange transport; Pankrats was hoping that Karpov would be able to provide a bus. No assistance was forthcoming, however; Karpov cited practical problems, telling Pankrats that the Peace Fund did not have any buses. This negative experience with the Peace Fund was similar to what Grushevaia had previously encountered. In the end, Vladislav Listev, a popular journalist and director general of the ORT television channel, stepped in and arranged a bus for the children.[220]

4.3.1.1 *The Role of the Diaspora*
The first people to take an interest in the Chernobyl children in Canada were members of the Belarusian diaspora. As early as 1989, the Canadian Relief Fund for Chernobyl Victims in Belarus (CRFCV) was founded by the translator and artist Ivonka Survilla, together with other members of the diaspora.[221] According to Survilla, the motivation came from reports in the Belorussian press through which she learnt that the BSSR had been affected by the legacy of the accident after all; she had previously assumed that only Ukraine was involved – as the information she had received from the Canadian health ministry had led her to believe.[222] The fact that she learnt about the true state of affairs from the Belorussian media rather than Western institutions and the Western press effectively meant that Soviet glasnost had extended as far as North America.

---

[219] Natasha Jones, 'A Reprieve from the Shadow of Chernobyl', *Langley Times*, 16 July 1994, 1.

[220] Pankrats, interviewed by Arndt.

[221] Survilla (Bel. also Survila) has been president of the Rada of the Belarusian Democratic Republic (Rada Belaruskai Narodnai Respubliki), the Belarusian government in exile, since 1997. For her biography: Ivonka Survila, *Daroha: Stouptsy – Kapenhahen – Paryzh – Madryd – Atava – Mensk* (n. p.: Radye Svaboda, 2008); Arndt, *Tschernobylkinder*, 281. For the Rada: Rada BNR, www.radabnr.org/ (2 March 2024).

[222] Survila, *Daroha*, 66.

Her first year of activity left Survilla feeling disappointed: major fund-raising successes had not materialized. Despite its best efforts, such as presentations at the 1990 conference of the Canadian Association of Slavists, the CRFCV had only managed to raise CA$5,000.[223] In an effort to generate greater interest in the Chernobyl children among the Canadian public, Survilla invited two representatives of the 'For the Children of Chernobyl' fund – Gennadii Grushevoi and Iouri Pankrats – to join her in Toronto.[224] This brought results: Grushevoi was interviewed by the *Ottawa Citizen* in April 1991, and the resulting attention was one of the reasons why, only a few months later, the CRFCV was in a position to invite its first children to Canada for six weeks. In the first three weeks after the interview was published, Survilla recalls, seventy families had expressed an interest in hosting Chernobyl children from the BSSR, before Grushevoi himself ordered a pause because he could not send more children at such short notice.[225]

The number of Belarusian children that went to Canada with the help of the 'For the Children of Chernobyl' fund and the CRFCV doubled between 1992 (110 children) and 1994 (250 children). In 1996 and 1997, more than 600 children flew to Canada at the invitation of the CRFCV.[226] In the mid-1990s, the CRFCV was employing a full-time member of staff and forty-two sections had been formed, in almost every province of the country, to organize children's visits. This involved collecting donations, identifying host families, drawing up activity programmes, and finding doctors who were prepared to carry out medical examinations at no cost;[227] in many cases, such doctors were themselves members of the diaspora.[228]

Survilla was gratified when her labours began to bear fruit: the children's visits meant that 'all Canada', as she put it, had learnt about Belarus and how the republic had been affected by Chernobyl. It was not just the humanitarian aspect that mattered to her: 'It was a great source of satisfaction for me . . . Canadians went to Belarus and returned enthused by the hospitality of our people.'[229] She still saw herself as an ambassador, as representing the country that she had left behind as a child and never

---

[223] Ibid., 84.
[224] Pankrats, interviewed by Arndt.
[225] Survila, *Daroha*, 83–5.
[226] Ibid., 86.
[227] Ibid.
[228] Private Archive Pankrats, Tom Goving, Letter to Grushevoi and Pankrats, 5 February 1994.
[229] Survila, *Daroha*, 86.

visited again since then. Bringing that country closer to the people of North America, making them aware of its existence, was one of the CRFCV's main concerns. Members of the group in London, Ontario, summarized their goals as follows in a letter to Pankrats and Grushevoi:

The purpose of the stay in London is to provide the children with rest and recreation and, while they are here, with an introduction to Canadian life and the English language. We also appreciate the opportunity to give Canadian children and their families direct knowledge of the people of Belarus through contact with the visiting children.[230]

According to its own figures, the fund secured several million dollars' worth of aid and obtained the backing of Canadian state institutions in its efforts to support the Chernobyl children. In 1992, the foreign ministry sent three aircraft loaded with medication to Belarus at the CRFCV's request.[231]

Canadian Aid for Chernobyl, a successor organization to the CRFCV founded in 1998, remains actively involved in arranging trips to Canada for Chernobyl children today. It continues to use a radiation symbol modified to depict a flower as a symbol – the logo of the 'For the Children of Chernobyl' fund.

4.3.1.2 *Children of Chernobyl Northwest*
Nadya Neal Hinson also learnt about the plight of the Chernobyl children through the media – in her case, Western sources.[232] In her autobiography, she describes how she watched a television documentary about a group of Chernobyl children travelling abroad together with her mother, who had breast cancer.[233] She had been unable to note down the address

---

[230] Private Archive Pankrats, Goving, Letter to Grushevoi and Pankrats.

[231] Survila, *Daroha*, 87.

[232] Nadya Neal Hinson has had this name since 2003. Prior to her Orthodox baptism in 1994, she used the first name 'Nancy'; up to 2003, she sometimes combined her maiden name 'Neal' with the surname of her then-husband, 'Oldenettel'. She also used her mother's Slovak maiden name, 'Wolchova'. All these variants appear in the sources.

[233] Neal, *Silent Clouds*, 228. Neal Hinson revised her autobiography several times. The first version, *Snapshots of the Clouds*, was published in 1996; a second followed in 2006 as *The Butterfly's Effect*. The third and so far last version is *On Silent Clouds of Butterfly Wings*, published in 2008. The work cannot really be considered an autobiography in the conventional sense of the word; Neal Hinson herself acknowledges having woven fictional elements into the narrative: Neal, *Silent Clouds*, n. p. On autobiographies as 'acts of social communication' and 'biographical illusions', see Volker Depkat, 'Autobiographie und die soziale Konstruktion von Wirklichkeit', *Geschichte und Gesellschaft* 29 (2003), 441–76, here 446.

displayed at the end of the programme in time, so she contacted the television channel for help. The address she was given was that of the 'For the Children of Chernobyl' fund in Minsk. Neal Hinson's interested had been piqued, and she duly wrote a letter to the fund early in November 1990. Gennadii Grushevoi's reply reached her almost half a year later, in April 1991. He was very direct in asking her to help host Chernobyl children for rehabilitation. According to her emotionally charged autobiography, Neal Hinson read the letter out to her mother on her deathbed. Her mother died shortly after that, and Neal Hinson saw it as her duty to take up the cause of the Chernobyl children.[234] She used her inheritance from her parents as starting capital to fund her work.

Neal Hinson explained that the approach Grushevoi took with his fund was a crucial factor in her decision to collaborate with it: 'It is not the only children's fund by any means but unique in its approach and commitment to international understanding and cooperation between people.' She was also willing to accept that things might not always be straightforward:

I have spoken for hours on the subject of the children's health and the history of the Fund. If I didn't believe in it, I couldn't possibly ask someone that I don't even know to believe in it. I believe in the ideal even if sometimes the methodology leaves something to be desired. This is not my country. I don't feel that it is my place to judge how they have to get things done in their own turf. Still, there are often many strained moments of disagreements on which children should travel out or stay behind.[235]

4.3.1.3 Nadya Neal Hinson: From Hanford to Chernobyl and Back

Neal Hinson repeatedly drew connections between her activism and her biography, in which nuclear landscapes and Eastern Europe both played a role. Even her birthplace – Moscow, Idaho – seemed a sign of a special affinity. Her father worked at the top-secret Hanford plutonium production complex.[236] In her autobiography, she describes how she spent weekends fishing with him on the Columbia river; they were oblivious to the fact that it had been contaminated by operations and accidents at the site.[237] Pressure from former workers at Hanford and opponents of nuclear power forced the Department of Energy to release documents

---

[234] Neal, *Silent Clouds*, 207–10.

[235] Ibid., 304.

[236] On the origins of the facility, and life and work there, see Brown, *Plutopia*.

[237] Ibid. On the Columbia river, see also the magisterial Richard White, *The Organic Machine*, 6th ed. (New York: Hill and Wang, 2001).

detailing the environmental consequences of plutonium production there. Between 2,000 and 5,000 Hanford 'downwinders' subsequently took legal action against the state for health damage that they believed had been caused by operations at the site.[238]

Neal Hinson found a further connection with the plight of the Chernobyl children in the fact that her own mother was the daughter of Slovakian immigrants. This was, she wrote, why she had always taken a particular interest in Eastern Europe and felt a special affinity with the region.[239] She also felt that this 'feeling of being Slavic' was not really at home in the United States.[240] Her mother's family background was kept quiet for years, and contact with her Slovakian relatives avoided, so as not to jeopardize her father's employment at Hanford; this only added to Neal Hinson's vague sense of longing for the unknown in the form of Eastern Europe.[241]

It is hardly surprising that, when she visited the Soviet Union for the first time, shortly before the 1991 August Coup, she noted in her journal that 'Nothing felt foreign to me – actually I felt very much at home.' The sense of familiarity did not, however, prevent her from seeing things from a critical distance. The picture she drew of conditions during this time of upheaval was not a rosy one: this first encounter with her Slavic roots unfolded against the backdrop of a Latvian kolkhoz and the grim conditions of life and work there.[242] Her journal reflects how coming from a family of immigrants from Eastern Europe – even in the second generation – could make it easier to connect with people by establishing 'kindred feeling'.[243]

Neal Hinson went to Minsk on 9 August 1991, ten days before the attempted coup in Moscow.[244] Once at the office of the 'For the Children of Chernobyl' fund, she was struck by the fax machine that had printed

[238] Tom Carpenter, interviewed by Melanie Arndt, 5 April 2016. Seattle. 'Downwinders' originated as a term for residents exposed to radiation from the Nevada Test Site; it was subsequently extended to apply to any group exposed to increased radiation due to weapons testing, uranium mining, or nuclear accidents. For the United States, see Valerie Kuletz, *The Tainted Desert: Environmental Ruin in the American West* (New York: Routledge, 1998); Masco, *The Nuclear Borderlands*; Brown, *Plutopia*; Zaretsky, *Radiation Nation*.
[239] Neal, *Snapshots*, 141.
[240] Neal, *Silent Clouds*, 260.
[241] Ibid., 68.
[242] Ibid., 221.
[243] Ibid., 222.
[244] Ibid., 224–8.

news of her arrival in the East – it seemed 'a little out of place in its modest environment', she observed, not least because a whole desk had been set aside for it.[245] She was struck by the 'high level of energy' that met her: 'It was obvious that things were getting done and that people were anxious to be part of it.' The visit also brought home the fact that there was a political dimension to the fund's work: a critical newspaper article about Gennadii Grushevoi had recently been published, eliciting a response of 'anger and humor mixed together' in the fund. Grushevoi had, she observed in her autobiography, become 'something of a hero and villain at the same time'.[246]

Contrary to what the fund had hoped, Neal Hinson's first visit did not result in Chernobyl children being invited to the United States. The airfare between Minsk or Moscow and Seattle was still too expensive at this point. This did not change until 1994, when Aeroflot started flying to Seattle and offered discounted tickets for the scheme.[247] The first thirty-one Chernobyl children flew to Seattle for six weeks in 1994; they were followed by two further groups, consisting of a total of eighty-five children, for seven weeks each in 1995. More and more children went to the American North-West in the next four years. They recuperated, received dental treatment, and in some case even had operations (e.g. to remove thyroid glands that had been damaged by radiation).[248] In addition, Children of Chernobyl Northwest – which Neal Hinson had founded in 1994 'to extend an international hand of kindness and goodwill'[249] – gathered two tonnes of humanitarian aid and had it shipped to Belarus.[250]

Neal Hinson spent six years organizing and funding Chernobyl children trips for the 'For the Children of Chernobyl' fund in Minsk, which she saw as her 'parent organization'.[251] As the end of the 1990s approached, however, she found herself facing an increasingly murky political situation, as well as increasing rumours that Lukashenka was going to impose a ban on such trips. In view of this, Neal Hinson felt that she had no alternative but to stop inviting children. She found confirmation that this was the right decision in the fact that the fund's fourth

---

[245] Ibid., 225.
[246] Ibid., 226.
[247] Neal, *Snapshots*, 141.
[248] Ibid.
[249] Neal, 'Introduction', in: *Silent Clouds*, n. p.
[250] Neal, *Silent Clouds*, 337.
[251] Neal, 'Introduction', in: *Silent Clouds*, n. p.; Nadya Neal Hinson to Melanie Arndt. Facebook chat, 21 June 2017.

'World after Chernobyl' congress (at which she gave a presentation on styles of leadership) was held in Poland in 1998 – Grushevoi had, she said, moved it there because he feared harassment by the authorities if it were to take place in Belarus.[252]

Long before the first children were invited and Children of Chernobyl Northwest founded, Neal Hinson's Chernobyl children activism had begun with speaking tours in which she drew attention to their plight and solicited donations for them. These engagements took her as far as Alaska; her autobiography presents this as a positive time, for 'the Chernobyl situation fascinated people'. This fascination, however, was sometimes a ghoulish one, accompanied by peculiar assumptions: 'They wanted to know if children were born with three arms and two heads. It probably would have been easier for us if they had.'[253] In most cases, the suffering of the Chernobyl children was not immediately obvious to the eye. They might be a bit pale, but that was a long way off from the obvious signs of need that many Americans knew from pictures of starving children in Africa, for example; it was thus all the more important to explain their situation to the general public in order to win over potential donors and host families.[254]

Given concerns about possible corruption, the best way to demonstrate that the Belarusian contacts of Children of Chernobyl Northwest were trustworthy and legitimate was for representatives of the 'For the Children of Chernobyl' fund to appear in North America in person. Ivonka Survilla had demonstrated this on the continent's east coast in 1991; two years later, Neal Hinson invited Grushevoi and Pankrats in similar fashion to join her for a series of talks along the west coast (Fig. 4.8).[255] Some of the travel costs for Grushevoi and Pankrats were covered by Neal Hinson's alma mater, the Jackson School of International Studies in Seattle; the rest was paid by the Women's Federation for World Peace (the women's organization of Sun Miung Moon's Unification Church, founded by Moon's wife, Hak Ja Han, in 1992). Neal Hinson is an adherent of the Moon movement and a member of the Women's Federation.[256] She had gone to Moscow on behalf of the Federation in 1991 to make preparations for the 'Women's Role in World

---

[252] Neal Hinsen to Arndt, 21 June 2017.
[253] Neal, *Silent Clouds*, 258.
[254] Ibid.
[255] Ibid., 293.
[256] For a balanced treatment of the Moon movement, see Friedmann Eißler, *Vereinigungskirche (Moon-Bewegung)*, April 2016, www.ezw-berlin.de/html/3_3065.p hp (2 May 2024).

FIGURE 4.8 From left to right: Nadya Neal Hinson, Iouri Pankrats, Gennadii Grushevoi, and Washington State Senator Peter von Reichbauer, who presented them with a certificate of recognition from the King County Council. King County, 1994. Courtesy of Iouri Pankrats.

Peace' conference that was to be held there the following year. Her work in support of the Chernobyl children was closely bound up with her membership of the Unification Church, but she underlined that Children of Chernobyl Northwest was not a religious organization.[257] Instead, she hoped to find new audiences for her Chernobyl children activism in the Moon movement[258] – which indeed came to pass in the form of Mark Boitano, a real-estate agent and Republican politician in New Mexico.

Even though he was committed to anti-Communism (a cause that was firmly embedded in the Unification Church's teachings right up to the end of the Cold War), Boitano agreed to join Neal Hinson on a visit to Minsk. On his return, he founded an organization called Friends for Friends and invited numerous children to New Mexico in the following years. He also got involved in Belarus himself, sometimes quite distinctively: at Christmas

[257] Neal, *Snapshots*, 142.
[258] Ibid.

1995, for instance, he and Iouri Pankrats turned up unannounced at 250 Minsk apartments to distribute gifts from the United States.[259] Boitano took up the cause of Chernobyl for humanitarian, religious, and political reasons. His efforts were, in Pankrats's view, a crucial factor in his election to the New Mexico Senate in 1997; Boitano had, he recalls, openly asked to be integrated into the Chernobyl aid effort in order to increase his appeal to voters.[260] Supporting the children of the former enemy in the East could clearly have its political benefits, even for a Republican on the periphery of the United States.[261] Whether or not Boitano's voters might have been particularly well disposed towards such actions because they themselves lived in 'nuclear borderlands' is a matter of speculation – but it is certainly plausible.[262]

For Neal Hinson, the draw of spirituality and the desire to confront her own fears were the main motivations behind her charitable work.[263] Self-discovery and helping others were not mutually exclusive.[264] Neal Hinson is convinced that it is only trying to change society for the better that can lead to a spiritually fulfilling life.[265] Her autobiography is replete with reflection on her fears, which stemmed primarily from misfortunes her family and she herself had suffered, and her attempts to confront them by helping others.[266] She first became active in this respect after her father's suicide.[267] US society, however, was unable to provide her with the opportunities she needed to live out her longing for spirituality. Even when the peace movement emerged in the 1970s, she was disappointed: in her eyes, 'the hippies confused drug use and sexual liberation with innate spirituality'.[268] She found much deeper fulfilment

---

[259] Pankrats, interviewed by Arndt.

[260] Mark Boitano continued as senator until 2013, when he did not stand for re-election.

[261] Pankrats, interviewed by Arndt.

[262] The first atomic bomb was developed and detonated in New Mexico as part of the Manhattan Project. On life in the state's 'nuclear borderlands', see the illuminating Masco, *The Nuclear Borderlands*.

[263] Neal, *Silent Clouds*, 153.

[264] See Pascal Eitler, '"Alternative" Religion. Subjektivierungspraktiken und Politisierungsstrategien im "New Age" (Westdeutschland 1970–1990)' in Sven Reichardt, Detlef Siegfried (eds.), *Das alternative Milieu: Antibürgerlicher Lebensstil und linke Politik in der Bundesrepublik Deutschland und Europa 1968–1983* (Göttingen: Wallstein, 2010), 335–52, here 347. Lingelbach, *Spenden und Sammeln*.

[265] Neal, *Silent Clouds*, 155.

[266] Ibid., 156.

[267] Neal, *Snapshots*, 70; *Silent Clouds*, 245.

[268] Ibid., 155.

in the Moon movement and the (former) Soviet Union: 'What our counterparts in the former Soviet Union lack in material needs, we lack in spiritual satisfaction.'[269] Eastern Europe was a spiritual space for Neal Hinson. This may have been due to the obvious boom in religion(s) during the transformations that accompanied the fall of the Soviet Union;[270] it may also have been a consequence of her individual perspective as an American on a spiritual quest. Her Orthodox baptism in Belarus in May 1994 was a high point in her spiritual journey; she took the name 'Nadezhda' (the short form of which is 'Nadia') – the name of an angel, as she pointed out.[271]

Neal Hinson's motivations in the West were underpinned by a sense of moral duty resulting from the horrors of the Second World War – just like those of most people in the East who became involved in Chernobyl aid work. In a Chernobyl Children Northwest press release, she wrote:

As we near the 50th anniversary of Hiroshima, it is timely to remember that it will soon be ten years after the nuclear disaster of this generation that is at least 40 times greater than the bomb dropped in 1945. History has a tendency to be repeated unless we learn from the mistakes and correct them. The 80 children who arrive at SeaTac airport on July 5th at 11 a.m. can be our teachers.[272]

For Neal Hinson, hosting a Chernobyl child was an act of peace,[273] a way to acquire knowledge that was vital and true: 'One family in the Pacific Northwest 'adopts' one child from the Chernobyl area and both of their worlds are made larger. True knowledge comes from experience. There is no amount of textbook research which can compare to the direct exchange of cultures'.[274] As well as warning against repeating the mistakes of the past and calling for solidarity, Neal Hinson – like Ivonka Survilla before her – wanted her work to inform and open people's eyes. That, she wrote, is why she started to set her down memories on paper. She sees recording what she experienced as a service both to her children and to her country, where 'children know very little about the world wars, Vietnam, Hanford or Chernobyl. They don't ask. When they get older – they will.'[275] In contrast to Carter and Christensen, whose Chernobyl

---

[269] Neal, 'Introduction' in *Silent Clouds*, n. p.
[270] See Melissa L. Caldwell, *Living Faithfully in an Unjust World: Compassionate Care in Russia* (Oakland, CA: University of California Press, 2017).
[271] Neal Hinson to Arndt, Facebook chat, 21 June 2017.
[272] Neal, *Silent Clouds*, 337.
[273] HA, Center for Civil Society International, Box 48–4 (1994).
[274] Neal, *Snapshots*, 7.
[275] Ibid., 5.

children book was aimed at readers in Eastern Europe in the first instance, Neal Hinson was writing against the 'loss of community – loss of concern for our neighbor, distrust in the motivation of our governments and laws, protectionism, false identification, dysfunction, denial' – in the United States.[276] But she too wanted to give the people of Belarus an outsider's perspective – as seen through 'the eyes of a stranger, a visitor from another planet' – that would be of benefit to the children most of all.[277]

Neal Hinson's engagement with the Chernobyl accident awakened a new interest in the nuclear legacy of her own backyard: the Hanford plutonium production facility.[278] She began to reach out to others who were trying to come to terms with this past.[279] This would seem to confirm that other people's radioactivity is more interesting at first than one's own, as Liudmila Zhirina, a biologist and Chernobyl children activist from Briansk, suggested. The late 1980s and early 1990s had seen the first US 'environmental delegations' visiting the Soviet Union to lend their support to the nascent environmental movement there;[280] Zhirina had been struck by the fact that US interest in the consequences of Chernobyl was accompanied by a glaring lack of awareness about the country's own nuclear legacy.[281] Neal Hinson also connected Gennadii Grushevoi with the Hanford Technical Steering Committee.[282] The committee's chair, Mary Lou Blazek, from the Oregon Department of Energy, even flew to Minsk in 1996 to take part in one of the 'World after Chernobyl' congresses organized by the 'For the Children of Chernobyl' fund.[283]

4.3.2 'Seeking the Christ Child among the Children of Chernobyl'

Early developments outside the network of the 'For the Children of Chernobyl' fund were also marked by the fact that it was primarily religious groups – now Protestant and evangelical ones in particular – that took an interest in the plight of the Chernobyl children. They too set about finding out more about these children and exploring ways of

---

[276] Ibid., 4.
[277] Ibid.
[278] Ibid., 7.
[279] Cf. UWLSC, Hanford Litigation Office Records, acc. 5911–001, Box 33, Folder 'Children of Chernobyl Northwest, USA / Neal, Nancy – The Women's Federation for World Peace'.
[280] See Arndt, *Nostalgic Bonfires*.
[281] Liudmila Zhirina, interviewed by Melanie Arndt, 21 October 2014. Online.
[282] HA, Center for Civil Society International, Box 48–4.
[283] Ibid. On these congresses, see Section 4.5.

helping them.[284] At some point, almost all of these groups came into contact with the fund, collaborated with it on individual projects, or made use of its know-how. They were never, however, associated with it in the way that 'Friends across the Sea' or Chernobyl Children Northwest were.

4.3.2.1 Bringing Christmas to the Children of Chernobyl: CitiHope International

Paul Moore, a Nazarene pastor in New York, set up CitiHope International to help the Chernobyl children in 1990. Between its foundation and 2017, the organization had provided humanitarian aid worth more than US$1 billion for various projects in thirty-six countries.[285] The Chernobyl disaster turned this urban ministry in New York City into a presence on the global humanitarian stage. Before starting CitiHope International, Moore had been producing *CitiHope Radio Ministry*, a weekly radio show soliciting donations for people in need in the city, together with another Nazarene, Michael Christensen.[286] After reading about Chernobyl children in the BSSR in the *New York Times*, Moore got Christensen to look into how it might be possible to help these far-away children in the Soviet Union. This was quite a challenge: Moore's partner 'wasn't even sure how to pronounce the name of the Soviet republic or how to find it on the map'.[287]

From the outset, the Nazarene initiative collaborated with state officials. In this respect, it was very different from initiatives that had long-term partnerships with the 'For the Children of Chernobyl' fund.[288] In August 1990, Moore and Christensen met Aleksandr Vasilev, counsellor to the Belorussian ambassador to the United Nations, and sounded him out about ways of helping the children. Vasilev was not the least bit bothered by the fact that the pair turned up in their priests' collars and were perfectly open about their religious motivations: 'Do you think there

---

[284] In Western Europe, particularly Italy, Catholic groups were actively involved in supporting the Chernobyl children as well.

[285] Paul Moore, *My CitiHope Story: Invitation*, www.youtube.com/watch?v=6ifoBL-dK5E (4 May 2024).

[286] CitiHope, *History*, https://web.archive.org/web/20160317011743/http://citihope.org/history/ (4 May 2024). Christensen later moved to San Francisco, where he founded the Golden Gate Community Church and directed the United Methodist AIDS Project. Carter, Christensen, *Children of Chernobyl*, 6.

[287] Ibid.

[288] The fund's records, however, do indicate that it worked with CitiHope, at least for a time. Cf. HA, Center for Civil Society International, Box 48–4, 18 August 1993.

is any way we could come to Byelorussia this Christmas to bring presents to children in the hospitals and orphanages in the spirit of the Christ child?', Moore asked.[289] Despite his previously 'reserved expression', this Soviet representative 'seemed to be deeply touched' by the suggestion. He was apparently on the verge of tears as he recalled his own childhood, when there had still been presents on 7 January (Christmas in the Russian Orthodox Church). Vasilev expressed a tentative hope that it might be possible to mark the festival in some way, but he also struck a note of caution – he acknowledged that 'Christmas gifts would be a most welcome gesture', but explained that the children needed something else much more urgently: 'medicine and medical supplies'. He also asked whether the two men could set off for the BSSR sooner: 'Many will not live until Christmas.'[290]

Moore and Christensen were up for it. Ten days later, a 'fact-finding team' of three – Moore, his son Paul Moore, Jr (aged seventeen at the time), and Christensen – were on their way to Minsk to see the situation for themselves.[291] With them they brought 'ten bulky bags of medicine, medical supplies, and toys' donated by listeners to CitiHope's radio broadcasts – and bibles.[292] The official invitation required by the authorities had been provided by Metropolitan Filaret, head of the Belorussian Orthodox Church. 'Nothing can prepare a child of the Cold War for that first encounter with the USSR. Nothing', Moore noted in his journal as they entered the East.[293] When Moore and Christensen looked round Brest while the bogies on their train were being swapped, however, everything seemed unremarkable: 'For the most part, they were ignored. No KGB agents lurked as they walked around, and no Red Army soldiers confiscated their cameras.'[294]

After arriving in Minsk, they met Vladimir Lipskii, chairman of the Belorussian Children's Fund, in his office in the heart of the city. The windows of the building on the corner of Ulitsa Kommunisticheskaia (Communist Street) and the wide and imposing Prospekt Nezavisimosti (Independence Avenue) had a view of the eternal flame and obelisk commemorating the dead of the Second World War on Victory Square. Lipskii, a man of 'warm smiles and broad, theatrical gestures', used this

---

[289] Carter, Christensen, *Children of Chernobyl*, 7.
[290] Ibid., 7–8.
[291] Carter, Christensen, *Children of Chernobyl*, XXI; Moore, *My CitiHope Story*.
[292] Carter, Christensen, *Children of Chernobyl*, 9, 12.
[293] Quoted in ibid., 11.
[294] Ibid., 14.

first meeting to tell the delegation about the letters the fund received from distraught parents in the 'Zone'. 'All our children are the children of Chernobyl', he is recalled as saying – and that they were 'not too proud to ask for help'.[295]

The visitors gained an insight into Lipskii's own motivations over dinner, when he told them about his experiences of childhood in the Soviet Union under Nazi occupation. Having his own childhood taken away from him, he told them, had meant he 'could identify with the needs of other children who had been deprived, too: first orphans, and then these children of Chernobyl'. Earlier that day, he had explained to the Americans why children were still living in contaminated regions: it simply was 'not possible to find new homes' for all of them: 'We want to, but we cannot. The task is too great.' Instead, he said, the children had to be taught 'how to live with radiation'.[296]

Lipskii hoped that CitiHope would be able to help create opportunities for thousands of children from the BSSR to recuperate in the United States. In the days that followed, the trio of Americans toured various healthcare facilities chosen by Lipskii, speaking to doctors in the process. The Americans were impressed by the level of medical knowledge they encountered; the problem was that drugs and modern equipment were in short supply, they realized. The Soviet doctors were also keen to engage in exchange with their Western counterparts in order to learn about cutting-edge therapies and approaches to treatment. All the conversations took place through an interpreter: none of the three Americans spoke Russian, and their hosts had only a limited knowledge of English.[297]

That was different in this respect in the city's Hospital No. 1, where a paediatric oncologist by the name of Olga Aleinikova had close to eighty seriously ill young patients in her care. Aleinikova – who spoke English almost fluently, 'with a British accent'[298] – and her hospital were to leave a lasting impression on the visitors. The Americans were met with a horrifying scene: the usual shortage of medicines and equipment made itself felt with particular urgency here because the children were so sick and in such great need of help. The hospital had 'no working x-ray equipment, no lab of its own'. There was also not enough methotrexate,

---

[295] Ibid., 17.
[296] Ibid., 18–19.
[297] Ibid., 17–19. Cf. also the recordings in CitiHope, *Belarus 1990: CitiHope's Journey Begins*, 2015, www.youtube.com/watch?v=X3ZorA_V-hI (4 May 2024).
[298] Carter, Christensen, *Children of Chernobyl*, 20.

an established leukaemia treatment in the West, where it had helped achieve a cure rate of 85 per cent in childhood leukaemia patients. The equivalent figure in the BSSR was 15 per cent – and the number of patients was increasing: in 1986, Aleinikova's hospital had had to treat forty children; there were almost twice as many in 1990, and the numbers were still rising. Yet the hospital was understaffed and in urgent need of renovation: 'From time to time, chunks of ceiling plaster and paint chips dropped onto the beds of the children.'[299]

Just What the Doctor Ordered: Olga Aleinikova
Olga Aleinikova became a key figure in efforts to help Chernobyl children with cancer in the early 1990s. In many ways, she was the epitome of the ideal socialist woman: emancipated, highly qualified, resolute, ambitious, and a mother. Work had 'always come first' for her and her husband.[300] Now a professor, she has received numerous honours and distinctions at home and abroad for her achievements. With her abilities and assertiveness, she was able to convince many more visitors after Moore and his team to support her work. She had acquired some of her knowledge in the West, for instance when she went to Frankfurt am Main in 1989 and spent two and a half months in 'training facilities', as she told the CitiHope trio:

We are at least twenty years behind the West … It was a shock for me. Not your standard of living, but your hospitals and your treatments and what you could do for your children. This was my first time in those hospitals, but what about all the other Soviet doctors who go to those conferences and congresses in the West? Why didn't they tell me what they saw? I was angry![301]

On her return from Frankfurt, Aleinikova set about trying to bring her hospital up to the standards of Western facilities. She secured foreign donations,[302] as well as sending colleagues (nurses and psychologists) for training abroad, for instance through Children's Hospice International in the United States.[303] The new political situation made such actions possible, and she sought to take advantage of this in bringing

[299] Ibid., 21–2.
[300] Tat'iana Shakhanovich, 'Vrach i mnogodetnaia mama Ol'ga Aleinikova: Poslerodovaia depressiia – u tekh, kto vyros v teplichnykh usloviiakh', *Komsomol'skaia Pravda*, 5 July 2016', www.kp.by/daily/26550/3567450/ (23 August 2017).
[301] Carter, Christensen, *Children of Chernobyl*, 22.
[302] A German telethon apparently raised over DM3 million in support of Aleinikova's 'modern hospital dream'. Carter, Christensen, *Children of Chernobyl*, 21.
[303] Private Archive Pankrats, Jane P. Conkey, 'Children's Hospice International, Letter to Aleinikova'. 28 April 1994.

methods she had encountered abroad to her own country. Although she is not religious herself, after seeing Christian volunteers helping sick children in German and Swiss hospitals, she persuaded Orthodox nuns to visit her own hospital to work with the children and comfort their parents. She also made religious festivals and rituals part of everyday life at the hospital.[304] The enlistment of the nuns also led to 'a sort of art therapy', in the course of which the children produced artwork that Aleinikova, in turn, made use of for fund-raising purposes.[305]

In addition to her medical and organizational talents, Aleinikova was also able to put her intercultural skills to good use. As Michelle Carter, who visited the hospital shortly after the CitiHope trio and went on to support Aleinikova for many years, recalled:

She also seemed to understand and even anticipate the American need to ask endless, sometimes pointed questions in an effort to comprehend the scope of the tragedy. She was neither offended nor bored, and she made each guest feel as though his or her efforts for the children would alone determine whether they lived or died.[306]

Aleinikova worked with suffering, with children who were in many cases terminally ill, on a daily basis, and this too gave her a distinctive air of authenticity. Visitors were won over by what she had to say; the interest in helping was so great that foreign groups of all kinds were practically queuing up to visit the hospital in the early 1990s. The pragmatic Aleinikova did not lose any sleep over the fact that her seriously ill young patients were constantly being exposed to curious, often horrified gazes: 'Privacy is something my children cannot afford. The world needs to know about my hospital and my children because I need the world's help.'[307]

No visit by foreign humanitarian initiatives at this time was complete without dinner with the metropolitan. The CitiHope trio were among the first to experience this lavish affair, complete with caviar. As they were eating, the guests – who included representatives of the Belorussian Children's Fund and other state institutions – discussed the new role of the Church during perestroika, as well as how the Americans might be able to help the Chernobyl children and the Children's Fund. The exchanges

---

[304] Carter, Christensen, *Children of Chernobyl*, 34. The nuns, however, sometimes combined this with religious moralizing (e.g. by framing the children's illnesses as a consequence of their parents' sins). Personal observation of the author, Christmas 1996.

[305] See also Section 5.1.3.

[306] Carter, Christensen, *Children of Chernobyl*, 33.

[307] Ibid., 21.

were underpinned by a sense of common ground: 'We are one!',
Metropolitan Filaret said, by which he meant all the initiatives and
institutions that had made the Chernobyl children's cause their
own.[308] The opulence of the Orthodox Church and the 'magnetism' of
the metropolitan meant that Christensen, unlike Moore, was not really
at ease during the encounter, according to the account that he co-
authored. He had also been 'embarrassed that the metropolitan obviously saw them as something more than ordinary American clergymen
with decidedly limited resources on a journey of personal discovery' –
concerns that Moore seemed not to share.[309]

'It Will Be Thanksgiving Day in Chernobyl'

Aleinikova was probably speaking in jest when she told Moore that 1,000
bottles of methotrexate would be enough to make her 'believe in a God of
miracles',[310] but that was motivation enough for the Americans. On their
return to the United States, they were to discover that getting hold of the
drug was not entirely straightforward. The pharmaceutical companies
would only deal with specific agencies, so Moore had to start networking –
and thereby managed to find a way of obtaining the vials for US$10 each,
rather than the US$200 retail price that Aleinikova had to pay in the
BSSR.[311] Moore launched an emotional appeal in one of his radio broadcasts, wherein he even played a recording of Aleinikova herself. Before
long, CitiHope had raised the funds it needed. Ultimately, with the help of
Interchurch Medical Assistance, it was able to obtain not 1,000 vials of
methotrexate, but almost twice as many, as well as other medications and
vitamins, with a total value of around US$500,000.[312]

Rather than waiting for Christmas in order to pass on the donations as
planned, CitiHope went for a 'typically American "do-it-now"' approach
and sent a team to Minsk for Thanksgiving instead.[313] This time, Moore's
wife, Sharon, went to the BSSR together with three other women to meet
Aleinikova and hand over the medicine. The delivery of the methotrexate
appears to have been a moving, emotional occasion that saw both Moore
and Aleinikova in tears. The American team also had some more exotic
gifts with them: frozen turkeys. The United Parcel Service had donated

[308] Ibid., 28.
[309] Ibid., 28–9.
[310] Ibid., 23.
[311] Ibid., 48.
[312] Ibid., 49.
[313] Ibid.

them so that CitiHope could celebrate Thanksgiving with the Chernobyl children. The US media picked up on the initiative, which became 'Citihope's first national publicity': the *San Francisco Examiner* published a story headlined 'It Will Be Thanksgiving Day in Chernobyl', which was picked up by a news agency.[314] The turkeys turned out to present an unexpected practical challenge in the BSSR: they were far too big to fit inside Soviet ovens. What really mattered to CitiHope, of course, was not so much getting the perfect roast as the symbolic value of the gift: 'Of course, the feast was more than just a meal.' For the 240 children at an orphanage and boarding school who enjoyed the meat,[315] the Americans reasoned, this was the first food in four years that they could be sure was not contaminated with radiation, thus making the occasion all the more special – above all as a 'psychological boost, beyond the nutritional value of the meal'.[316] The children expressed their gratitude by singing American folk songs, and Sharon Moore told them about the first Thanksgiving. The CitiHope team could be sure the message had got through when Lipskii told the children 'This is our first Thanksgiving in the Soviet Union. When you grow up and take power in this country, you must also have a Thanksgiving so that your children will know how to give thanks to God for food and medicine.'[317]

After Thanksgiving, the four women boarded a propeller aircraft 'of questionable safety' and set off for the Chernobyl exclusion zone as the first Americans to be allowed into it – or so they were told. The women may have had no reason to doubt this, or they may simply not have been troubled by the fact that it was not actually the case.[318] At any event, they got to see the evacuated villages at first hand. When they stopped 'at one of the small cities on the perimeter of the zone, Sharon met a scientist who handed her a bootleg videotape of the secret liquidation efforts conducted in the first few hours after the explosion'. CitiHope later used it alongside its own recordings 'to help tell the story of the children of Chernobyl'.[319] The team also visited an orphanage in Homel; what they saw there left

---

[314] Ibid., 51.

[315] The following film account reports instead that there were 175 children: CitiHope, *Chernobyl's Children in Crisis: CitiHope Responds*, 2015, www.youtube.com/watch?v=qZxZECyXZj4 (4 May 2024).

[316] Carter, Christensen, *Children of Chernobyl*, 51.

[317] Ibid., 51–2.

[318] The women were far from being the first Americans to enter the area; one of the most prominent individuals to do so before them was the bone marrow transplant specialist Robert Peter Gale. See Chapter 2.

[319] Carter, Christensen, *Children of Chernobyl*, 52.

them 'emotionally drained' and feeling 'powerless'.[320] Just as the first CitiHope team had done, the women turned to religious rituals to cope with their helplessness: back at Aleinikova's cancer centre in Minsk, for instance, Sharon Moore prayed for the children, anointed them, and placed pictures of Jesus on bedside tables.[321]

After the women returned to the United States, CitiHope held talks on its premises near Times Square in order to publicize its visits and raise further donations. Petr Kravchenko, the BSSR minister of foreign affairs, spoke at one of these events. Only a few days earlier he had proclaimed the dawn of a new age of solidarity in a speech on 'Chernobyl – a twentieth-century Golgotha' at the UN General Assembly.[322] Kravchenko, who clearly saw the work of CitiHope as a pioneering part of this new era, expressed his gratitude to the 'courageous American women who ventured beyond the ocean to the unknown world on a mission of understanding'. He set this in the context of political transformation in Eastern Europe: the Berlin Wall had fallen a year earlier, he said, but 'that which prevented mutual understanding and compassion fell as well': 'We start to understand when we see each other better.'[323]

If Nadya Neal Hinson had felt that the Chernobyl children needed an icon, CitiHope supplied one with the image of a thin, pale girl who was unmistakably ill with cancer. Her name was Natasha Ptushko; the first CitiHope team had met her at Aleinikova's hospital, where Paul Moore, Jr, had fallen in love with her. This 'withering white rose', as Michael Christensen described her, came to epitomize the Chernobyl children for CitiHope: 'the children of Chernobyl now had a face for the Americans'.[324] In this case, however, help came too late: Natasha Ptushko died in February 1992, just after an invitation for her to come to New York for a bone marrow transplant had been arranged.[325] The way she was used in publicity demonstrates once again the importance of individual fates – 'the power of one', as the *New York Times* journalist Nicholas Kristof has called it – in humanitarian aid work.[326] Eliciting

---

[320] Ibid.

[321] Ibid., 53–4.

[322] Petr Kravchenko, *Belarus' na perelome: Diplomaticheskii proryv v mir* (Minsk: BIP-S Plius, 2009), 311–19.

[323] Carter, Christensen, *Children of Chernobyl*, 54–5.

[324] Ibid., 25.

[325] Ibid., 115.

[326] Nicholas D. Kristof, 'The Power of One' in Scott Slovic, Paul Slovic (eds.), *Numbers and Nerves: Information, Emotion, and Meaning in a World of Data* (Corvallis, OR: Oregon State University Press, 2015), 85–8.

compassion and fostering a readiness to make donations meant replacing an anonymous mass with individual names and stories, thereby giving suffering a human face. It is interesting to note in this context that the Chernobyl children whose plight was highlighted were often pretty, vulnerable-looking girls; representations of them were often indebted to 'traditional' values.

After bringing Thanksgiving to 'Chernobyl', CitiHope continued its work in the form of Project Magi. In this case, the wise men came not from the East but from the West, and they included the Moore family, 'who had become celebrities of sorts in Byelorussia' by this point. The CitiHope party was 'following a star of hope and seeking the Christ child among the children of Chernobyl': they planned to deliver their gifts on Orthodox Christmas Day, which would be celebrated legally again for the first time that year (1991) on 7 January.[327]

In the course of their journey, CitiHope's representatives visited medical facilities, state agencies, and a collective farm where people were living and working, and food was being produced, despite high levels of background radiation. The Americans kept waiting for someone to mention Chernobyl as they toured the farm; eventually, Michael Christensen raised the matter directly. The chairman of the farm explained in response that the radiation levels in food produced there were within limits – but he also told them that they did not give any of the food they grew to their own children, many of whom were already suffering from thyroid problems, headaches, and nosebleeds. Any illusions that remained on the part of the Americans came to an end when a number of parents from the farm appeared at the end of the tour and started imploring the group to help their children.[328]

The group evidently found it hard to realize their missionizing objective in the face of such conditions. As Sharon Moore said of their tour of the farm: 'We came here to talk about God's love, but you can't talk about God's love when people are starving. They can't hear over the rumble of their own stomachs.'[329] But they kept going. At one of the collective's schools, after the nativity scene they had brought as a gift had been

---

[327] Ibid., 62. In the 1930s, Stalin had banned the Christian festival on 7 January and replaced it with the secular tradition of New Year on 31 December.
[328] Ibid., 90.
[329] Ibid., 91.

unpacked, Paul Moore told the children a version of the Christmas story specially adapted for the collective:

Then came the magi to the farm with their gifts – uncontaminated food, toys, and medicine – for baby Jesus. And all the children on the collective farm sang for joy and danced around the tree that was especially decorated for the occasion. The children from the farm village of Vu Pokaliubichi were not forgotten by the God of the universe, who so loved the world and the children of the village that he sent his only son to be born in a barn in their midst.[330]

Moore, who took his religious role and missionizing intentions particularly seriously in Project Magi, found great satisfaction in having introduced the children to the story for the very first time.[331] He must have been even more satisfied when, during an improvised midnight Eucharist on Christmas Eve, Lipskii and another figure from the Belorussian Children's Fund 'declared their new faith in God, and Paul invited them to partake of the elements of communion'. It is not known whether the two men's conversion was a lasting one, but it was, in any event, later understood by 'these American Christians' as 'a moment of immense importance that would help sustain them and their Byelorussian partners through many struggles and misunderstandings in the months ahead'.[332] In the late Soviet Union, religious motivations were no longer a barrier to collaboration, even when they were openly professed. Quite the opposite, in fact – state officials could even see them as an important asset amid the social turmoil. As Vasilii Kazakov, the BSSR minister of health, apparently said to Moore and those who accompanied him on his third visit to the BSSR: 'We are trying to cure the body; you can cure the soul. We will succeed when doctors and priests work together.'[333] This openness toward religiosity on the part of the state accompanied a resurgence in public interest in Christianity and a renaissance for the Churches.[334]

On Christmas Day in the Orthodox calendar in 1991, Paul Moore received the 'highest civilian award' that the BSSR could bestow. The distinction was remarkable not just because it was being conferred on a priest who made no secret of his missionary intentions, but also because 'other international relief organizations had shipped far more food and

---

[330] Vu Pokaliubichi is probably the village of Pokoliubichi (Bel. Pakaliubichy), north-east of Homel.
[331] Ibid., 92.
[332] Ibid., 97.
[333] Ibid., 85.
[334] See Caldwell, *Living Faithfully*.

medicine than CitiHope'. The occasion, however, also gave pause for thought because of the possibility that Nikolai Dementei (Bel. Mikalai Dzemiantsei), the beleaguered chairman of the BSSR Supreme Soviet, might be taking the opportunity to improve his own image by associating himself with CitiHope. Such concerns about being politically exploited were very likely justified at a time when the state was faced with the growing popularity of the Belarusian Popular Front and other opposition movements.[335] It was in its own interests to demonstrate openness and a readiness to change, not least because its opponents were far ahead of it in terms of international connections and religious 'rebirth' was a central theme of the Popular Front.[336] The state also made sure that Moore's wife, Sharon, was not left out: Lipskii appointed her to the board of directors of the Belorussian Children's Fund on the same day.[337]

Sharon, who had trained as a signer, represented the Children's Fund and CitiHope onstage a few months later. She performed with the Zorachka children's folk ensemble at 'Musicians of the World for Chernobyl Children', a charity concert in Minsk's Dinamo Stadium to mark the fifth anniversary of the disaster.[338] Despite a decidedly chilly 5°C, 20,000 people gathered to watch the concert live; it was also broadcast on television. Practically all the big names in Soviet pop and rock music sang for the Chernobyl children: from DDT, to Laskovyi Mai and Mashina Vremeni, to Nautilus Pompilius and Igor Talkov. Instead of Pink Floyd, as Sharon Moore was later to recall mistakenly,[339] however, the West was largely represented by older folk music bands and B-list stars – China Crisis and Lindisfarne from Great Britain, for example, and the Canadian country singer Lesley Schatz.

Foreign initiatives had to negotiate attempts to push commercial interests as well as political ones. In many cases, these had just as little – if anything – to do with the Chernobyl children. Ordering 5,0000 straw stars from a secondary school of art in Minsk might just about still count

[335] Carter, Christensen, *Children of Chernobyl*, 97–9.
[336] Cf. different documents in NARB, f. 4p, op. 157.
[337] Carter, Christensen, *Children of Chernobyl*, 99.
[338] The first charity concert for those impacted by the disaster took place in Moscow on 30 May 1986. The pop singer Alla Pugacheva was instrumental in the initiative; she also performed for 'liquidators' in Chornobyl soon after that. Philip Taubman, 'Moscow Rock Concert Aids Chernobyl Victims', *New York Times*, 31 May 1986, https://bit.ly/3ZYkezp (4 May 2024).
[339] Carter, Christensen, *Children of Chernobyl*, 121; A. Leitman, 'Rok-n-roll na vozdusiakh', *Molodoi Dal'nevostochnik*, 1 June 1991, http://sovr.narod.ru/articles/91015.html (4 May 2024).

as a humanitarian gesture, but the same could hardly be said about the idea of promoting the work of an artist whom Lipskii had befriended. This did not stop Paul Moore from organizing an exhibition in New York. It was, however, a flop: the bare-breasted Madonnas evidently were not to American tastes, and none of the paintings were sold.[340]

The Zorachka children's ensemble was among those keen to go to the United States. CitiHope representatives had seen the group perform at Christmas 1990 and were immediately won over. Encouraged by an insistent Lipskii, they had, before the day was out, agreed to bring the group to the United States in order 'to carry the message of the children of Chernobyl to the American public' – after all, Zorachka's singers and dancers included children from Mazyr (Ru. Mozyr), one of the worst-contaminated cities.[341] The plan was that the tour would cover New York, New Jersey, and Pennsylvania, and raise funds for a mobile clinic, among other things. The expectations that CitiHope had raised on all sides ultimately resulted in bitter disappointment, however. CitiHope had been busy working on publicity and finding sponsors, but the ensemble never made it to the United States as a group. The Belorussian Children's Fund, which was due to pay for the flight tickets, was unable to do so.[342]

CitiHope also helped to arrange respite in the United States. This began in 1991, when a group of ten Chernobyl children from the village of Vetka flew to San Francisco to stay with Sonoma County families as part of Project Fresh Air. Constance and Clifford McClain from Petaluma – two prominent and widely travelled figures in Californian civil society[343] – were behind the group's visit. The McClains had already been trying to invite Chernobyl children for over a year. They had initially had Ukrainian children in mind, but nothing seemed to be happening on that front. When Michael Christensen in San Francisco came to their attention through a newspaper article about CitiHope's 'Chernobyl' Thanksgiving, they contacted him and expressed an interest in helping bring Chernobyl

---

[340] Carter, Christensen, *Children of Chernobyl*, 79–81.

[341] Ibid., 96.

[342] Ibid., 118–19.

[343] Connie McClain (1920–2013) was a political scientist, her husband Cliff (1923–2013) a professional baseball player and, later, coach. They became acquainted with the effects of the Chernobyl accident while travelling in the Soviet Union, and decided they wanted to bring children to California. See Julie Johnson, 'Constance McClain', *The Press Democrat*, 25 February 2013, https://bit.ly/49Ko3Zi (4 May 2024); 'Clifford McClain', *The Press Democrat*, 15 November 2013, www.legacy.com/obituaries/press democrat/obituary.aspx?pid=168002879 (4 May 2024).

children from the BSSR to the United States. At that point, however, the Soviet Union was only allowing sick children to go to the United States for treatment or therapeutic purposes, which the McClains felt would be too much for them to handle. It was thus six more months before the first Chernobyl children arrived in California in June 1991. There was no advance warning when they did appear, either. The flight had been arranged at short notice, and the Belorussian partner organization (Grushevoi's 'For the Children of Chernobyl' fund) had not been able to inform the McClains about it in time – so the children ended up at San Francisco airport with nobody to meet them.[344]

The McClains' efforts in hosting the children from Vetka marked the start of the Chernobyl Children's Project. It was to facilitate visits to California for Belarusian children for almost twenty years – until one day in August 2008, when the sixteen-year-old Tatstsiana Kazyra decided she did not want to go back to Belarus. In a tragic irony, visits to the United States for Chernobyl children thus came to an end in the same place that they began.[345]

At around the same time as the group from Vetka, Project Fresh Air was also giving children from Narouvlia 'a dose of Americana' on the east coast of the United States.[346] In the case of this group, Lipskii's Children's Fund was the partner organization in the BSSR – and not everything went as planned for this cohort either. Paul Moore had just resigned as an associate pastor at St George's Episcopal Church in Manhattan in order to concentrate on his work for CitiHope and the Chernobyl children.[347] Like Grushevoi's organization, Lipskii's state fund had to operate on an ad hoc basis and had therefore been unable to provide a firm arrival date. The resultant uncertainty had been too much for the planned host families, who had withdrawn at short notice. The Moores stepped in to take their place.[348]

Project Fresh Air also brought six more children to the United States in summer 1991. They were in remission from various forms of cancer (leukaemia, Hodgkin's disease, thyroid cancer), and went to the Sierra Nevada Mountains to stay at the American Cancer Society's specialist Camp Sunshine Dreams, which had been running since 1988.[349]

---

[344] See Carter, Christensen, *Children of Chernobyl*, 173.
[345] See Section 5.5.2.
[346] Carter, Christensen, *Children of Chernobyl*, 175.
[347] Ibid., 176.
[348] Ibid.; see also Section 5.4.1.
[349] Carter, Christensen, *Children of Chernobyl*, 177–80.

The groups of Chernobyl children may have been small to begin with, but other CitiHope initiatives were taking on increasingly impressive dimensions. In summer and autumn 1991, in an effort overseen by CitiHope, twenty-two American NGOs assembled 130 tonnes of medicine, medical equipment, and baby food for the BSSR's Chernobyl children – a quantity that rendered the costs of transporting it to Minsk prohibitive. Attempts to secure a contribution of US$55,000 from what was at this point still the Soviet republic were unsuccessful, despite Kravchenko's best efforts; it appears that Gorbachev himself refused to give his approval. In the end, the donations made their way to the BSSR – via Kyiv – with the help of Donald Kendall, the former boss of Pepsi Cola, and the Children of Chornobyl Relief Fund, a Ukrainian diaspora organization (Fig. 4.9). When the goods finally arrived, some medicines and vitamins had already passed their expiry date and had to be disposed of.[350]

In order to prevent such situations from arising in future, Stanislau Shushkevich (Ru. Stanislav Shushkevich), the new chairman of the Belarusian Supreme Soviet, put forward an agreement between CitiHope, the Belarusian Children's Fund, and the Belarusian government to the effect that the Belarusian state would facilitate the distribution of humanitarian aid – overseen by CitiHope. The Belarusian government demonstrated its commitment by giving CitiHope a warehouse of its own to use as a storage facility for donations.[351] At this meeting with Moore in November 1991, the new head of state adopted a tone similar to that of his predecessor, Dementei, earlier the same year. Shushkevich, a nuclear physicist by background, reportedly underlined how important it was for the people of Belarus – and the children in particular – that CitiHope bring religious values to the country in this time of upheaval, not just aid:

We need to know God. We've been raised not to believe in God, to believe only in the state. Therefore our children have no spiritual roots. They don't know right from wrong. They have no moral guidance, and we have paid a terrible price for this. You must show us the way to know God and help bring spiritual life to our country.[352]

The Soviet system saw itself as setting high ethical standards for its children, as was reflected not least in the rules of the Pioneer organization. Shushkevich's words amounted to a rejection of that self-image.[353]

---

[350] Tamara Tershakovets, 'Mria Airplane Takes Off for CCRF's Thanksgiving Airlift', *The Ukrainian Weekly*, 1 December 1991, 3, 363.
[351] Carter, Christensen, *Children of Chernobyl*, 188.
[352] Ibid., 188.
[353] See Section 1.4.2.

FIGURE 4.9 Vova Malofienko with his mother and the doctor treating him in front of the Antonov An-225 Mriia, which transported tonnes of relief supplies.

Soon after this, a CitiHope delegation experienced the dawn of the new political era at first hand, as the end of the Soviet Union played out on television in Minsk. They were joined by Michelle Carter, who was accompanying them in order to pass on donations for Aleinikova's paediatric cancer unit in person.[354]

After the delegation returned, CitiHope began planning its next big project. President George Bush had announced a major food relief programme for the Soviet Union in May 1991; Paul Moore consequently asked the US Department of Agriculture for US$2.4 million to support his most recent idea. Moore proposed 'to deliver 3,500 metric tonnes of flour, rice, sugar, cooking oil, powdered milk, dried fruit, and baby formula to six cities in Belarus' and 'to use "ordinary American volunteers" to distribute food parcels to 175,000 families and 63 institutions'. The Department of Agriculture agreed to the sum, and Moore approached a former Navy

[354] Ibid.

officer he knew for help with the logistics. This was Operation Hope Express/Operation Nadezhda Express. Carter calculated that the aid was equivalent to 3,500 Toyota cars in weight. Accordingly, CitiHope recruited sixty-three Americans who were prepared to spend two weeks distributing food in Belarus – and to pay the not insignificant sum of US$1,200 to cover their own travel expenses. CitiHope described Hope Express as a form of 'citizen to citizen diplomacy'.[355] Given that the food was paid for by the US state, this was not completely accurate. What was genuinely new, though, was the fact that volunteers were distributing it.

CitiHope again attached considerable importance to symbolic actions in the context of this enterprise. Sharon Moore revelled in attaching an Operation Nadezhda Express poster to the military helicopter that flew the team and some of their cargo from what, during the Cold War, had been a secret military base near Minsk to Homel in the south of Belarus. They were accompanied by a film crew to record events. A certain amount of time and energy was needed to coordinate such enterprises. Despite his best efforts, Moore had been unable to get the Belarusian government to arrange a helicopter trip. However, he was not going to give up that easily, and he turned to the Belarusian Children's Fund instead. Aleksandr Truchan, a close colleague of Lipskii, still had connections with the military and former Party apparatus – and a helicopter soon materialized.[356] Michael Christensen subsequently described his thoughts during the flight in his journal: 'I couldn't help but focus on the irony of all this. Here we were, American volunteers and former enemies, in a military machine once used in the Afghan War, delivering life-saving medicine and humanitarian assistance at the edge of Chernobyl's Dead Zone. We really were on a wild frontier of international relations.'[357]

Moore attempted to pull off an unprecedented symbolic coup at the end of Operation Hope Express, when he came up with the idea of setting up a permanent headquarters for CitiHope in Belarus – in the building where the first congress of the Russian Social Democratic Workers' Party had been held, which was now home to a museum commemorating the origins of the Communist Party. On the one hand, Moore savoured the irony of the project; on the other hand, he also felt that there were fundamental similarities between the Christian faith and Communist ideology insofar as both were striving to achieve a more just world.

[355] Ibid., 197–202.
[356] Ibid., 201–2; cf. also the photograph in ibid., n. p.
[357] Ibid., 202.

Kravchenko attempted to dissuade him, but Moore was not to be put off. He eventually got his way, but the triumph was not to last for long. Moore had hardly left the country when angry voices of protest appeared. Signatures opposing the desecration of the sacred site were collected, and posters appeared declaring 'Yankee, go home!', 'Minister of Culture, resign!', or 'Give us back our history!' The Belarusian minister of culture yielded to the pressure and cancelled the lease agreement. In order not to alienate CitiHope, 'all parties agreed to a permanent exhibition in one room of the museum about the children of Chernobyl and the work of CitiHope International'.[358]

4.3.2.2 *The United Church of Christ*
Soon after the CitiHope 'fact-finding team', twelve members of the United Church of Christ (UCC) set off for the Soviet Union from the west coast of the United States in November 1990. These 'peace exchange delegates' had been invited by the Soviet Peace Fund, and included Michelle Carter, who was in her mid-forties and managing editor of the *San Mateo Times*. Unlike many others who became involved with the Chernobyl children, she could speak Russian and had been to the Soviet Union before, in a delegation that had visited to join in the celebrations marking the millennium of the Russian Orthodox Church in 1988 – a trip that had also been organized by the Soviet Peace Fund. Visiting the Soviet Union was of personal and professional interest to her.[359] The group was supposed to be going to Moscow, but due to the August Coup the Peace Fund changed the programme and sent the group on to Minsk instead – a decision that turned out to have consequences nobody could have foreseen, for it was in Minsk that Carter met Aleinikova, the paediatric oncologist: 'Everything changed after.'[360]

After returning to California, Carter wrote up a five-part report on her experiences at the children's cancer hospital for the *San Mateo Times*. Its publication was followed by several hundred dollars' worth of donations, and Carter founded the Children of Chernobyl Project. She collaborated closely with her church, the Congregational Church of Belmont (part of the UCC) – an approach that had three crucial benefits. First, in legal terms, it meant that donations were tax-deductible and thus a more attractive proposition. Second, it gave Carter access to a pool of

[358] Ibid., 203–5.
[359] Ibid., 30–1.
[360] Carter interviewed by Arndt.

committed volunteers from the outset. Third, it meant that she did not have to use donations to help cover overhead costs. She concentrated her efforts exclusively on supporting Aleinikova and her children's cancer hospital.[361]

In contrast to CitiHope, with which Carter initially collaborated, the UCC did not have any missionary aims. Instead, she said, the church saw its work as creating social impact: 'Our church is a social action church. So we see this [Chernobyl aid] as social action. This is our role. This is one of the things we do.' They had a rule of 'three Bs' for their work: 'No bibles, bullets, or booze.'[362] The partnership between CitiHope and the Children of Chernobyl Project was pragmatically motivated rather than a reflection of shared mindsets. Carter needed a way to bring the donations she had assembled to the BSSR. She was wary of CitiHope's conservative, evangelical roots, but she was also taken by Christensen's charitable work, which included AIDS patients and the homeless in San Francisco. Christensen's suggestion that she join him on a visit and see for herself how CitiHope operated sealed things. CitiHope's connection with Interchurch Medical Assistance made it possible to get hold of methotrexate; the medicine purchased with the US$7,000 she had raised in California was waiting in CitiHope's New York offices when she arrived there to join CitiHope on the way to Minsk in June 1991.[363] The book *Children of Chernobyl* was later to emerge as a result of this joint trip.[364]

It was, however, not long before Carter's relatively matter-of-fact and cautious approach started to clash with Moore's more 'flamboyant' manner and missionary zeal. There was too great a gulf between the liberal views of the UCC and the conservative Church of the Nazarene. Women and men are ordained in the UCC, which also actively supports LGBT rights, whereas the Church of the Nazarene is wedded to deeply patriarchal ideas of the world and society. Carter also felt that Moore's practically endless stream of new ideas – 'it just kept getting bigger' – was overshadowing the real work to be done. As a result, she parted company with CitiHope after the trip.[365]

Carter was also unsure what to make of the Orthodox Church, at least in the form it presented itself at the residence of Metropolitan Filaret. She

---

[361] Ibid.
[362] Ibid.
[363] Carter, Christensen, *Children of Chernobyl*, 126–9.
[364] In practice, the book was written by Carter. Carter, interviewed by Arndt.
[365] Ibid.

was uncomfortable with the opulence on display; such wealth would, she felt, be better spent helping those in need.[366] Carter realized that the concept of charity in the (former) Soviet Union differed from that in the United States. Private aid, of the kind she knew from the United States and considered universal, was not to be found in the BSSR. 'Identify a need, and somewhere an American or an American church would raise money for it' – but that was not the case here. Carter explained this in terms of the country's history: 'The Soviet system had sanitized the notion of charity – and even the idea of unmet need – from the culture.' It was perfectly obvious by autumn 1990 that the Soviet state 'could no longer provide for a whole panorama of needs. What surprised the Americans was that the Soviets no longer knew how to respond as individuals.'[367] Time and again on her travels, Carter recalls, Belarusians asked her in bewilderment why Americans were giving money for children they had never seen, and why they cared about their plight at all in the first place. It was, she said, often too much for her (post-)Soviet interlocutors to grasp that it was simply a 'western principle that's just ingrained in us'.[368] She was repeatedly told that charity had been the concern of the state in the Soviet Union: 'The state took care of all that. And if the state didn't take care of it, it probably didn't need taking care of.'[369] The state's inability to help children with cancer meant that Carter felt she had to act. She spent fifteen years obtaining medicine and medical equipment for the treatment of Belarusian children in Aleinikova's care. At first, she had to arrange for the donations to be taken to Belarus herself; it was not until the end of the 1990s that suppliers were able to ship to the hospital directly, after which Carter simply settled the bills. In the early 2000s, responsibility for funding the methotrexate passed to the Belarusian state.[370]

Carter's Children of Chernobyl Project also provided Aleinikova with supplies and equipment for her research into what was causing her young patients' cancers. In addition, it covered her membership dues for the American Society of Hematology and subscriptions to medical journals that Aleinikova would otherwise not have been able to access or could not afford herself. Aleinikova made frequent visits to California to tell the Congregational Church of Belmont about her work. Some donors and

[366] Carter, Christensen, *Children of Chernobyl*, 43.
[367] Ibid., 43.
[368] Carter interviewed by Arndt.
[369] Ibid.
[370] Ibid.

church members went to Belarus as well. Carter describes the close friend-ship she developed with Aleinikova, despite the latter's Soviet back-ground. She sees the oncologist as a 'soul sister'. Conversely, it was through Aleinikova that Carter was presented with an award for her work on the twentieth anniversary of the disaster in 2006.[371]

According to Carter, the Children of Chernobyl Project had raised around US$250,000 by 2012. This was supplemented by numerous dona-tions of supplies from private individuals, businesses, and medical facil-ities. Hospitals had to discard consumables such as cannulas or catheters on a regular basis because standards had changed, rendering stockpiled supplies useless. It was often financially expedient for vendors to donate products that had become redundant rather than disposing of them, not least because donations could be used to reduce tax liability. Generous hospital staff also recycled disposable items, such as stainless steel scalpels or syringes, so that they could be donated to medical facilities in the former Soviet Union.[372]

Carter's initiative supported other efforts to help the Chernobyl chil-dren on a sporadic basis – for instance, by arranging accommodation for groups attached to other organizations on the way to their host families. This was how Carter crossed paths with a girl called Tatsiana Khvitsko, who broke her journey in California while on the way to Kansas City with other children with physical disabilities.[373]

Alongside her involvement in Chernobyl aid, in her capacity as a journalist Carter took an active interest in supporting the democratization of post-Soviet society. In 1995, she began working with journalists across Russia to help them adapt to the freedom of the press.[374] She remembers the *Cheliabinskii rabochii* (*Cheliabinsk Worker*) newspaper in Cheliabinsk particularly well: the editor's investigations into the environmental conse-quences of plutonium processing at the Maiak facility, including the Kyshtym accident of 1957, left a lasting impression on her.[375] Looking back, Carter described her humanitarian work as a source of great personal enrichment: 'I became involved in people's lives that I would have not in any other fashion. So for me it was a life-changing experience.'[376]

---

[371] Ibid.
[372] Ibid; Carter, Christensen, *Children of Chernobyl*, 107–8.
[373] See Section 5.6.2.
[374] See Michelle Carter, *From under the Russian Snow* (Fairfield, CA: Bedazzled Ink Publishing, 2017).
[375] Carter, interviewed by Arndt.
[376] Ibid.

4.4 NEW NETWORKS IN THE WEST

As the collaboration between CitiHope and the Children of Chernobyl Project shows, humanitarian work to support the Chernobyl children not only led to the formation of new networks among those offering their assistance; those networks could also, at least for limited periods, demonstrate a remarkable capacity to accommodate different ways of seeing and doing things. The unprecedented situation in the context of the fall of the Soviet Union meant that all involved had no alternative but to exchange knowledge in all manner of areas and make connections with other individuals and organizations. This was the only way that local partners in the East could be found, trips for children facilitated, and hard-to-source supplies obtained and delivered to the places – not always readily accessible – where they were needed. The resultant alliances spanned the whole of North America: when the Center for US–USSR Initiatives gave a slideshow with pictures from a trip to the East – the one that led to the press conference with which this chapter began – in Sacramento, California, in winter 1991, it did so not to further its own cause but to solicit donations for CitiHope.[377]

As the case of the Belarusian 'For the Children of Chernobyl' fund has shown, approaches in East and West were similar in this respect. The alliances themselves may have existed for varying lengths of time, but they still left a lasting mark on society. In many cases, they resulted in the various groupings involved developing a conscious sense of their own identity that had not been there before. This seems to have had particularly positive implications for the diasporas. Awareness of the existence of the Belarusian diaspora in Canada, for instance, clearly increased because of the humanitarian response to Chernobyl. The level of networking between women, and the extent of their involvement, is striking. Alongside this empowerment, the networking also resulted in a general professionalization of activism in organizations of all sizes. The first major networking event for the various US initiatives working with Chernobyl children took place at the Capitol in Washington on 11 April 1991. The idea came from Robert Miller, professor of music at the University of Connecticut and director of volunteer services at Paul Newman's Hole in the Wall Camp, six months after the first Ukrainian Chernobyl children with cancer had gone to the camp for respite.[378]

---

[377] HA, Francis U. Macy, Box 4, 13 November 1991.
[378] Carter, Christensen, *Children of Chernobyl*, 134–42.

Representatives of the various initiatives, including CitiHope's Michael Christensen, used the event to discuss their experiences and, above all, the problems they had run up against. There was a place for forthright criticism as well. Miller, for instance, cautioned against underestimating just how sick some Chernobyl children, particularly those with cancer, were. He was adamant about the reality that goodwill on its own was not enough to help the children: 'Though well-meaning, some of the host groups ... seem to be unaware of the special care and facilities that the children may need. They are unaware, for instance, that some children are at risk of becoming very ill very quickly.'[379] He wanted to introduce medical screening for the children before they left the Soviet Union, in order to establish which children were actually in a fit state to travel. He also advocated 'the sharing of pharmacological data among medical professionals at the various camps'. These sentiments were echoed by Molly Schwenn, a paediatric oncologist from the Floating Hospital for Infants and Children in Boston. She had helped care for the Chernobyl children at the Hole in the Wall Camp and arranged for Vova Malofienko to be treated at her hospital. She related how the Ukrainian children and their accompanying doctor had turned up at the camp with non-existent English and incomplete medical records. As a result, the US doctors had struggled to work out even what the children's illnesses were, never mind the treatment they had been receiving at home, how that treatment should continue, and what drugs they had been getting. Piecing that information together had been 'real detective work', Schwenn said. As the camp came to an end, she had been particularly concerned about the prospects facing two children when they returned home. She drew up a detailed treatment plan for one boy and gave it to the accompanying doctor to take back to the BSSR. She arranged treatment in Boston for the second child, Vova, with the consent of his parents. The city's Ukrainian community helped her fly his mother over for support.[380]

At the networking event in Washington, Miller also addressed the moral dilemma of having to send children who came to the United States for treatment or respite back to the Soviet Union at the end of their stay – in full knowledge that they were returning to medical care that would fall short of US standards. Very few were able to stay in the United States for longer in the way that Vova did. Chernobyl initiatives had to learn to deal with this problem, Miller felt, suggesting that sharing experiences would be a first

[379] Ibid., 135.
[380] Ibid., 137–40.

step in that direction. As well as pooling their knowledge and experience, delegates addressed the mental health needs of those impacted by the disaster. Bill Walsh, the spokesman for Project Hope, identified a 'Chernobyl of the spirit, a deep and widespread despair over the tragedy that has befallen their land'. Chernobyl initiatives had to bear the importance of psychological help in mind, Walsh pointed out, not just the more immediately obvious medical effects of the disaster.[381] Christensen's impressions confirmed that 'each delegate had encountered that sense of despair in the families affected by Chernobyl' – and that, by showing they cared, aid organizations offered a way out of it: 'It appeared that the kind of hope they could offer really did help. Knowing that the people in the rest of the world cared and were working to provide help for their children seemed to loosen the grip that Chernobyl had on people's souls.'[382] The event also facilitated networking of a very practical nature: 'A speaker would say, "We need this", and someone in the audience would stand up and say, "We've got it"', Christensen recalled of the energizing atmosphere.[383] New practical links were only one of the benefits of the event for CitiHope – it also represented a significant endorsement of the organization's work. It was clear that CitiHope, although it operated on a grassroots level, could hold its own with major established organizations such as AmeriCares and Project Hope. Besides, Christensen was the only delegate from an initiative that focused solely on the BSSR; all the others were still concentrating on Ukraine.[384] As successful as this networking event was, it was not until 1998 that an umbrella organization for Chernobyl initiatives in the United States was formed – the Children of Chernobyl United States Alliance, which represented organizations from over twenty states.[385]

For all the early enthusiasm, collaboration between the various parties involved in the United States did not necessarily result in the lessons that were learnt being retained. Carter is quite clear about the fact that attempts to reinvent the wheel were all too common.[386] She was contacted repeatedly with requests for advice by people who wanted to help Chernobyl children; but, she says, their goals were so ambitious that many of them did not want to listen to what she had to say. This was particularly so when it came to collaboration with state institutions,

[381] Ibid., 141–2.
[382] Ibid., 142.
[383] Ibid., 136.
[384] Ibid., 140.
[385] On the Children of Chernobyl United States Alliance, see Section 5.5.2.
[386] Carter, interviewed by Arndt.

something that Carter supported in principle on the grounds that such institutions were, ultimately, part of the system in which the doctors and children were living: 'If you really want to be helpful, you've got to work with the system the way the system is. You're not going to be able to change it. You've got to realize it.'[387]

Around the tenth anniversary of the disaster, many organizations were adjusting their priorities, and the first signs of processes that would lead to the Chernobyl aid movement petering out were becoming apparent.[388] Some initiatives, such as CitiHope, had already largely abandoned the idea of being 'rescuers' and were now pursuing the far more modest goal of providing assistance for limited periods of time in order to help at least some parts of the affected populations deal with the legacy of the disaster. The turning point for CitiHope occurred in April 1991, when Svetlana, a ten-year-old girl with leukaemia, was brought to the United States. She had come, accompanied by her mother, for treatment in place of Natasha Ptushko following the latter's death. However, it soon became clear that she was weaker and sicker than had been thought – exactly the kind of situation that Robert Miller cautioned against at the networking meeting in Washington. Rather than being treated as an outpatient as planned, she would have to stay in hospital the whole time. Nobody had anticipated such a scenario or the costs it entailed. The New York doctor waived his fees, and the hospital agreed to reduce its charges, but that still left a bill for US$80,000 for Svetlana's bone marrow transplant, which CitiHope settled two years later. This was 'the first part of a broad learning curve' as CitiHope found itself having to deal with an increasing number of such advocacy cases; Sharon Moore 'learned that an emotional "rescue" response had to be replaced with "wise facilitation"'. In practice this meant, for example, making sure that verbal agreements had been set down in writing before a child left for the United States. The most important insight for Sharon Moore, though, was the need to stop taking on responsibility for everything or trying to be a 'rescuer'.[389]

This recognition was closely connected to growing reflection on the moral issues surrounding children going abroad and selecting individuals for medical treatment; this had also figured at the 1991 Capitol meeting. The moral pressures felt by some activists stemmed in part from the fact that it was only possible to help a fraction of all the Chernobyl children,

---

[387] Ibid.
[388] See HA, Center for Civil Society International, Box 48–4.
[389] Carter, Christensen, *Children of Chernobyl*, 132.

and in part from the fact that there were also so many children in their own country who needed help.[390]

4.5 KNOWLEDGE TRANSFER

Starting in 1992, the 'For the Children of Chernobyl' fund held 'World after Chernobyl' congresses every two years. The congresses provided a distinctive forum for networking and knowledge transfer. Specialists from home and abroad presented their work on the medical, environmental, and social effects of the disaster to a broad, international audience that was largely drawn from civil society. The fund attached particular importance to giving 'home-grown' experts who had grown up in the Soviet Union a platform because their voices were often overlooked in an international context. After Aliaksandr Lukashenka took office, it became harder and harder for them to pursue their research and present their findings in their own country, too. The Belarusian nuclear physicist Vasil Nestsiarenka (Ru. Vasilii Nesterrenko), the Moscow biologist Elena Burlakova, the Minsk radiation biologist Evgenii Konoplia, the Belarusian sociologist Evgenii Babosov, and the Belarusian doctor Iury Bandazheuski all presented at the congresses several times.[391] Their papers addressed the national and global consequences of the explosion. Smaller workgroups had presentations on various aspects of international humanitarian collaboration to kick-start discussion. At the 1994 congress, for instance, Grushevoi advocated collaboration on the basis of partnership. He underlined that the fund stood for partnership, not paternalism, and had no time for projects that simply deposited humanitarian aid and left; only collaboration between equals could help people to help themselves, he told participants.[392]

The 'For the Children of Chernobyl' fund complemented the presentations by organizing a comprehensive programme of other activities aimed at giving delegates an insight into the reality of life in post-disaster and post-Soviet Belarus. This typically included visits to hospitals such as Aleinikova's

---

[390] Ibid., 133.

[391] Iury Bandazheuski, rector of Homel Medical University from 1990 to 2001, was arrested in 1999 and sentenced to eight years in 2001 after being accused of accepting bribes from students' parents. Human rights organizations denounced the arrest as a politically motivated response to the fact that Bandazheuski's work on the effects of Chernobyl, particularly low-level radiation, did not support the official position. He was released from imprisonment in 2005.

[392] Author's own records, Vladimir Bogdnanov, '"Die Welt nach Tschernobyl 8 Jahre danach"', 16 April 1994.

children's cancer hospital, research centres such as Nestsiarenka's Institute of Radiation Safety, facilities for children such as orphanages for children with disabilities, and state and non-state Chernobyl organizations. The fund also offered excursions to contaminated areas.

Delegates and the organizing committees did not shy away from expressing political views, either. They demanded support for democratization in the countries worst affected by the fallout, and for the development of renewable energy sources. According to the programme of the 1996 congress, this was the only way to stop being held hostage by the nuclear policies of major states.[393] Papers that framed the Chernobyl movement as part of worldwide opposition to nuclear power set out a clear mission for organizations involved in Chernobyl aid at home and abroad.[394] Before the construction of a nuclear power plant in Belarus became a certainty, delegates were discussing alternatives to the planned project with foreign experts and those affected by it.[395] As well as natural scientists, social scientists and humanities scholars also had a voice: David Marples, Canadian political scientist and friend of the Grushevois, for instance, gave a presentation on nuclear power.

On the fourth day of the 1996 congress, shortly before the tenth anniversary of the disaster, delegates passed a resolution, 'The World after Chernobyl is the World without Nuclear Threat to Mankind', aimed at national state institutions, international organizations, and the world community (Fig. 4.10). It held that the use of nuclear power was morally and scientific indefensible, and that instead of continuing to rely on it, Belarus should start introducing renewable energy and use that alongside traditional energy sources. The resolution also wanted energy-saving measures to be introduced and firmly established in society. The resolution called on the Belarusian government to develop a national programme for supporting small power stations for the immediate future up to 2010, and on the international community to pressurize the IAEA to remove support for nuclear power from its statute.[396]

In 1994, Grushevoi still hoped that that year's congress would provide an impetus for aid projects on the state level; civil society would not, he

[393] Author's own records, BBF, *Prahrama. The Third International Congress "The World After Chernobyl", Minsk 25–29 sakavika 1996*, March 1996.

[394] Author's own records, BBF, 2. *Kongress "Die Welt nach Tschernobyl", Programm*. 1994.

[395] Author's own records, Vladimir Bogdnanov, '"Die Welt nach Tschernobyl 8 Jahre danach"', 16 April 1994.

[396] Author's own records, 'Participants of the 3rd International Congress "World After Chernobyl", Draft Resolution "World After Chernobyl is the World without Nuclear Threat to Mankind"', 28 March 1996.

III МІЖНАРОДНЫ КАНГРЭС
"СВЕТ ПАСЛЯ ЧАРНОБЫЛЯ"

INTERNATIONAL CONGRESS "THE WORLD AFTER CHERNOBYL"

MINSK, 1996

ПРАГРАМА

THE THIRD INTERNATIONAL CONGRESS
"THE WORLD AFTER CHERNOBYL"

3.INTERNATIONALER KONGREß
"DIE WELT NACH TSCHERNOBYL"

Рэспубліка Беларусь, Мінск,
25-29 сакавіка 1996 года

FIGURE 4.10 Programme of the Third International Congress "The World after Chernobyl". April 1996.

believed, be able to alleviate the legacy of the disaster on its own.[397] Up to and including 1996, state representatives were involved in the organization

---

[397] Author's own records, Vladimir Bogdnanov, '"Die Welt nach Tschernobyl 8 Jahre danach"', 16 April 1994.

of the congresses, as well as participating as speakers – for example, Inesa Drabysheuskaia (Ru. Inessa Drobyshevskaia), minister of health and a qualified gynaecologist and radiologist, who came from the Homel oblast. Meetings with high-ranking political figures also took place, such as with Stanislau Shushekvich, chairman of the Supreme Soviet, in 1992, and with the mayor of Minsk in 1994.[398] By 1998, things had changed, and the fund found that it had no choice but to hold its fourth congress abroad – in Poland – for political reasons. This was a reflection of how much the situation had deteriorated following the KGB's audit of the fund in 1997 and Grushevoi's flight abroad. Accordingly, foreign partner organizations made a special effort to publicize the event, as a letter from Nadya Neal Hinson to supporters of Chernobyl Children Northwest demonstrates. Taking part in that year's congress was, she wrote, an important way 'to support and apply pressure on the Belarusian government, to keep doors open for the children to travel, for aid to be fairly distributed and to encourage those inside the republic who cannot speak for themselves right now'.[399]

Delegates to the congresses returned from their travels with knowledge that they passed on to initiatives and the wider public in their home countries in circulars, newsletters, presentations, and personal conversations.[400] The programmes and papers were translated and printed in several languages, and the congresses came to play an important role in spreading awareness outside the immediate circles of Chernobyl activists, as well as opening up a different perspective in the form of insights from Belarusian, Ukrainian, and Russian researchers, whose ideas were otherwise not at all easy to find. In the process, the congresses helped to create a better understanding of the post-Soviet space in more general terms. Activists were repeatedly met with questions in their home countries such as 'Where in the heck is Belarus?'[401] – and now they could answer them properly. This ultimately helped to break down the mental image of homogeneous blocs carried over from the Cold War and foster an awareness of how varied the post-Soviet space really was. The knowledge acquired at the congresses further helped activists to reassure themselves and others that their assistance really was needed and was making a meaningful difference in reducing suffering, particularly among children.

---

[398] Survila, *Daroha*, 94.
[399] Neal, *Silent Clouds*, 366–7.
[400] Cf. HA, Center for Civil Society International, Box 48–4, Oct. 1996.
[401] Ibid.

Alongside information transfer, generating knowledge, and establishing the ground rules for collaboration (especially with the 'For the Children of Chernobyl' fund), the congresses also served other purposes that were only indirectly related to Chernobyl activism. Belarusian exiles such as Ivonka Survilla from Canada and her husband, Janka, who had decided to support the Chernobyl children from afar, for instance, used the first congress in 1992 as an opportunity to return for the first time to the place that they still saw as home.[402]

CONCLUSION

In an article on charitable organizations in Russia published in the early 1990s, Anne White described 1989–91 as the 'best years of the independent welfare movement'.[403] This contemporary assessment is certainly a fair reflection of the sense of a new beginning and the extent of social mobilization in Russia. But the opening up and subsequent collapse of the Soviet Union also released an unprecedented wave of humanitarian activism in civil society in the West. The Chernobyl children presented countless individuals and initiatives in the West with new ways to get involved, both at home and in the uncharted territory of their former enemies. The adventure of working on the 'wild frontier of international relations', as one of the pioneers, the Nazarene Michael Christensen, put it,[404] held a special appeal for a great many people. For a time, everything seemed possible in both East and West. The world had come a little bit closer together: Western NGOs saw the Chernobyl children as 'children on the other side of our small earth' who self-evidently deserved their attention.[405] The media had a crucial role in this process: they closely followed efforts to support the Chernobyl children from the very beginning, and thus played a significant part in the emergence and consolidation of the transnational aid effort.

---

[402] Survila, *Daroha*, 94. The extent to which they had lost touch with reality in the country can be seen in their disappointment that Belarusian speakers did not deliver their presentations in Belarusian, and that there was no interpreting into the mother tongue that was so important to them. Instead, they were forced to choose between a Russian or English option. In subsequent years there was simultaneous interpretation into Belarusian as well, even though Russian remained the main language in everyday use in Belarus up to the end of the 1990s.

[403] White, 'Charity', 797.

[404] Quoted in Carter, Christensen, *Children of Chernobyl*, 202.

[405] HA, Center for Civil Society International, Box 48–4.

Once it became clear that the Soviet Union was not in a position to cover the treatment and recuperation needs of the Chernobyl children – particularly where high-quality options were concerned – the Soviet leadership decided to ask other countries to host children for limited periods. Initially, this was confined to seriously ill children going to socialist Cuba, but it was not long before the state was also sending children to the United States – its former enemy – whether they were ill or not. In order to fund these trips, the state turned to new fund-raising methods, such as telemarathons, that were imported from the West. Parallel to the state measures, which were soon at the limit of their financial and organizational capacities, non-state organizations – above all, the 'For the Children of Chernobyl' fund – began advocating for the Chernobyl children in the West. They were so successful in this that they were soon handling almost the entire effort to send children away for recuperation.

As this chapter has shown, it was generally religious groups that took the first steps. They brought with them Christian ideals and traditions that even high-ranking representatives of the collapsing state could accept and appreciate, as a means of boosting the morale of a population that was experiencing increasing hardship or as a way of reinforcing their own position. It soon became apparent, however, that the newly won freedoms in the East also brought new challenges of their own as well. Cracks were appearing everywhere and had a profound effect on individual lives. The revelations of *katastroika* left many people bitterly feeling that everything before had been grounded in a lie.[406] Moving forward meant finding new sources of meaning and new ways of compensating for the lost certainties. In Belarus, the cause of the Chernobyl children fulfilled such a role for functionaries and critics of the state alike.

Those in the East derived affirmation, recognition, and new perspectives from the unprecedented foreign interest in helping the Chernobyl children. But that same attention also tested the capabilities and experience of all concerned, including both the 'For the Children of Chernobyl' fund itself and its ever-growing network of foreign partner organizations. Improvisation and decision-making on the fly were the only option in many cases; there were no guaranteed recipes for success. The West had never been involved in the (former) Soviet Union like this before. The double-crisis situation – post-Chernobyl and post-socialism – was new for

---

[406] I am alluding here to Stephen Kotkin's formulation: 'All previous life was revealed a lie.' Kotkin, *Armageddon Averted*, 70.

everyone, including organizations that had many years of experience in humanitarian aid work.

In the first ten years, there was a pragmatic focus on getting things done. Humanitarian efforts largely unfolded through ad hoc decisions – verbal agreements and spontaneous actions.[407] The material context – a lack of modern means of communication (faxes, computers, etc.) and a laborious infrastructural context (e.g. when travelling to Belarus) – often posed a further hindrance. Learning, negotiation, and compromise were necessary on every level. Charismatic figures such as Paul Moore or the Grushevois sought to lay down the rules of the game. The Belarusians were fighting on two fronts at the same time. They were trying, on the one hand, to help the Chernobyl children and, on the other hand, to drive the democratization of the country forward – aid efforts by parties other than the state were always political, even if not openly so. The main source of friction between aid organizations was the question of relationships with the state. The 'For the Children of Chernobyl' fund rejected collaboration with the state outright and demanded that its foreign partners do likewise. The fund saw its existence as constantly under threat: if the idea of repression by Soviet Party and state organs was the Grushevois' main worry in the early period of their work, this dovetailed almost seamlessly with later fears of being the target of conspiracies that even extended into the West. This perception of imperilment came to a head with Lukashenka's moves towards an authoritarian police state, which the fund read as initiating a direct attack on it. It was not alone in this interpretation of events. Practically all the non-state organizations that did not profess loyalty to Lukashenka found themselves in a 'constant struggle for survival' from the mid-1990s onwards because of these – real or perceived – threats.[408]

---

[407] On the concept of adhocracy, see Dunn, 'Chaos of Humanitarian Aid'.
[408] Bekus, *Struggle over Identity*, 114.

5

The Chernobyl Children

'THAT FEELING OF FLYING'

In an interview for *Her Life* magazine, Tatsiana Khvitsko recalled 'that feeling of flying' when she ran for the first time. It was still a long way from there to her first marathon in 2018, but getting there was nothing in comparison to that first run – for Khvitsko had no lower legs and only four fully developed fingers when she was born in Minsk in 1990. 'I'm flying because I have no feet', as she put it. A Florida firm provided her with her first running prostheses in 2011, when she had already been living permanently in the United States for three years.[1] For eleven years prior to that, she had spent each summer there, turning from a 'feisty' girl whose plight was emblematic of the 'Chernobyl babies' into an impressive 'Chernobyl woman'.[2] As well as being held in high regard for her sporting achievements as a runner and bodybuilder,[3] she has also become a sought-after motivational coach and a source of inspiration for numerous children and adults.[4]

Olga O., who was born near Minsk, was aged seven when she first went to the United States as a Chernobyl child. She too has made the country her home. She was diagnosed with alopecia (hair loss) when she was three.

---

[1] Quoted in Cindy McDermott, 'Tatsiana Khvitsko: "The Joy of Running Makes Me Unstoppable!"'. *Her Life Magazine*, 2017, https://bit.ly/3ZYuuHT (4 May 2024).

[2] Ibid.; Sophia Rosenbaum, 'Chernobyl Woman Born without Legs is Competitive Bodybuilder', *New York Post*, 14 October 2015, https://bit.ly/49O8gvJ (4 May 2024).

[3] Jay Akbar, 'Woman Born with Only Four Fingers and No Legs after Exposure to Chernobyl Radiation while in the Womb Defies the Odds to Become a Bodybuilder', *Daily Mail Online*, 14 October 2015, https://bit.ly/3P9uJcM (4 May 2024).

[4] I would like to thank Michelle Carter for bringing Khvitsko to my attention.

US specialists later confirmed the diagnosis; the condition was irreversible. At the age of eight, during her second visit to Washington State, she was given a permanent hair replacement. Robert Sorbo, an Associated Press photographer, captured the scene at the Men's Hair Club with eloquent images that reached audiences far beyond the usual circles of Chernobyl children organizations; one particularly iconic picture was disseminated all over the world in various newspapers.[5] Today, only a few people know that this confident young woman wears a wig.[6]

This chapter is dedicated to the 'Chernobyl children' themselves. That label will probably remain attached to them for the rest of their lives; thus, the chapter is concerned with the individuals subsumed by the term, and considers the ways in which being defined as such has affected how people see themselves. What connects the Chernobyl children with the disaster, and what do they themselves associate with it? The chapter also examines the children's experiences abroad and the impact the visits had on their lives as they grew older; the political and social upheaval in the former Soviet Union makes its presence felt here. Throughout, the Chernobyl children's own accounts are set in dialogue with the perspectives of the many other parties with which they were – whether they consciously noticed it or not – confronted.[7] The children's own families, their foreign host families, the (mostly female) adults who accompanied them, representatives of the organizations that invited them, those who contributed financial support, and journalists all played a role in constructing the Chernobyl child.

5.1 WHO ARE THE CHERNOBYL CHILDREN?

When the champion boxer Wladimir Klitschko said that all Ukrainians were 'children of Chernobyl',[8] he was echoing the reasoning that many aid initiatives had adopted from the mid-1990s onwards. Essentially, any child from the republics of Belarus and Ukraine could be a Chernobyl

---

[5] Cf. AP Images, 'Chernobyl Aftermath', https://bit.ly/3P4TUxf (6 November 2019).

[6] Ol'ga O. to Melanie Arndt. Email, 23 October 2017.

[7] The accounts of the Chernobyl children document how they see themselves in retrospect; this does not, of course, necessarily always reflect how they perceived things as children. On the methodological challenges of doing oral history with contemporary witnesses to events who have since grown up into adults, see Martina Winkler, 'Kindheitsgeschichte: Version 1.0', *Docupedia-Zeitgeschichte*, 17 October 2016, http://docupedia.de/zg/Winkler_kindh eitsgeschichte_v1_de_2016 (4 May 2024).

[8] See Chapter 1.

child. It was generally much less common for children from the Russian Federation to be classed as such: there were certainly sufficient irradiated landscapes there, but the country's geographical dimensions and the size of its population meant that a much smaller proportion of children overall were affected than was the case in Belarus and Ukraine.[9]

Despite this broad understanding of what constituted a Chernobyl child, the children were also conceptually divided into various groups in their home and host countries on the basis of how much they were (or were thought to be) suffering and in need of help. On the one hand, the resultant hierarchies were useful when it came to making a convincing case to donors and state authorities. On the other hand, they also served a purpose for those involved in the humanitarian response to Chernobyl, helping them to better understand what they were trying to achieve and define the scope of their work more clearly. These hierarchies were based on geographical provenance, social factors, and health needs. A flyer for the North American public from the Belarusian 'For the Children of Chernobyl' fund explained that it was

giving priority to children from contaminated areas of Belarus who come from low-income families, from large families and from families of evacuees, invalids and 'liquidators'. These are the children who would have little opportunity for recuperation without the assistance of our Fund. Although the children appear to be healthy, their immune system has been weakened as a result of the lack of vitamins in food and ongoing exposure to low levels of radiation.[10]

The fund thus decided who was able to travel as a Chernobyl child on the basis of a wide range of factors. The formulation 'giving priority' generally entailed a degree of flexibility that also allowed other children to be included. It is notable that the fund felt it had to explain that children's outward appearance was not indicative of how badly they needed help; this was necessitated by the fact that the children who went abroad generally looked very different from the images of the Chernobyl children propagated in the media and informational material, which largely involved children who had physical deformities or were seriously ill.

---

[9] The definition of 'Chernobyl children' had a much narrower scope where Russia was concerned, and was generally limited to children from the Briansk oblast, where the Chernobyl fallout was at its worst.

[10] Private Archive Pankrats, 'The Belarusian Charitable Fund "For The Children of Chernobyl"'.

Activists all had in common the fact that they attached an ethical significance to the children, giving them the role of 'natural' bringers of peace, messengers from a better future, 'citizen diplomats'. As Neal Hinson put it, 'Children ... are often our most ardent peace seekers.'[11] This idealization put a huge amount of pressure on the children and the organizations working with them to fulfil expectations that they would never be able to meet. It did not help that such glorification of the children often glossed over substantial differences in how childhood was conceived. The idea of children as part of the collective was still predominant in Eastern Europe during this period of upheaval, whereas adults in the West were increasingly treating children as independent individuals.[12] At times, however, the reverse held true where perceptions of the Chernobyl children were concerned; I shall return to this point later.

5.1.1 'True' Chernobyl Children: Children from Contaminated Areas

Practically all initiatives displayed a preference for 'directly affected' children in their work. The criteria listed by the 'For the Children of Chernobyl' fund, for instance, included children's geographical provenance ('from contaminated areas') and the status of their parents ('from families of ... invalids and "liquidators"'). The fund also treated children from socially disadvantaged families ('from low-income families, from large families') along similar lines. According to figures for 1990 to 1993 produced by the fund, the vast majority of the 38,884 children who went abroad with the help of its network during that time came from the two southern oblasts of Homel and Mahileu, which were particularly badly affected by the radioactive fallout.[13]

Even so, far from all 'directly affected' children made regular trips abroad as Chernobyl children, not least because the schemes began too late for those who were already teenagers at the time of the disaster. The story of Alena Oginets is a representative example of this. She was already seventeen when she went to Bavaria in summer 1990 on her first and only trip abroad as a Chernobyl child. Her visit was organized by the Ukrainian Chernobyl Youth Fund. The fund's president, Oleksandr Bozhko, who had recently returned from Cuba, accompanied the group, which was

---

[11] Neal, *Snapshots*, 44.
[12] See Chapter 1.
[13] Private Archive Pankrats, BBF, 'Number of Children of the Republic of Belarus, Who Travelled Abroad in 1990–93'. (1994).

comprised of children of *OLBshniki* ('liquidators' who had either died or were seriously ill).[14] Oginets recalled in a Skype interview that she had been very reluctant to go abroad to a foreign country and a family of strangers whose language she did not speak – yet she had been impressed and won over by the genuine interest in her plight that she encountered and the kindness displayed by her host family.[15] When schemes for sending children up to the age of eighteen abroad began in earnest in 1991, Oginets was already too old to participate. She had been to Artek, and then to a sanatorium in Transcarpathia, but that was it. Such trips were so popular that they had to be allocated at random, and Olga's luck had, she explained, run out. Despite being a Prypiat evacuee and daughter of a 'liquidator', and thus among the 'most affected' Chernobyl children, she was barely able to benefit from the opportunities to recuperate in other countries; for the most part, younger Chernobyl children spent a greater amount of time abroad.[16]

Oginets and her husband, who also comes from Prypiat, later tried to obtain one of the free recuperation trips to which they were entitled by virtue of being officially classed as 'affected'. However, there were not enough places to accommodate all those who were eligible. Those who wanted to apply had to face hours of queuing; Oginets and her husband once took turns waiting in line through the night in order to get on the list as early as possible. The effort was worth it – they got to go to Kherson on the Dnipro delta – but they soon decided not to apply again; Oginets's husband had to work in the 'zone' in summer, so they had little time for travelling anyway.[17]

In the early years of the schemes for sending children abroad, it was not uncommon for potential host families to be concerned about the idea of hosting those 'most affected': radiation was just as badly understood in the West as in the East. Some Western host families were afraid that taking in a Chernobyl child might literally mean bringing Chernobyl into their homes as well. Organizations often had to explain to hosts that children from contaminated areas were not a health hazard. As the North American flyer circulated by the 'For the Children of Chernobyl' fund

[14] *OLB* stands for *ostraia luchevaia bolezn'* ('acute radiation sickness').

[15] Oginets, interviewed by Arndt; Oginets and her husband later visited the Bavarian family again on their own initiative, but at some point they stopped keeping in touch.

[16] This impression is based on my own research and the statistical information available (which is sparse and does not come close to covering all the Chernobyl children; cf. Gatal'skaia, Zaitseva, *Sotsial'no-psikhologicheskii analiz*).

[17] Oginets, interviewed by Arndt.

explained: 'These children pose no health risk to other children. What they need is wholesome food, fresh and clean air, and a little love. Experience has shown that these recuperative visits have been very beneficial to the children.'[18]

5.1.2 'By Far the Saddest': Children from Orphanages

'Of all the Chernobyl children they were by far the saddest', Nadya Neal Hinson wrote of children from orphanages.[19] The appeals of the Lenin Children's Fund and its successor organizations had done nothing to change the – in some cases shocking – conditions in residential facilities for children with disabilities, particularly outside the larger towns and cities. They had, however, at least helped to make sure that such homes were no longer completely out of sight and out of mind. It became commonplace for all the young residents of such facilities to be described as Chernobyl children from the 1990s onwards; this was probably both a reflection of desperation and an attempt to get the attention of foreign donors.

Like Tatsiana Khvitsko, the children in these homes were generally not orphans in the strict sense of the word: their parents were still alive but had, because of the children's disabilities, chosen not to take them home when they were born. After the BNF's Chernobyl children committee managed to get the Slauharad children's home relocated in autumn 1989,[20] (social) orphans were the object of ongoing attention. Even so, children with disabilities did not travel abroad until much later than other Chernobyl children. Those with significant physical or cognitive disabilities, in particular, were almost entirely excluded throughout the lifetime of the schemes. Host families felt unable to cope with the special arrangements that might be needed; the additional costs and workload this entailed were beyond what most groups could contemplate.

Neal Hinson wanted to bring children with minor disabilities, at least, to Washington State. In 1994, for instance, two children with speech impediments came to Seattle. She later visited them in 'Velyeka' (probably Vileika, around 100 km north-west of Minsk) following their return to Belarus, turning up unannounced at the facility in December 1994 in order to 'see the reality ... without the usual "special" attention given to foreign

---

[18] Private Archive Pankrats, 'The Belarusian Charitable Fund "For The Children of Chernobyl"'.
[19] Neal, *Snapshots*, 116.
[20] See Section 4.2.1.

guests'.[21] She was shocked by the destitution and despondency there. There was not enough food, neither for the children nor for the staff. The institution's funding from the state – which had run out in October – was not sufficient to enable anyone to live or work there with dignity. The director, as Neal Hinson put it, 'apologized with resignation' for conditions that he felt powerless to change.[22] Only a small number of those working in such homes were suitably qualified; there was little social recognition for their work. Even if the *sanitarki* (the women who cared for the children) tried to keep things clean, the facilities were filled with an unpleasant smell of urine, *kompot* (a drink made from dried fruit cooked in water), overcooked buckwheat gruel, and disinfectant.[23]

Given the widespread impoverishment in the children's homes – and in Belarus as a whole – it came as little surprise to activists such as Neal Hinson that humanitarian aid from abroad did not always go straight to the children and that staff at the facilities thought of themselves and their own families before anyone else.[24] Numerous organizations and individuals made similar observations, but the tolerance articulated by Neal Hinson was far from representative. The extent to which people took advantage of the opportunity to benefit for themselves varied considerably. In some cases, well-meaning foreign donors were presented with requests that had nothing to do with the realities of the situation – for example, for video cameras in a facility where even the most basic needs were lacking.[25] In others, donations simply disappeared.

The 'power of one'[26] – the plight of a single child whose individual circumstances could be changed for the better – played a crucial role in efforts to help children in residential facilities. If a personal connection developed, it was not uncommon for people who came from abroad to offer their help to subsequently feel responsible for a child's future for years, if not for the rest of their own lives. They took heart from quick and uncomplicated early successes, such as the happiness and gratitude that children displayed when given toys. The misery, neglect, and hopelessness inside the homes were such that, in the first instance, it did not take much to make a tangible difference to the plight of children there. The lives of

---

[21] Neal, *Snapshots*, 117.
[22] Ibid.
[23] Personal observations of the author in various Belarusian children's homes between 1996 and 1999; see also Phillips, 'No Invalids'.
[24] Neal, *Snapshots*, 117.
[25] Personal observation of the author, spring 1997.
[26] Kristof, 'The Power of One'.

many Chernobyl children – including some of the 'saddest' among them, as Neal Hinson put it – took a turn for the better in the guise of adoption. Chernobyl children initiatives in Ireland and Italy were particularly active when it came to adopting children with disabilities, and it was in those countries that the greatest success stories were to be found, starting in the mid-1990s. Achievements also came in other ways, too, such as when foreign initiatives helped to improve children's lives closer to home.

Nonetheless, children often had to be left to their fate in residential facilities without it being possible to offer them hope for the future or any prospect of a rapid improvement in their circumstances. As easy as the initial successes may have been, 'saving' the children turned out to be a very different proposition. Neal Hinson wrote that she 'felt like the worst kind of traitor' when she said goodbye to one of the orphans that she visited at the children's home in Belarus after he had returned from recuperation in Seattle:

We had healed him, gave him a home and brought him back to this. Sometimes kindness really is cruelty in disguise. I think about him now, wanting him to have Christmas presents, a dog, a future and yet I know that I must walk away without looking back. Will he remember all of this and hate America someday? We promised to ship clothing and protein powder but would Denis ever get any of it? Certainly, he would never know it was from us.[27]

It took a long time for perceptions of people with disabilities in Belarusian society to start changing. The impetus often came from foreign initiatives, as in the case of the Belarusian Aid Association for Children and Young People with Disabilities (Beloruskaia assotsiaciia pomoshchi detiam-invalidam i molodym invalidam).[28] It originated in the context of international efforts to help the Chernobyl children in the mid-1990s. Later, USAID and the United Nations funded various inclusion programmes. As the change in attitudes became cemented, invoking the idea of 'Chernobyl children' became increasingly unnecessary. This can be seen from the first Miss Wheelchair contest, held in Warsaw in 2017, where two Belarusian women competed for the title in the final round.[29] Neither was a 'true'

---

[27] Neal, *Snapshots*, 117.

[28] The association's early origins are reflected in the fact that its name in both Russian and Belarusian includes the word *invalid*, the usual term for a disabled person in the Soviet Union. Organizations founded later, such as Different – Equal (Raznye ravnye), avoid the term almost entirely because it is considered discriminatory by European and North American groups. Cf. the websites of two organizations: http://belapdi.org (1 March 2024); http://www.rrby.org (1 March 2024).

[29] See The Only One Foundation, *About Us*, https://bit.ly/3VPl3Z3 (4 May 2024); and, on one of the Belarusian finalists, USAID, *Changing Societal Perceptions of People*

Chernobyl child, but they would probably never have participated in a beauty contest for women with disabilities if it had not been for the international effort to support the Chernobyl children.

5.1.3 'Every Child . . . at Risk'

The sociologist Vladimir Lupandin presented a scathing assessment of the early opportunities to recuperate abroad (which at this point were still being arranged by the state) in the *Literaturnaia gazeta* in December 1991. 'The recuperation trips are pointless', he wrote of 'the trips to Sochi, to Cuba': 'It is the children of trade representatives, of apparatchiks that travel, even if they are in no need of help.'[30] This idea – that the only people who benefited were offspring of the *nomenklatura* who did not really need help or who were simply taking advantage of naive foreigners and their goodwill – is as old as the practice of sending Chernobyl children away itself. There is no denying that Party and state functionaries, as well as parents who were well connected in other ways, could use their influence to ensure that a child got one of the coveted places on the list of those travelling. This was particularly true for trips to the United States or other Western countries, which in the early days were far from commonplace. There is, however, insufficient evidence to determine how widespread such manipulation really was; the recollections of those involved in organizing the trips and of the children who went on them are not in themselves enough to allow clear conclusions to be drawn. Nor should the simple fact that a child came from a functionary family lead to assumptions about the extent to which they were or were not 'affected' by Chernobyl and in need of recuperation (either at home or abroad).[31] Neal Hinson was very clear that 'every child in Belarus was at risk'.[32] Like other activists, she was already arguing early on that it was not just children from evacuated areas that should be taken abroad. All children, she wrote, were at risk because of the nutritional situation. She also observed that children in well-known contaminated areas had the attention of numerous aid organizations, whereas those in less 'visible' areas were disadvantaged when it came to getting places on trips abroad or

---

with Disabilities in Belarus: Art and Media Put the Spotlight on Talent and Potential, www.usaid.gov/node/226636 (4 May 2024).

[30] Podgoronikov, 'My ot prirody ne sposobny k demokratii?': For 'every child . . . at risk': Neal, *Snapshots*, 115.

[31] See Section 4.2.1.

[32] Neal, *Snapshots*, 115.

receiving humanitarian aid in their own country.[33] Neal Hinson's Chernobyl Children Northwest initiative did, in fact, extend its work to the Minsk oblast – for example, the *atomgrad* of Druzhny, Olga O.'s home town – in recognition of this. In such cases, there was not much left separating Chernobyl aid from wider efforts to help children in under-developed countries in general.

Given that every child was essentially at risk, there were circumstances in which the label 'Chernobyl child' could become a sought-after status. In a manner indebted to the place of competition in Soviet society, getting to go abroad as a Chernobyl child became a mark of distinction: for the best in maths, the winner of an art competition, or the most engaging folk-dance group. Aid organizations took advantage of this for marketing purposes. The best or most touching drawings by Chernobyl children could be auctioned to raise money for Chernobyl aid or the children's educational institutions.[34] Chernobyl children with a gift for music or dance brought authenticity to fund-raising events abroad, attracting more donations as a result. These star Chernobyl children themselves, who were meant to represent not just the disaster but a land of talent, had mixed feelings about the experience. The dancer Svetlana Bodrunova looked back on her travels as a Chernobyl child not as 'having fun abroad' but as a demanding 'performance tour'.[35]

Children's creative responses to what they had experienced – most of all, particularly impressive and/or expressive drawings, paintings, and writing – figured prominently in national and international efforts to help the Chernobyl children. In Germany, *Der Spiegel* published the first children's drawings as early as May 1990 – Irina Grushevaia had brought them with her.[36] Viola, an NGO run by the biologist Liudmila Zhirina in the Russian city of Briansk, organized painting and writing competitions for Chernobyl children for ten years. The idea here was not primarily to choose children to be sent abroad; instead, Zhirina wanted to present children's perspectives on the disaster. Common themes included activities that were not possible in a post-Chernobyl childhood: playing outdoors, gathering mushrooms, or tapping birch sap in the forest in spring. The texts and images also document other experiences of loss – for example, grandparents who were

---

[33] Ibid.

[34] Cf. Private Archive Pankrats, Alla Stulova, 'Kleny v Chausakh pokhozhi na kanadskie' *Vechernii Minsk*, 12 December 1995.

[35] Bodrunova, 'Chernobyl in the Eyes'.

[36] 'Reaktorkatastrophe: Retter nicht in Sicht', *Der Spiegel*, 7 May 1990.

hardly known because they chose to remain in the 'zone' and thus could be visited only rarely, if at all. In other cases, the sense of loneliness is palpable. Monsters and dark, malevolent creatures figure in many pictures, symbolizing radiation, danger, and death. The short accompanying texts also tell of fathers who had become alcoholics because vodka was vaunted as a means of warding off the effects of radiation.[37]

Artwork by the children was not so much an 'authentic' document of individual experience as – particularly if it was chosen for publication – a reflection of what could be said and depicted where the disaster was concerned, and of the memories of it that society deemed acceptable and wanted to cultivate. This was particularly pertinent when the children's work was being used to decide who got to go abroad to recuperate. It is also likely that it led to children's memories of the disaster becoming aligned.

5.1.4　The Selection Process

Host families had a say when it came to decisions about the kind of children who came to stay with them. They generally wanted someone whose age and gender would be well matched to their housing situation and existing children. Such preferences could present a challenge. Rather than being able to work through waiting lists sequentially, organizations – almost always under pressure of time – had to actively pick out suitable candidates. Preferences regarding age and gender, on the one hand, and the expectation that a child would be genuinely in need, on the other, could be mutually exclusive – something that was not always easy to explain to host families.[38] Conversely, the Chernobyl children's own families were often unhappy because of the lack of clarity about how the process worked – they felt they had been treated unfairly if other children received permission to travel even though their names had been added to the list later.

On occasion, the trips could also be used akin to currency in bargaining processes of the kind that were common in Soviet society. In a newspaper interview on the twentieth anniversary of the disaster, Gennadii Grushevoi related how he himself had taken advantage of this. The children from the Slauharad home were eventually moved

---

[37] Ludmila S. Zhirina, *Chernobyl through the Eyes of Children: A Compilation of Children's Drawings and Tales* (n. p., 2013).
[38] Diane Dillon, interviewed by Melanie Arndt, 10 June 2015. San Rafael, CA.

permanently to Bialynichi (Ru. Belynichi). Their new facility, though, was in a poor state of repair. There was water in the cellar for four years, but nothing was being done about it. When it became clear that enquiries were leading nowhere, Grushevoi turned to senior engineers at three enterprises and proposed a deal: 'Let's do it like this: 120 children of workers at your enterprises go abroad this summer for humanitarian reasons, and you mobilize parents to fix the home at no cost.'[39] Within a month and a half, Grushevoi recalled, the basement was dry – and stayed that way. The parents would not even accept money to pay for the materials used in the work.[40]

The Belarusian organizations tried to find ways of accommodating the wishes of host families. They reconciled themselves pragmatically with the fact that they could not live up to all the expectations placed on them, however frustrating this was.[41] Neal Hinson also recalls 'strained moments and disagreements' when it came to making decisions about which children were to travel and which ones were not; the weight of the responsibilities associated with the selection process was not, she felt, always properly appreciated.[42] Problems could arise even if a host family's wish for a child 'in particular need' was met: hosts did not always understand what taking in a socially disadvantaged child entailed. There could be considerable frustration if children turned out to have behavioural issues (e.g. because they found it hard to follow rules), yet the Belarusian organizations were sometimes concerned to facilitate trips for precisely such children, not least in the hope of improving their behaviour.[43]

For all the challenges it entailed, many host parents were adamant that they wanted to host a child in genuine need, rather than a privileged one. This did not, however, necessarily have to mean getting incontrovertible evidence of hardship – personal impressions were often enough. After hosting a girl who came across as 'kind of a princess' because she was 'a little privileged at home', or at least behaved that way, the McClellands in California decided 'If we're going to do this again, I'd rather get somebody that really needs [it].' The next child they hosted met these expectations: 'The boy that we had desperately needed help.' He hardly had any other clothes with him and seemed very modest. He had a Matchbox car that

[39] Quoted in Gennadii Grushevoi, 'Esli by Chernobyl' sluchilsia segodnia, my by tozhe ne uznali pravdu', *Khartyia '97*, 27 April 2006.

[40] Ibid.

[41] Neal, *Snapshots*, 127.

[42] Neal, *Silent Clouds*, 304.

[43] Irina Narkevich, interviewed by Melanie Arndt, 2 September 2010. Minsk.

had lost all its windows, but he cherished it nonetheless: 'That was his prized treasure to bring with him on the plane. And I just thought – yeah', Julie McClelland recalled.[44]

The expectations – which could amount to a sense of entitlement – on the part of the organizations that invited the children did not just pose practical problems for their counterparts in the children's home countries; they increasingly became a source of frustration as well. In April 2006, Grushevoi complained about a Canadian organization that was unhappy because his fund had again sent a child from relatively well-off circumstances rather than one from a disadvantaged background, as had been requested. Grushevoi held Cold War stereotypes to blame for the Canadians' behaviour:

We simply cannot escape from how things were before Chernobyl! I have been explaining for sixteen years that the children we send abroad are those whose health needs to be protected, not the ones that need to get to wear your jeans. We are sending these children for the simple reason that, even now, radiation is still a hazard. But the world is governed by the principle of needy/not needy.[45]

This kind of thinking, an exasperated Grushevoi went on, explained why 'a man living in Canada' could decide that a family of five from the Brahin area did not qualify as being in need.[46]

5.2 BETTER TO RECUPERATE AT HOME OR ABROAD?

Starting in the mid-1990s, a debate about the health benefits and moral risks of going abroad erupted in the wake of remarks by President Lukashenka. Critics spoke of 'recuperation tourism' (*ozdorovitelnyi turizm*) and a 'Chernobyl children industry'.[47] Some also claimed that the Chernobyl children were subjected to a 'culture shock' when visiting other countries, and this too became a source of contention.[48] These controversies reflected

---

[44] Debra Zapata, Julie McClelland, interviewed by Melanie Arndt, 11 April 2016. Santa Rosa, CA.

[45] Quoted in Grushevoi, 'Esli by Chernobyl' sluchilsia segodnia'.

[46] Ibid.

[47] 'Ozdorovlenie "chernobyl'skikh detei"'; Stefano Lorenzetto, 'L'industria dei bimbi di Chernobyl', *Il Giornale*, 30 September 2006, www.ilgiornale.it/news/l-industria-dei-bimbi-chernobyl.html (4 May 2024).

[48] On the concept of culture shock, see Arndt, *Tschernobylkinder*, 552–3; R. Michael Paige (ed.), *Education for the Intercultural Experience* (Yarmouth, ME: Intercultural Press Inc, 1993); Paul Pedersen, *The Five Stages of Culture Shock: Critical Incidents around the World* (Westport, CT: Greenwood Press, 1995).

how society negotiated the changes it was undergoing in a post-socialist country where authoritarian structures had returned and efforts were being made to reinstate state hegemony.

5.2.1 Tomatoes with Smetana: The Culture Shock Argument

Critics of recuperation abroad worried that it was psychologically too much for the children, who were often still very young. They argued that spending four or more weeks in completely unfamiliar surroundings, with no knowledge of the language and culture of the host country, entailed more stress than it did respite and might even have a negative impact on the children's health. These opponents of sending children abroad argued that they would be better off recuperating in their own country, where they would be spared the burden of havening to adapt to unknown surroundings and situations and spared the burden of homesickness.

On a political level, meanwhile, there was a concern that the children might be being morally corrupted by the liberal and consumerist Western lifestyle. President Lukashenka expressed such concerns repeatedly.[49] Sviatlana Aleksievich's *Chernobyl Prayer*, a compendium of witness interviews rendered as literary monologues, also includes voices that touch on the culture shock theory. A teacher, for instance, observes on the one hand that nobody talks about Chernobyl with the children at school – people only talk about it with them in other countries. On the other hand, he describes how children who are sent abroad seem to lose the ability to act independently; they 'become spectators', as he puts it: 'They just look, instead of living' (precisely the opposite of what the foreign organizations were actually trying to achieve). Their lives are defined by anticipation of the next trip and the gifts it will bring, he says, before explaining how he tries to snap them out of this state of passivity and help them rediscover their agency in order to move on, just as he himself had managed to do after the siege of Leningrad.[50]

The culture shock debate overlooked the fact that for some considerable time, the system simply did not have the capacity to enable all eligible children to recuperate in their own country. Even as late as November 2008, Grushevoi felt that recuperation to an appropriate

---

[49] See Section 4.2.1.
[50] Svetlana Alexievich, *Chernobyl Prayer: A Chronicle of the Future*, tr. Anna Gunin and Arch Tait. (London: Penguin Books, 2016), 134–5.

standard was not possible in Belarus.[51] This may have been something of an exaggeration: facilities such as the Nadezhda recuperation centre or (financed by the state Belarusian Children's Fund) the 'Raduga Nadezhdy' respite programme for children with cancer had been around for quite some time. Nonetheless, Grushevoi's remarks were indicative of a widespread view among Belarusians. Confidence in the options available at home was lacking; recuperation abroad seemed far more attractive. Alongside material considerations, increasing significance was attached to the broadening of cultural horizons, as reflected not least by the Chernobyl children interviewed for this book. They echoed Grushevoi's assessment: even if they were positively disposed in principle to the idea of recuperation in uncontaminated locations in their home countries, they were not confident that it would be of the same standard as recuperation abroad. In the case of many Chernobyl children, however, this was not a view grounded in personal experience: they were not familiar with what was available in their own country precisely because they had been sent abroad. The context was very different, too, because children would often stay with host families abroad, rather than in the shared accommodation that was the norm at home. At any event, those interviewed agreed that experiencing a new country and a new culture was the deciding factor.

The criticism of sending children abroad, as well as the claims of culture shock, were notorious. Grushevoi's line was that sending the children abroad was important because they were his last hope for the country. He conceded that they might behave just like their parents to start with, 'trying to appropriate chocolate here, trainers there'.[52] But he also reasoned that the children could not be made worse, only better; many, he argued, had already changed, and this in turn was influencing their parents. 'If only you knew how many children out in the countryside speak a foreign language thanks to our work', he enthused in a newspaper interview. His fund, he said, had initiated a 'revolution' as early as 1991, when the parents of Chernobyl children had called for foreign languages to be given more attention at school. Grushevoi recognized that many people had strong feelings about the risk of culture shock – but if there had been one, in his view, it could only be seen as a good thing. Even in the 1990s, he observed, village children had not

[51] Gennadii Grushevoi, 'Dlia ozdorovleniia detei v Belarusi "poka net podkhodiashchikh uslovii"', *Telegraf.by*, 14 November 2008.
[52] Quoted in Grushevoi, 'Esli by Chernobyl' sluchilsia segodnia'.

known how to use a toothbrush; even basic skills like this had been acquired abroad. That is what the trips were all about, in his eyes: 'a quantitative accumulation that after a certain amount of time becomes qualitative in nature'. Far more was at stake here for Grushevoi than simply brushing one's teeth properly: he hoped that Belarusian society would gradually develop for the better, which for him meant democratization most of all.[53]

The comments of the Chernobyl children are more or less in line with Grushevoi's views. Like him, for the most part, they associate 'culture shock' with positive experiences, particularly in retrospect. For the second and third generations of Chernobyl children in particular (who went abroad for the first time around the turn of the new millennium), the concept was meaningless. Individuals who described themselves as open-minded and ultimately remained abroad had particularly fond memories of their encounters with a new culture. The words of Olga O. are emblematic of this: 'All my life, I have found it easier than many other people I know to get used to new things. I liked American culture much better than my own because the Americans are far more discreet, don't ask about things that aren't necessary, and are a lot more friendly as well.'[54] The open-mindedness and goodwill encountered abroad is a recurring theme for Chernobyl children. Cuba, the United States, Australia, Germany, Italy – no matter where they went, the Chernobyl children enthusiastically recall smiling people on the street and the good-natured atmosphere in their host families. Those who attempted to import such easy-going ways into everyday life back home, however, were in for an unpleasant surprise: 'When I got back to Belarus, I decided to smile at everyone on the street. They looked at me as if I had gone mad', Tatstsiana Kazyra recalls.[55] It was not the done thing to be overly friendly in public in the former Soviet Union: being on one's guard and the fear of doing something wrong were too deeply ingrained, the everyday strains and hardships of Soviet and post-Soviet life too great a burden.

Olga O. did not agree with the theory that younger children were particularly vulnerable to culture shock during their time abroad. She said that she found it particularly easy to cross-cultural boundaries precisely

[53] Ibid.
[54] Ol'ga O., interviewed by Melanie Arndt, October 2017. Online.
[55] Ales' Siry, 'Tat'iana Kozyro, geroinia mezhdunarodnogo skandala v 2008 godu, rodila rebenka i zhivet v Borisove', *Tut.by*, 5 February 2014, https://bit.ly/3P95TKl (4 May 2024).

because she was only seven when she made her first trip – she wasn't aware of the borders, as such, in the first place.[56] Even so, positive experiences could still be accompanied by homesickness and distressing misunderstandings. Olga O. recounted how on one occasion she ran crying to her room in Seattle when there was ice cream with a red fruit sauce for dessert: she had immediately assumed it was tomatoes with smetana, one of her favourite dishes back in Belarus, and been overcome by a wave of homesickness. Only years later did she learn what it really was.[57]

Tatiana Chulitskaia, who holds a doctorate in political science, rejected the idea that spending time abroad is damaging for young people – and, indirectly, the nation – because it prevents the next generation from envisaging a future in their own country: 'I think that exactly the opposite is the case: the more experience, the better it is in a cultural sense – and that includes Belarus. Because this experience of communication can't be had inside Belarus. And seeing that there are other ways of living – that's a really good thing.'[58] Chulitskaia also took a favourable view of the adoption of children from orphanages. 'It's not as if Belarus is disadvantaged by this, is it? It causes the nation no harm', she said, adding that it was actually a good thing in terms of the whole person: the Belarusian facilities saw to it that the children had a healthy upbringing, she felt, but there was still much better support for their development abroad.[59]

Iryna Shaparava (Ru. Irina Shaporova) from Homel, who was born in 1981, understood 'culture shock' completely differently. When asked about the idea, she did not even connect it with the decade during which she went abroad twice every year: first as a Chernobyl child herself and then as a chaperone for groups of Chernobyl children. Instead, Shaparava associated 'culture shock' with the changes that her own country witnessed after the breakup of the Soviet Union; this included the distress she felt when the universal school uniform was abolished. Her trips abroad had a practical benefit in this respect: 'Even today, I'm still wearing things that I was given to grow into back then!' She could not understand the hype surrounding the newest fashion trends at that time, which she recalls as one of 'collective hysteria'.[60]

---

[56] Ol'ga O., interviewed by Arndt.
[57] Ibid.
[58] Tatiana Chulitskaia, interviewed by Melanie Arndt, 10 September 2010. Vilnius.
[59] Ibid.
[60] Ibid.

Grushevoi had no time for parents who instructed their children to come back with as many presents as possible; in his eyes, they were reducing themselves and the people of Belarus to beggars when they sent their children abroad with lists of requests ('we'd like a tape recorder, trainers, etc.'). He was in no doubt that many people had used the children's trips to benefit themselves, ignoring the fact that the children were being sent abroad not 'because of a bit of bread or some trainers' but to cleanse their bodies of radionuclides.[61] Tatiana Chulitskaia, however, did not share the idea that trips abroad were primarily about experiencing material prosperity. Her first trip as a Chernobyl child saw her visiting Poland, Belarus's western neighbour, in 1992. Looking back almost twenty years later, she recalled that she had not exactly been happy at exchanging the urban surroundings of Mahileu for the countryside and a host family with an income significantly below average – but she also said that she now saw it as a particularly educational experience, pointing out how important it is to appreciate the fact that the world is not the same everywhere.[62]

According to a study by two Belarusian social psychologists, Galina Gatalskaia and Nina Zaitseva, in the eyes of the Chernobyl children, recuperation abroad helped them to improve their health, broaden their horizons, and make important decisions affecting their lives.[63] In addition, Gatalskaia and Zaitseva argue, the trips sharpened national consciousness (*natsionalnoe samosoznanie*): becoming aware of a Belarusian 'us' distinct from 'others', they report, had led 82 per cent of respondents to value their own national identity more.[64] Participants in the study seemed optimistic about the future of their homeland. They had a positive outlook regarding the Belarusian economy (including apprenticeships and jobs for young people), as well as the state's desire and ability to improve healthcare and the environmental situation, look after the country's cultural heritage properly, and foster a love of the mother tongue and the homeland in the young generation. The study found that Chernobyl children who had been abroad at least four times were most likely to develop patriotic feelings for their

---

[61] Quoted in Grushevoi, 'Esli by Chernobyl' sluchilsia segodnia'.

[62] Chulitskaia, interviewed by Arndt.

[63] Gatal'skaia, Zaitseva, *Sotsial'no-psikhologicheskii analiz*. The authors of the study evaluated questionnaires and details from 258 individuals: 224 Chernobyl children and 34 accompanying adults. Unfortunately, it is not clear from the study whether and to what extent the questionnaire had an open-ended component; if there was one, it does not appear to have been substantial.

[64] Ibid., 50.

homeland. As one student observed: 'The trips made us realize how little we appreciate our own natural beauty and what we've got here.'[65] Gatalskaia and Zaitseva clearly did not pursue the reasons behind this change. Time spent outdoors may have played a role; several organizations recommended that host parents bring the children close to the 'simple pleasures' of nature rather than concentrating on shopping trips and the attractions of consumerist society.

Nadzeya Husakouskaya (Ru. Nadezhda Gusakovskaia) from Minsk did not find herself in a particularly patriotic mood during her one trip abroad as a Chernobyl child in 1992. Aged twelve at the time, in addition to gratitude, it was above all a sense of shame that she experienced. This was combined with a 'vague sense of "being different"' caused by the fact that the children had all their personal belongings taken away from them after they arrived in the host country. They were not even allowed to wear their own underpants, she said; instead, all the Chernobyl children had been outfitted in the same yellow shorts and checked blue shirts. She remembered how passers-by looked at them with pity and waved – albeit from a safe distance, as though they were infectious. For Husakouskaya, coming from Belarus was associated more than anything else with the feeling of being 'unclean and embarrassed'.[66]

A third of the chaperones surveyed by Gatalskaia and Zaitseva said they had encountered negative influences on the children. They felt that host families gave the children too much latitude and tended to spoil their guests. A further third had noticed children picturing themselves living in their host country in the future. Gatalskaia and Zaitseva describe such fancies as 'infantile designs' motivated by superficialities such as having an easier life; they argue that the children were glossing over the problems of other countries as well as the advantages of life in their own. According to chaperones, however, the desire to emigrate increased if children made multiple trips.[67]

According to Diane Dillon, who was both a host parent and the president of the Chernobyl Children's Project in California, many children did experience a culture shock – most of all as a consequence of the unprecedented choice available to consumers. Children in California in particular, she said, were overwhelmed by the fruit and vegetables on offer; she had recommended that host families wait until the children had

---

[65] Quoted in ibid., 53.
[66] Nadzeya Husakouskaya, interviewed by Melanie Arndt, 2 July 2019. Email.
[67] Gatal'skaia, Zaitseva, *Sotsial'no-psikhologicheskii analiz*, 52.

acclimatized to some degree before going shopping with them.[68] She also recalled how the culture shock had become apparent in remarks the children made about everyday life in the United States. They picked up on differences in social norms and expectations (e.g. the fact that women and pensioners drove cars).[69] Aleksandr Gorobchenko, a Chernobyl child from Vetka who later chaperoned groups himself and is now a psychologist in Minsk, confirmed that children in the early groups that went to the West were left 'in a state of shock' by the variety on offer. He remembers getting his photograph taken in front of supermarket shelves and feeling a great sense of helplessness. This did, however, get better with future visits, he observed.[70]

Neither the children nor the adults who accompanied them were immune to the shock of consumerism (which need not necessarily be a culture shock). Particularly in the early years of collaboration with foreign partners, the Belarusian adult chaperones were astonished at the extent to which the Soviet state had deceived them about the reality of life in the West. Dillon remembers how one woman had expressed her state of disbelief: 'You feel like, "Oh my gosh, everything that I was told is wrong."'[71] Remarks such as this must always be seen in terms of the context in which they were originally made. One party was reliant on the other for help; it is conceivable that some things were said with a view to meeting expectations and may therefore reveal more about the people to whom they were addressed than about those who made them. Dillon herself confirmed that the adults who accompanied the children tended to say what they thought their US counterparts wanted to hear. She observed that this was particularly the case in communication between Chernobyl children and their hosts when conversations were being interpreted by the adults: 'The interpreters will actually kind of almost change their [the children's] answers, too, which I found out later. And they'll give you the answers you want to hear.'[72]

Irina Narkevich reported that the adults who accompanied the children sometimes experienced a much stronger culture shock than the children themselves. When a same-sex couple were hosting a Chernobyl child, there were no problems until the Belarusian chaperones visited the family.

[68] Dillon, interviewed by Arndt.
[69] Ibid.
[70] Aleksandr Gorobchenko, interviewed by Melanie Arndt, 3 September 2010. Minsk.
[71] Dillon, interviewed by Arndt.
[72] Ibid.

The child was clearly not at all bothered by the couple's orientation, but the two interpreters had a lengthy discussion about whether they would have to take the child away from the family to avoid morally 'corrupting' them.[73] Even leaving President Lukashenka aside, conservatism was deeply rooted in Belarusian society as a whole, and the generally well-educated and widely travelled adults who went with the children were often no exception to this; homophobia was particularly marked.[74]

Fears that the Chernobyl children would all abandon their homeland to go abroad may have been widespread, but they were also blown out of proportion. If nothing else, as Narkevich noted, the concern was unfounded for the simple reason that most of the Chernobyl children never became sufficiently proficient in the language of their host country to work there. But if they equipped themselves 'to do something respectable and find a job in Germany or wherever, if it's a respectable job – why not?' That was globalization, after all, she felt, acknowledging that she herself had benefited professionally from her trips as a chaperone.[75] In emphasizing that it had to be a 'respectable job', Narkevich was alluding to an idea that became increasingly common from the mid-1990s onwards: it was claimed that young women in particular were returning home only to marry foreigners and move abroad or fall for phoney job offers that actually involved prostitution. What is true is that the number of Belarusian and Ukrainian women working in brothels and street prostitution in the West and in East Central Europe did increase in the 1990s, and that there were former 'Chernobyl girls' among them.[76] The 'For the Children of Chernobyl' fund set up an information programme in an attempt to address this.

Seen through Western eyes, the close-fitting fashion preferred by Belarusian and Ukrainian women in the 1990s could result in a culture shock of a very different kind for hosts, not least because it was at odds with stereotypes of hardship.[77] Particularly in the case of the young women who accompanied groups of children, neediness did not necessarily manifest itself in external appearances.

---

[73] Narkevich, interviewed by Arndt, 2 September 2010. Minsk.
[74] Ibid.
[75] Ibid.
[76] There are no statistics on this. See also Melanie Arndt, 'Ever Been to a Brothel?', *Plotki* (2001), 27–30.
[77] Cf. Narkevich, interviewed by Arndt.

5.2.2 Medical Aspects of Recuperation Abroad

Early in the 2000s, Belrad – an independent radiation research institute headed by Vasil Nestsiarenka that was itself involved in recuperation trips for children – studied the medical benefits of going abroad for recuperation as compared to recuperation within Belarus.[78] In doing so, it collaborated with foreign partners and state institutions,[79] above all the Zhdanovichy (Ru. Zhdanovichi) children's rehabilitation centre, which had been established on the site of a holiday facility belonging to the Ministry of Defence.

Nestsiarenka's team used the SICh whole-body counter (Fig. 3.2) to see whether recuperation led to a reduction in caesium-137 levels in children. With the consent of children and parents, they also studied the effect of a formulation that was made with high-dose apple pectin and intended to bind and excrete caesium-137 ingested with food and drink.[80] Their conclusions are unlikely to have been welcome news for those who insisted on the importance of recuperation abroad. Various tests showed that, at least where caesium-137 decorporation was concerned, it made no difference where the children went for recuperation – the effect was more or less the same: caesium-137 levels were reduced by around 15 to 20 per cent. A marked improvement on this was only achieved if children went abroad for at least two months and were given a combination of pectin and vitamins, as observed in tests on Chernobyl children who were sent to Canada or Spain: their caesium-137 levels dropped by an average of around 50 per cent.[81]

Nestsiarenka's team concluded in 2007 that the most effective protocol was to give the children the vitamin–pectin mix two to three times a year and send them to a country with 'clean' food for a month. While there, they were to continue taking the formulation – a puzzling recommendation,

---

[78] V. B. Nesterenko et al., *Sravnenie effektivnosti ozdorovleniia detei chernobyl'skikh regionov Belarusi v respublikanskikh sanatoriiakh i za rubezhom* (Minsk: Belrad, [2007]).

[79] On the German side, members of the Forschungszentrum Jülich took part in the studies and the Bundesministerium für Umweltschutz und Strahlensicherheit co-funded the experiments. See PSR/IPPNW Schweiz, Kinder von Tschernobyl Belarus/Frankreich, Weimarer Verein 'Hilfe für die Kinder von Tschernobyl': JANUN e.V. (eds.), *Können Pektinkuren die Kinder von Tschernobyl schützen? Programm Internationales Pektinhearing*, Hanover, 16 February 2007; Achim Riemann, V. B. Nesterenko, *Häufig gestellte Fragen zum Thema Pektin* [2007].

[80] Pectins bind heavy metals such as caesium-137 in the gastrointestinal tract; they are then naturally excreted. See V. B. Nesterenko, *Einwirkung der Radiation auf die Gesundheit von Kindern in Belarus 12 Jahre nach dem Tschernobylunfall* (n. p., 1998).

[81] Nesterenko et al., *Sravnenie*, 1; Nesterenko, *Einwirkung der Radiation*.

because pectin is only effective at preventing further incorporation of caesium-137, which was not generally likely to occur abroad anyway; Nestsiarenka did not explain this apparent contradiction.[82] In any event, the positive effects were limited. Pectin did not cleanse the body of radio-nuclides that had already been incorporated, such as strontium, which accumulates in bones. It is also worth noting that the Belrad studies did not cover the psychological effects of the different kinds of recuperation; at the time of writing, there are still no robust studies of the impact that trips had on the mental health of Chernobyl children.

5.3 EXPERIENCES ABROAD

Most Chernobyl children went abroad more than once (assuming they were young enough when their first trip was arranged to be able to make further trips before turning eighteen). In Gatalskaia and Zaitseva's study, only 16 per cent of those surveyed said they had made only one trip abroad as a Chernobyl child. Almost half (47 per cent) went abroad at least four times.[83] Most children stayed with host families. Only a small proportion stayed in group accommodation such as holiday camps or youth hostels; even then, most of them spent a few days, at least, with a host family.[84] The Chernobyl children in Gatalskaia and Zaitseva's study spent an average of four and a half months abroad in total. Hardly any Chernobyl children were later able to recall the names of the organizations that invited them; memories of towns and cities and families were much more persistent.[85]

5.3.1 New Worlds

Looking back after more than twenty-five years, Nadzeya Husakouskaya reflected on how being sent abroad as a Chernobyl child had affected her life: 'It [my trip] opened up a new world for me – made me realize that it's a big world, that there are different countries in it and people speak different languages, that you can travel and even live in another country. A month is a pretty long time, after all; it felt like a lifetime.'[86] Getting to know the 'new

---

[82] Nesterenko et al., *Sravnenie*.

[83] Gatal'skaia, Zaitseva, *Sotsial'no-psikhologicheskii analiz*, 52.

[84] According to Gatal'skaia, Zaitseva, *Sotsial'no-psikhologicheskii analiz*, 53, 89.5 per cent of children spent some or all of their trips with a host family.

[85] This is documented by Gatal'skaia, Zaitseva, *Sotsial'no-psikhologicheskii analiz*, 46, as well as by my own research.

[86] Husakouskaya, interviewed by Arndt.

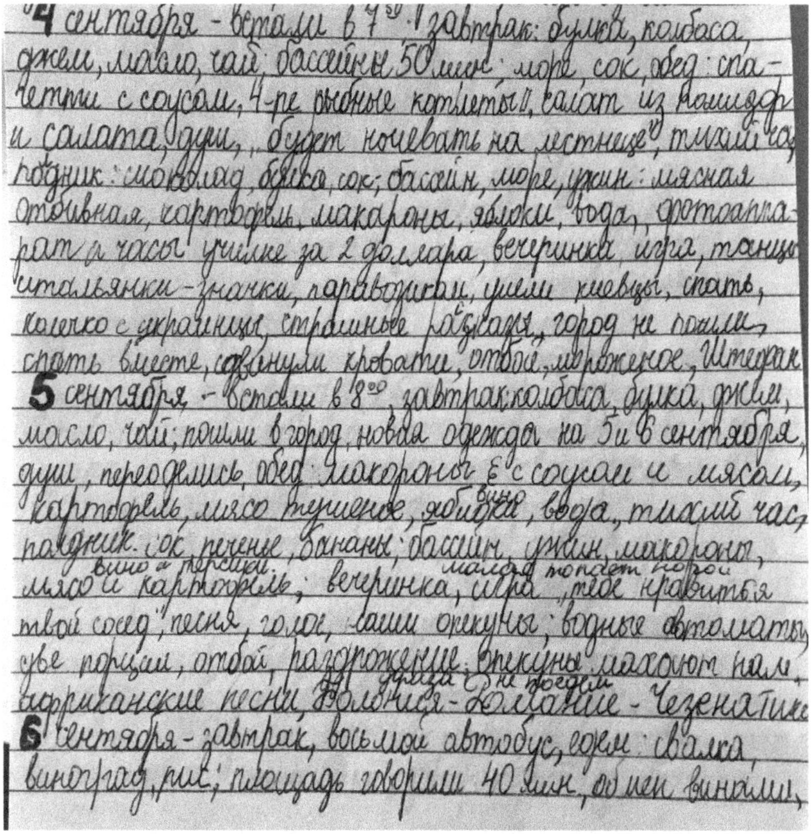

FIGURE 5.1 Extract from Nadzeya Husakouskaya's diary.

world' often started with everyday things such as unfamiliar food and new routines. A diary entry where Husakouskaya recorded in detail what meals and the daily routine consisted of is a striking example of this (Fig. 5.1).

In 2015, a journalist called Olga Bubich published a newspaper article on the Chernobyl children and the effects of being sent abroad. She was interested in what it was like for a child to return to their home village after spending several weeks in a world where showers and toilets were not simply 'marvels' belonging to the 'world of the imagination' (*mir fantastiki*).[87] Differences in the standard of hygiene figure repeatedly in

---

[87] Ol'ga Bubich, '"Chernobyl'tsy" za rubezhom: Pravo na druguiu zhizn', *Belarusskii Zhurnal*, 23 April 2016, https://bit.ly/4gGqO35 (4 May 2024).

the stories recounted by Chernobyl children, particularly when recalling their first impressions of their host country. One Chernobyl child, now a businessman in Lithuania, for instance, remembered how the self-flushing toilets at a cafe had left a lasting impression on him and his companions when they were on the way from the airport to their host families.[88] By way of explanation, he quickly added that he came from an 'extremely poor and degraded post-Soviet state'. This remark reflects a centuries-old narrative of backwardness that can manifest itself particularly starkly in the context of such basic aspects of everyday life as personal hygiene. Looking back on her time as a chaperone, Elena Pankratova of the 'For the Children of Chernobyl' fund noted that many children – especially in the first groups to go abroad – were simply not accustomed to using toilets and would surprise their host families by disappearing into the garden or behind bushes instead.[89]

Vivid anecdotes such as this should not obscure the fact that a large number of the Chernobyl children who went abroad came from towns and cities. Accounts indebted to the narrative of backwardness cannot accommodate experiences such as that of Tatiana Chulitskaia and her disappointment at ending up on a Polish farm on her first trip. Reactions such as Chulitskaia's may not lend themselves particularly well to the image of impoverished Chernobyl children, but her case was far from unique. Despite the novelty of encountering self-flushing toilets, one future businessman was also somewhat nonplussed at where he ended up: 'Well, I expected to come to a high-tech megapolis and was a bit surprised when ended up living in a country side [*sic*]. Still, I was surprised in a good sense.'[90]

According to Gatalskaia and Zaitseva's study, aspects of the natural world – particularly mountains and seas – were what the children liked best about their host countries. But they were also impressed by architectural richness and diversity, attractive and well-looked-after buildings and towns, and the various tourist sights. Favourite activities included getting up close with nature; going to swimming pools, concerts, exhibitions, and restaurants; and making shopping trips.[91]

---

[88] A 3 [anonymized], interviewed by Melanie Arndt, 5 March 2014. Online.
[89] Elena Pankratova, 'Grustnyi pessimizm' in Tamkovich, *Filasofiia dabryni*, 303–9, here 306.
[90] A 3, interviewed by Arndt.
[91] Gatal'skaia, Zaitseva, *Sotsial'no-psikhologichesky analiz*, 47.

What bothered the children most while they were abroad was the fact that their homeland was a completely unknown quantity for their hosts. They kept having to explain where Belarus is and point out that it is not part of Russia. Some children were also upset when Belarus – if it was actually recognized as a country in its own right at all – was reduced to the Chernobyl disaster. They felt that their country had more to offer than the wrecked power plant. Other negative aspects that the children identified included not being proficient in the local language, having trouble with unfamiliar food, being patronized by host families (particularly in Italy), and, sometimes, being viewed with scepticism and envy by other members of the host family. The relationship between a host family and the child they hosted could change as well: if a family invited the same Chernobyl child again, the fact that 'their' child had grown older at home in the meantime could, on occasion, result in distance between them, or even frustration, on the next visit. It did not help matters that many host families idealized 'their' Chernobyl children in their absence, which increased their sense of anticipation for the next visit and helped them justify the time and effort needed to arrange it.[92]

Most of the Chernobyl children surveyed by Gatalskaia and Zaitseva, however, stayed in touch with their host families for years, even after their last trips. Ninety per cent said they wanted to go back to the country they had visited as a Chernobyl child. Equally, by the time their stays came to an end, most children would be looking forward to going home to Belarus again; only a fifth would have preferred to remain abroad for longer there and then.[93]

5.3.2 Travelling with the Children: The Accompanying Adults

The children who went abroad were accompanied by adults who not only served as important points of contact for them, their host families, and the various organizations involved but also acted as interpreters and engaged in outreach work. The trips would have been impossible without fund-raising and outreach. In its proposition to potential partners in North America, the 'For the Children of Chernobyl' fund explicitly mentioned that each group of children would be accompanied by a volunteer who would be able to help with fund-raising efforts. The fund also added some suggestions about who best to approach: 'local businesses, Churches and

[92] Ibid., 48.
[93] Ibid., 49.

service organizations' had, it said, proved 'eager to assist' the Chernobyl children.[94] This example involves practical suggestions that originated in the East rather than the West – an exception to the rule in the 1990s.

The success of lobbying and advocacy for the trips, however, soon meant that too many groups were going abroad for all of them to be joined by an adult who had been suitably prepared for the role, let alone a member of the 'For the Children of Chernobyl' fund. As trips abroad became more commonplace, the adult chaperones became less well placed to inform those in the West about the legacy of the disaster. The pool of accompanying adults was initially made up primarily of (trainee) teachers and lecturers in foreign languages who were engaged through personal networks. A considerable portion of those who went on behalf of the 'For the Children of Chernobyl' fund were recruited by Grushevaia at the Linguistic University in Minsk, where she taught. Conveniently, language students were also able to act as interpreters.[95] According to Gatalskaia and Zaitseva's study, each adult would accompany groups of Chernobyl children abroad an average of eight times; this would mean spending a total of around nine months outside Belarus. It was predominantly women (62 per cent) with a university education who went with the groups; in most cases, they had pedagogical jobs.[96]

Chaperones were generally recompensed for looking after a group; the exact amount depended on the organization inviting the children. In the 1990s, the average figure was US$300 for trips to the United States and DM100–200 for Germany.[97] Particularly in the early days, the very prospect of visiting another country was incentive enough in itself; the allowances merely added to the appeal: in the 1990s, the fee could be equivalent to as much as twice the average monthly income in Belarus.[98] Even so, some still felt that the amounts on offer were humiliatingly low. It seemed to them that they were not being valued very highly in the context of average salaries in the West and the financial resources that – in their minds – inviting organizations had at their disposal.[99] In actual fact,

---

[94] Private Archive Pankrats, 'The Belarusian Charitable Fund "For The Children of Chernobyl"'.

[95] Tamkovich, *Filasofiia dabryni*, 239.

[96] Gatal'skaia, Zaitseva, *Sotsial'no-psikhologicheskii analiz*, 51.

[97] The allowance for accompanying children to the United States could later amount to up to US$800. Diane Dillon to Melanie Arndt. Chat, 30 November 2017.

[98] The average monthly income in Minsk in the mid-1990s was equivalent to around DM100–120.

[99] Tamkovich, *Filasofiia dabryni*, 306.

however, particularly in smaller initiatives, funding was often tight and people gave their time for free.

The vast majority of children were very appreciative of the adults who accompanied them. Their support made it easier to get used to new surroundings and adapt to unfamiliar cultures. The children also thought that their presence helped to ensure discipline was maintained in the groups.[100]

The accompanying adults, like the organizations that they worked for, adopted a discourse that framed their work for the Chernobyl children in terms of a triad of moral duty, a sense of responsibility, and self-sacrifice. Neal Hinson noted that 'families have broken up' because 'the adult chaperones . . . leave their families during the only time of year that they have vacation', spending the time with Chernobyl children instead. She felt they had no choice: 'it is something that must be done despite the personal hardship . . . something like getting refugees out of a war zone. Timing is the essence. Tomorrow may be too late.'[101] The narrative of self-sacrifice, however, elides the fact that the accompanying adults not only got to travel and practise their language skills while being recompensed for their services, they also drew benefits for their own children: the latter, as a rule, were able go abroad as Chernobyl children themselves as a result.

5.4 A HOME FROM HOME?

5.4.1 The Host Families

Families that wanted to host a Chernobyl child in the United States had to undergo a police background check and were visited by representatives of the local US organization who would form an impression of them and their homes.[102] This approach had been standard practice for several years in the context of international exchanges for schoolchildren, but it did not guarantee that everything would be straightforward. Members of the inviting organizations often had to step in at short notice if children

---

[100] This was stated by 78 per cent of those surveyed in Gatal'skaia, Zaitseva, *Sotsial'no-psikhologicheskii analiz*, 47.

[101] Neal, *Silent Clouds*, 304; a similar tendency was apparent in the interviews I conducted.

[102] See Neal, *Snapshots*, 142; HA, Newsletter 'For the children', vol. 1, no. 4; Dillon, interviewed by Arndt.

were struggling to fit in with their host families and accommodate them in their own homes instead.[103]

In terms of background and social status, the host parents that organizers collaborated with were a 'conglomerate of different people';[104] in the United States, though, they were largely the white middle classes. They were recruited through publicity in the local media and informally, by word of mouth. According to Diane Dillon, motivations were usually clear and uncomplicated: 'We felt like we were doing something good.' She saw nothing remarkable in hosting a child and looking after it alongside one's own children – she herself had five. Dillon, a Catholic, also recognized that religious motivations could play a role: 'that's kind of like maybe more of our faith'.[105]

Debra Zapata heard about the Chernobyl Children's Project through her sister-in-law, saw a video, and thought it would be a good idea to give her ten-year-old daughter a 'sister' for the summer. That was how Tatstsiana Kazyra came to her family as a Chernobyl child: 'I just love kids, always have loved kids. And my background was being a nanny, then running a daycare, then having children, then working in paediatrics. So I love kids. So having that chance to give something to another child, who is less fortunate, and also involve my children – that's why I was doing it.'[106]

Julie McClelland, Zapata's friend, recognized that there was a sense in which this was the less demanding way to make a difference: 'Here are kids that completely need help, they'll come to us – it's not like I needed to fly to Europe – and I could help.'[107]

The inviting organization usually covered all the costs the first time a Chernobyl child went abroad. If a family wanted to invite 'their' child again the following year, the organization would help with practical aspects – primarily visas and travel arrangements – but the family would have to pay the actual costs themselves. Entire extended families and social circles were mobilized to raise the necessary funds. When such children did return, they were generally allowed to join other Chernobyl children in group activities that local organizations arranged for them.

---

[103] Cf., e.g., Dillon interviewed by Arndt; Debra Zapata, Julie McClelland, interviewed by Melanie Arndt, 11 April 2016.
[104] Dillon, interviewed by Arndt, 10 June 2015.
[105] Ibid.
[106] Zapata, McClelland, interviewed by Arndt.
[107] Ibid.

The extensive support networks that host families had to nurture in order to help 'their' Chernobyl children make further visits also meant that awareness of the children's plight and background – and thus of the disaster in general – circulated more widely. Even if the knowledge involved often remained superficial – most people continued to refer to Belarus as Russia, for instance – minds were at least opened to some degree where what had once been enemy territory was concerned.

Some of the problems that the children experienced while staying with their host families can be put down to a certain amount of naivety. There was a good chance of trouble if the values and expectations of the host family were not aligned with what the Chernobyl child was used to – for example, when families with alternative and/or relatively well-off lifestyles took in 'particularly needy' village children. Disappointment all round was inevitable if there was no television in a household and the host parents thought that the child coming to them would be fascinated by English literature, when they had actually been looking forward to US television.[108] The bar was set very high when it came to the children's ability to fit in with expectations. As Pastor Paul Moore of CitiHope later reflected: 'We had arranged civic programs and receptions, but the kids weren't interested, and for the most part were uncooperative. They only wanted to swim and play – which we should have known.'[109]

After their initial experiences of going it alone, more and more initiatives and host parents started exchanging information with other organizations and host families. The increasing use of the Internet made it easier to circulate guidance for host families and supporters. This material distilled practical experience and imparted an awareness of the Chernobyl children's background and the needs to which it gave rise.[110] Above all, though, the guidance reflected the views of the activists behind it.

In the western United States, Nadya Neal Hinson became a popular expert on how to approach these children from a 'very, very different world'. With their paternalistic tone, her 'Host Family Guidelines' are typical of how Chernobyl children initiatives understood their role from the mid-1990s onward. She warned prospective host parents not to be misled by the fact that Chernobyl children looked so similar to North

---

[108] Ibid.

[109] Quoted in Carter, Christensen, *Children of Chernobyl*, 176.

[110] A particularly effective account of this can be found in one such text, entitled 'There's No Such Thing as "the" Chernobyl Child', produced by a German initiative. Cf. Eva Balke, '"Das" Tschernobylkind gibt es nicht', *Bundesarbeitsgemeinschaft 'Den Kindern von Tschernobyl'*, 16 April 2008, https://bit.ly/41ILmDG (4 May 2024).

American ones on the outside; the reality was that there were 'profound differences between the cultural values, way of life and perception of this immense part of the world and … North America'.[111] For Neal Hinson, ex-Soviet society differed fundamentally from that of North America because of its historically grounded and deeply rooted, typically Slavic religiosity and connection with nature. She also turned to a narrative according to which Belarusian history in particular consists of one long sequence of catastrophes that gathered pace in the twentieth century. She painted a grim picture of remote and impoverished villages that the Chernobyl children called home. Just like poverty, she highlighted the prevalence of alcoholism (which was indeed widespread, even if there is insufficient data to pin down exactly how much so). The picture she drew of the climate (which was, in reality, hardly different from that in Washington State) had more in common with the classic cliché of the frosty East than the actual state of affairs in Belarus with its rich harvests. There is every indication that Neal Hinson's main concern was to demonstrate the 'otherness' of the Chernobyl children as emphatically as possible.[112] This includes her recommendations about what to give the children to eat. She described Belarusian food as 'very bland by any culture's point of view', writing that 'there is little seasoning and sour cream is the universal dressing of choice', and that the children might not take immediately to other 'sauces or spices or even peanut butter'. As banal as some of these points may seem, they did have a practical purpose. Numerous host families and chaperones had seen how unfamiliar food and drink could become a problem.[113]

Overall, though, Neal Hinson stressed that Belarusian children learnt from an early age 'to have control and restraint' – something that set them apart from their American counterparts. The everyday lives of the Chernobyl children were subject to much closer control by grandmothers, mothers, and teachers. Neal Hinson was probably largely correct when she wrote that American children simply took what they wanted, in contrast to children with a Soviet upbringing. Politeness and modesty dictated that they would try to refuse things, even if they actually wanted them. This behaviour was probably also a consequence of the fact that so many things were hard to come by in the Soviet Union. The upshot of all

---

[111] Nancy Neal-Oldenettel, 'Belarusian Culture, History & Children: Host Family Guidelines', *Children of Chernobyl Foundation*, https://bit.ly/4gmFoyA (4 May 2024).
[112] Ibid.
[113] Pankratova, 'Grustny pessimizm', 306.

this was that it was not easy for Chernobyl children to articulate their needs and desires or decide things for themselves. The challenge this posed figures repeatedly in the recollections of the Chernobyl children. Neal Hinson thus had good reason to encourage host families to be patient in this respect. Her advice was to offer the children things repeatedly and 'insist that they take things they might obviously like'.[114]

Gifts were another ongoing talking point among the Chernobyl children and in discussions between host families and inviting organizations. If host parents showered Chernobyl children with presents such as gold jewellery, expensive branded fashion, and shoes, envy ensued on the part of those who got less to take home with them.[115] In response, the organizations involved increasingly advocated restraint – however great 'the desire to give everything' might be, as Neal Hinson put it. Pointing to the catastrophic state of post-Soviet healthcare, she also suggested giving the children vitamins and painkillers such as paracetamol and aspirin to take home for their families.

Many Chernobyl children expressed considerable gratitude towards their host families, particularly if they had been to them several times. They said that their hosts had opened up new worlds to them and looked after them very well during their visits – and, in some cases, beyond. The children were impressed by the kindness with which people treated one another, in contrast to what they were used to at home, particularly the speed with which they themselves were integrated into family life, school classes, and other social groups. The children's perceptions here may well have been influenced by what they had gone through in the course of evacuation or resettlement. Time and again, they had had to adapt to new surroundings and stand up for themselves as *Chernobyltsy* ('Chernobylers') or *svetliachki* ('glowworms' – a pejorative term for Chernobyl children). This had, for many of them, involved difficult, if not traumatic experiences: if they were 'innocent' and 'needy' Chernobyl children deserving of empathy during their time abroad, this was often a far cry from the everyday realities of life in their home countries.[116]

Looking back, many Chernobyl children praised the easy-going interactions in their host countries; this impression stemmed above all from the fact that relationships were relatively permissive and there were few constraints on their activities. In contrast to their home countries, where

[114] Neal-Oldenettel, 'Belarusian Culture, History & Children'.
[115] Cf. Svetlana Bodrunova, interviewed by Melanie Arndt, 2012. Online.
[116] Cf. Gorobchenko, interviewed by Arndt.

they were closely monitored, the children were given considerable latitude by their host families – there was a lot they could do. The Chernobyl children also noticed host parents giving them more attention than they gave to their own children. They were impressed by the relationships that parents had with each other and with their children. Children from socially disadvantaged families, in particular, emphasized this experience and modelled their own aspirations on it: they wanted to have families where everyone respected, helped, and looked after one another.[117]

Only a small number of Chernobyl children – merely 6 per cent in the group surveyed by Gatalskaia and Zaitseva – reported negative experiences with their host families.[118] These can usually be traced back to the feeling of being excessively patronized. As one of my interviewees put it, host families thought they knew best as regards what was good for the children.[119] As experience grew, some perspectives on the children did at least become less simplistic, as did the picture being drawn of their home countries. The 'Host Family Guidelines' that Neal Hinson compiled in the mid-1990s were still accessible on the website of the Children of Chernobyl Foundation (based in San Diego, California) in December 2017, but by then it also hosted a much more detailed account of the Chernobyl children's background, together with suggestions on how to look after them, based on guidance compiled by the Canadian Relief Fund for Chernobyl Victims in Belarus.[120]

A significant number of hosts saw helping the Chernobyl children as a form of personal development. Anne Culbertson, a host parent from Seattle who took in a Chernobyl child in 1995, overcame her fear of public speaking in order to give a presentation promoting the children's cause for her church.[121] Culbertson saw hosting a child as good 'for our health as well as theirs'.[122] Her whole family had become emotionally involved in the plight of Andrei, 'their' Chernobyl child, as shown by the following anecdote: after reading the story of Aladdin and the magic lamp to her

---

[117] Gatal'skaia, Zaitseva, *Sotsial'no-psikhologicheskii analiz*, 53; various interviews by the author.

[118] Gatal'skaia, Zaitseva, *Sotsial'no-psikhologicheskii analiz*, 53; various interviews by the author.

[119] Bodrunova, Gorobchenko, interviewed by Arndt.

[120] Children of Chernobyl Foundation San Diego, *Host Family Guidelines*, https://bit.ly/3 YCjQEX (4 May 2024). For another example of later guidelines, cf. Linda Walker et al., 'Guidelines for Host Families', *Chernobyl Children's Project (UK)*, https://bit.ly/3BPYxIt (4 May 2024).

[121] Neal, *Silent Clouds*, 351.

[122] Ibid., 354.

eight-year-old son, she had asked him what his three wishes would be; one of them was 'to go back in time and make the Chernobyl disaster never happen'.[123] The accident had been something that had 'happened on the other side of the world' before Andrei brought it home to her, she said: only through him had she understood that 'anywhere on Earth is our backyard – Chernobyl, Bosnia, Hiroshima, Hanford'.[124]

Many host families were struck by the positive impact that the visits had, not least on their own children. As Diane Dillon, a single mother, put it:

> I thought it was a great experience for my kids, because we are honestly probably one of the poorest people on the block, and so my kids always felt like all their friends were doing all these other things and their friends had this and their friends had that. But then when they saw these children who had absolutely nothing, ... I think my children got more out of it than even the children that came here.[125]

The shadow cast by the nuclear past closer to home was a very real presence for many host parents for Chernobyl Children Northwest. Becky Freeman, a teacher and Hanford downwinder, was given her leukaemia diagnosis the day after 'her' Chernobyl child flew home; after she died in 1996, Neal Hinson dedicated the Children's Visit Program for 1997 to her memory.[126] Interest in the Chernobyl children was considerable among Hanford downwinders, but clearly never took off in the other nuclear landscapes of the United States. There was a grassroots movement dominated by mothers in the area around Three Mile Island,[127] but it concentrated its efforts on addressing the legacy of that accident. Chernobyl figured primarily as a contrastive point of reference there, even if important figures in the Three Mile Island movement did visit the site of the Ukrainian accident.[128]

5.4.2 The Chernobyl Children's Families

'There is no doubt that it affected my whole life, not just my own, but that of my family, my father, mother, and sister – and the people we knew, too, of course', Aleksandr from Mahileu explained when looking back on his

---

[123] Ibid., 352.

[124] Ibid., 352, 354.

[125] Dillon, interviewed by Arndt.

[126] Neal, *Silent Clouds*, 354, 359.

[127] Cf. Stamos (previously Osborn), interviewed by Arndt; Kinney, interviewed by Arndt.

[128] The politician and anti-nuclear activist Eugene Stilp, for example, went to Ukraine in 1996. Stilp, interviewed by Arndt; Arndt, *Nostalgic Bonfires*.

time as a Chernobyl child from the age of thirty-one.[129] He first went abroad as a Chernobyl child in 1990, to Germany. His experience resembled what Neal Hinson had described as 'almost like being in a Star Trek movie':[130] 'It was an explosion, a shock, nothing but positive emotions, like being in a different world ... I'll never forget that ... I simply liked everything, I wanted to stay there. Everything through rose-tinted glasses, absolutely everything.' Aleksandr had a hard time readapting to life with his own family after returning home. He recalls that they were simply not able to match the 'love, attention, tenderness, you name it' he had experienced with his host family. Repeated clashes with his parents ensued – something he came to regret as an adult, but all he wanted at the time was to embrace the lifestyle he had glimpsed in Germany. Aleksandr travelled a lot as a young adult and spent a year and a half in Prague. Living there alone was not easy, and he returned to Mahileu. Aleksandr said in his interview that he no longer wanted to live in the West, despite the fact that life was more comfortable there and the opportunities greater; as a child, he had not realized that there were more important things.[131]

Olga O. – the young girl with alopecia – had a very different experience. From her second visit onwards she was accompanied by her sister, and the whole family took an enthusiastic interest in their time abroad: 'It was an exciting experience for the whole family, and aside from missing each other, we enjoyed the new experiences. My parents were thrilled to see our vacation pictures, hear our stories about food, travel etc., as well I had more hope that somebody will find a cure for my illness.'[132] This account seems to gel best with Olga's annual visits, which always saw her returning to her own family after several weeks. Matters were not so straightforward when she left Belarus to study in the United States at the age of sixteen, funded by her host family. She did not see her own family again until nine years later, and recalled the 'emotional struggle, arguments, misunderstandings' that ensued because of how she had changed:

I forgot how it feels to be loved and cared for on a daily basis, so I have developed some form of defense mechanism, not sure from what though ... I am saddened that I have developed this type of defense mechanism after living w/out my family for so long. I somehow think it is all related.[133]

[129] Aleksandr, interviewed by Melanie Arndt, 6 September 2010. Mahileu.
[130] Neal-Oldenettel, *Belarusian Culture, History & Children*.
[131] Aleksandr, interviewed by Arndt.
[132] Ol'ga O., to Melanie Arndt. Email, 7 November 2017.
[133] Ibid.

Olga has hardly any contact with anyone in Belarus now, least of all with former schoolmates there. The negative memories are too uncomfortable, her old homeland too distant:

I have adapted to my life in the US so well, this is my home now and I have a whole new set of friends and memories here. Perhaps, I am blocking out people from my 'Belarus past' from my life to avoid mentioning of my hair problem, but I don't think I would benefit in any way from staying in touch with most of them.[134]

Alena Oginets from Prypiat, whose father was only fifty when he died, wished she could talk to him about the accident: 'If it were possible to turn back the clock, if he were still alive, ... and to ask about everything, I'd give anything for that.'[135] He was posthumously decorated for exceptional courage. Oginets recounted that when her mother tried to secure a 'fair' pension for her 'liquidator' husband, her efforts fell on deaf ears. Even the people her father taught at work, she said, had ended up earning more than him, who paid with his life. She also pointed out the pain caused to her mother by the fact that many people were able to get their hands on privileges as Chernobyl 'victims', even though they had hardly been touched by the disaster, if at all.

5.5 GOODBYE, AMERICA!

5.5.1 'Girl from Chernobyl Refuses to Go Back': Vika Moroz

Viktoriia (Vika) Moroz grew up in a children's home almost 100 km north-west of Minsk.[136] Her mother, an alcoholic, had lost custody of her, and her father did not want to look after her. Two years after Vika was born, her mother died in a traffic accident. Her elder brother, Aleksei, was also living in the home; Aleksei had a different father, but he took no interest in his child either.

Vika, who had learning difficulties, was aged seven when she went to Italy as a Chernobyl child; it was to change her life forever. From then on, she spent all her summer and Christmas holidays with a young couple located not far from Genoa. In summer 2006 – by which time she was ten – her host parents found bruises on her body. They decided

---

[134] Ibid.
[135] Oginets, interviewed by Arndt.
[136] The quotation in the section title is from John M. Glionna, 'Girl from Chernobyl Refuses to Go Back', *The Los Angeles Times*, 8 September 2008, https://bit.ly/4fspdwI (4 May 2024).

there and then that Vika should not return to her children's home in Belarus and disappeared with her. Almost three weeks later, the Italian police managed to track down the family; Viktoriia was flown back to Minsk, escorted by Belarusian doctors. By this point in time, foreign families had adopted hundreds of Belarusian children, particularly orphans with minor physical disabilities,[137] and this only added to the diplomatic controversy and media interest surrounding the case of Vika Moroz.[138] Her trips to Italy had not been facilitated by the 'For the Children of Chernobyl' fund, but Gennadii Grushevoi had something to say anyway. As with the 1997 coach crash,[139] he feared that the affair might have negative implications for the practice of sending children abroad as a whole.[140] He accused the organization that had arranged her trip of having made 'serious errors', while also criticizing the response of the Belarusian authorities as 'inhuman, even criminal'.[141]

Lukashenka took a personal interest in the affair. He gave instructions for conditions at Vika's children's home, and the allegation that she had been subjected to physical abuse there, to be investigated carefully. In an interview, he ruled out a ban on trips abroad for children, but did reaffirm the pledge he had made two years earlier to make sure that children could only be sent abroad subject to 'cast-iron controls'.[142] Lukashenka also returned to an old argument: the money spent on funding trips abroad would be better invested in the country that had suffered the most as a result of the Chernobyl disaster.[143]

The profound fissures that ran through post-Soviet society became apparent in the public discourse surrounding the Vika Moroz affair in

---

[137] RIA Novosti, 'Ministr Rad'kov o dal'neishei sud'be Viki Moroz', *Tut.by*, 2 October 2006, https://bit.ly/3VKkWoQ (4 May 2024).

[138] Tatiana Korovenkova, 'Apelliatsionnyi sud Genui prigovoril k vos'mi mesiatsam tiur'my suprugov Dzhusto-Bornakhin i vsekh, prichastnykh k ukryvatel'stvu belorusskoi siroty Viki Moroz', *BelaPAN*, 22 March 2019.

[139] See Section 4.2.1.

[140] In January 1997, a government minister had threated to ban the 'For the Children of Chernobyl' fund from sending children abroad because Elena Popravko, who was fourteen by that time, had been with her German host family since 1992 and had not returned to Belarus. Author's own records: 'Tschernobyl-Rundbrief 2/97 "Solidarität ist gefragt"', 1–2.

[141] Aleksandr Tomkovich, *Vzroslye igry – detiam*, 30 October 2008, https://bit.ly/4frtQXY (4 May 2024).

[142] See Section 4.2.1.

[143] 'Lukashenko rasporiadilsia proverit' v kakikh usloviiakh nakhodilas' Vika Moroz v belorusskom internate', *Newsru.com*, 7 October 2006, https://www.newsru.com/world/07oct2006/tobecntnd.html (4 May 2024).

Belarus. 'Put your homeland first, then yourself!', supporters of her forced return declared in good Soviet style. In a letter to the newspaper *Trud*, Vera Sushchenia underlined how important and proper it was that she herself had been brought up as a patriot in the Soviet Union. Young people now, on the other hand, thought only of themselves in her eyes – and that was why an example had to be set with Vika Moroz. For Sushchenia, the welcome that children received in other countries represented a threat to Belarus. Even if she did not refer to it directly, her concerns are familiar from the culture shock debate:

I would ban all trips to other countries for Belorussian children for good. What do they see there? A completely different way of life, all kinds of excesses. How are the children supposed to go back to their homes or boarding schools after all these experiences of natural beauty, tasty treats, indulgences? ... They all want to emigrate to warm countries, to lie on the beach and eat all manner of fruits.[144]

Who, Vera Sushchenia asked, would be left to support society if everyone wanted an easy life abroad? She was worried that one day there would be nobody to care for the elderly, to work in the factories, to milk the cows, adding sarcastically: 'Are we going to see Italians coming here to work then?'

The opposing view was also common, however. *Trud* also featured Polina Nikishina from Minsk, who said people should be 'eternally grateful to the foreign families that are prepared to take in Belorussian children and edify them, provide for them, and help them recover'. Belarusian families, she said, were not in a position to adopt other children because it was already hard enough for them to meet the needs of their own ones. As a critic of Vika's forced repatriation, Nikishina asked 'Why can't we bring ourselves to believe a child when she speaks of her fears about the orphanage – and accept instead without question the assurances ... on our infallible television that nothing bad could possibly happen to the child at the facility?'[145] Nikishina told of her own nephew, who had been raped as a ten-year-old by older children at a boarding school for children with visual impairments. For her, a state that asserted absolute power over its citizens was unacceptable. The way Vika had been brought back was nothing short of 'deportation' in her eyes. The girl had found happiness, she said, only for it to be replaced by an uncertain future after being destroyed by officials and diplomats acting as if they were 'holier than

---

[144] Quoted in Valentina Taran, 'Viktoriia Moroz: goriachii spor vokrug problemy', *Trud*, 12 October 2006.
[145] Quoted in ibid.

the Pope'.[146] The gulf between the views expressed by Sushchenia and Nikishina is characteristic of the debates that accompanied the return of Viktoriia Moroz to Belarus: the plight of a single individual provided a point of reference with which to negotiate fundamental questions facing post-Soviet society.

Some time after her return, the home placed Viktoriia Moroz in the care of a foster-family in Zhodzina (Ru. Zhodino). The foster mother, who looked after children's educational needs at the home, had clearly been 'asked' to take Viktoriia in. The foster family encouraged her to stay in touch with her host family in Italy. When Viktoriia tried to visit them with her foster mother in 2009, however, they got no further than the airport in Minsk: they were unable to travel because their paperwork was found not to be in order.[147]

5.5.2 'Girl from Chernobyl Refuses to Go Back': Tatstsiana Kazyra

As Grushevoi put it, the state 'banged the table and stamped its feet' again when, two years later, a sixteen-year-old called Tatstsiana Kazyra chose not to go back to Belarus in August 2008.[148] This was her ninth successive summer with her host family north of San Francisco; it was also due to be her last such visit as a Chernobyl child. She would have been studying for her school-leaving exams the next summer, and after that she would have been too old to be eligible. On the day she was due to fly home, she never turned up at the airport. What happened next reminded many of those involved of a Cold War thriller. There were no fatalities, of course – but it set in motion a chain of events that ultimately meant the end of trips to the United States for Belarusian Chernobyl children.

Tatstsiana Kazyra was born in 1991. Various accounts of her background exist, but they all make one thing clear: hers was not a sheltered upbringing. Both her parents had issues with alcohol; for a time, she lived in a residential facility, for a time with her grandmother.[149] She was eight when she first went to the United States. Debra Zapata, her host mother, vividly remembers how they met for the first time at San Francisco airport: Tatstsiana had arrived with 'almost nothing'; it had been obvious how

---

[146] Quoted in ibid.
[147] Svetlana Belous, 'Ukradennaia mechta Viki Moroz. Kak slozhilas' zhizn' siroty, kotoruiu rasluchili s ital'ianskoi sem'ei', *Tut.by*, 5 June 2014, https://bit.ly/3ZE5OD2 (4 May 2024).
[148] Aleksandr Tomkovich, 'Vzroslye igry – detiam'.
[149] For more on Kozyro, see Arndt, *Tschernobylkinder*, 384–95.

poor – and how excited – she was. With her red Russian shoes, she was the 'the cutest thing ever', Zapata recalled.[150] The first summer had proven difficult, Zapata said, because Tatstsiana had been a really 'wild' child and found it hard to adapt to the family rules. Yet, over the years, she developed a deeply personal bond with her host family; she was particularly close to her host parents' daughter, who was two years older than her. Zapata and McClelland also described how, with time, Tatstsiana had come to speak more and more openly about problems back home in Belarus and the fears associated with them. Even if her shaky English meant that communication was not always entirely straightforward, what the Zapatas heard – which went as far as threats on her grandmother's life – was more than enough to give them cause for concern. They decided to give Tatstsiana the option of staying in the United States.[151] Neither the host family nor the lawyer they appointed considered extending her stay legally, even though the family knew of precedents for this and Tatstsiana's visa was valid until the end of the year. The reasons for this will have to remain an open question. It is likely that they did not trust the authorities and consequently decided to go it alone.

This decision resulted in chaos for the rest of the group: children in tears, afraid that they were never going to see their parents again, before being sent back to their host families after twelve hours of uncertainty at the airport; worried parents in Belarus; questioning by border protection and the security services, not least the FBI. Rumours that the Zapatas had kidnapped Kazyra only made matters worse. The two Belarusian translators refused to leave without the girl, fearing prosecution by the Belarusian state. Their position remained unchanged even when a representative of the Belarusian embassy in Washington who had been hurriedly flown in provided written assurance that they would not be held responsible for Tatstsiana's whereabouts. The police's hands were tied because Tatstsiana's visa was still valid. They kept the Zapatas' home under guard, but this seemed more in order to protect those inside from angry host parents, representatives of the Chernobyl Children's Project – the Californian organization that had invited the group – and a growing number of journalists. On the advice of Julie McClelland's father, who had served in the Air Force, the Zapatas sequestered themselves in their house. They were afraid Tatstsiana

---

[150] Zapata, McClelland, interviewed by Arndt.
[151] Ibid.

might be removed. It was a traumatic situation, Debra Zapata recalled: they had all been afraid, and were too scared to leave the house.[152]

None of the efforts to persuade Tatstsiana to return home led anywhere. The representative from the Belarusian embassy even promised to repair her grandmother's ramshackle wooden house and pay for Tatstsiana's education, but to no effect. Only when his mobile telephone rang during the meeting and Tatstsiana's grandmother was on the other end of the line did Debra Zapata think that Tatstsiana seemed afraid.[153] Just as it had done with Vika Moroz, the Belarusian state insisted that Kazyra had to go back, particularly given that she was not an orphan and still a minor.[154]

Behind the scenes, attempts at mediation were under way via all possible channels. Meditsina i Chernobyl (Medicine and Chernobyl), the Belarusian partner organization involved, even contacted Arnold Schwarzenegger, the governor of California.[155] Concern was also growing among Chernobyl children initiatives in Canada and Europe because of fears that the future of their programmes was at risk. After almost a week of turmoil and uncertainty, the group flew back to Belarus, without Tatstsiana Kazyra but with the older of the two accompanying adults; her younger colleague preferred to remain in the United States until the situation with Kazyra was resolved. Pavel Shidlovskii, an official at the Belarusian Ministry of Foreign Affairs, also went to California in a further attempt to persuade Tatstsiana to return.[156] He too failed to achieve anything despite several meetings, and described the situation as a 'dangerous precedent'.[157]

On 14 August 2008, three days after the rest of the group had arrived back in their home country, Belarus indefinitely suspended schemes for sending Chernobyl children to the United States. The director of the Department for Humanitarian Activities, Aleksandr Koliada, announced that children would be allowed to go to the United States again only if the

[152] Ibid.
[153] Ibid.
[154] Glionna, 'Girl from Chernobyl'.
[155] Ol'ga Korelina, '"Prezident pozhimal mne ruku i blagodaril": Perevodchiki vstrech na vysshem urovne rasskazali o svoei rabote', *Tut.by*, 19 February 2015, https://bit.ly/3Z OSEmK (4 May 2024).
[156] Tatiana Korovenkova, 'Taniu Kozyro vernuli. A drugikh ne vypustiat', *Naviny.by*, 27 November 2008, https://bit.ly/49M5syY (4 May 2024).
[157] Quoted in Laura Norton, 'Despite Incentives from her Government, Girl in Petaluma Intents on Staying Here', *The Press Democrat*, 23 August 2008, www.pressdemocrat.c om/news/2203140-181/despite-incentives-from-her-government (4 May 2024).

United States was prepared to enter into an intergovernmental agreement setting out the conditions for recuperation trips.[158] The Children of Chernobyl United States Alliance, the umbrella organization for such initiatives in the United States, was dismayed by the suspension and condemned the Zapatas for their actions. Cecelia Calhoun, the alliance's Belarus liaison, made clear that 'we offer a health respite. We're not trying to rescue these children from any family condition in their home country.' 'This family knew the rules', she added: 'They made a conscious decision to break them. And it's harming the program.'[159] In Calhoun's eyes, the Zapatas had inflicted misery on more than 1,000 children who were now worrying that they might never get to see their host families again.[160] To this day, the Zapatas themselves are convinced that they did nothing illegal or wrong. They felt completely misunderstood, Debra Zapata explained, and had 'never thought ... the programme' would be stopped.[161] Zapata also said she knew that 'a lot of people hate us'– as a torrent of threatening phone calls and abuse had made abundantly clear.[162]

The diplomatic controversy rumbled on in the background, Chernobyl children initiatives in Europe and North America were up in arms, and the Zapatas were spending considerable sums of money on their lawyer and his Russian-speaking assistant, but things were starting to calm down on a more mundane level – at least on the face of it. The teenage girl from Belarus was not the object of so much attention now, and the new normal began to set in for her. She once again became increasingly unhappy with the strict rules in this religious family, and started to rebel, observing that she had been allowed to go wherever she wanted in Belarus – including the disco. The Zapatas were not so easy going. At one point, Tatstsiana voiced the idea of going back to Belarus after all. It is impossible to say whether this change of heart was connected to further attempts at persuasion by the Belarusian state. Kazyra herself said that homesickness was the deciding factor. In November 2008, she asked her grandmother to tell the Belarusian authorities that she wanted to return. They quickly sprang

[158] Vladislav Kagan, 'Gud-bai, Amerika', *Novye Izvestiia*, 19 September 2008; BelaPAN, 'Belarus' priostanavlivaet programmy ozdorovleniia belorusskikh detei v SShA', *Naviny. by*, 14 August 2008, https://naviny.by/rubrics/society/2008/08/14/ic_news_116_295692 (4 May 2024).

[159] Quoted in Glionna, 'Girl from Chernobyl'.

[160] Norton, 'Despite Incentives'.

[161] Zapata, McClelland, interviewed by Arndt.

[162] Ibid.

FIGURE 5.2 Tatstsiana Kazyra outside her grandmother's wooden house.

into action and arranged for her to be flown back to Belarus. Her sudden departure left her host family full of questions; 'their daughter', as they call her, had not even said goodbye.[163]

Kazyra was visited by numerous journalists at her grandmother's home. One of them came from the Russian broadcaster NTV, which branded her return an 'unexpected foreign-policy success'.[164] The images that BelaPAN, the Belarusian state news agency, circulated of Kazyra and her home, though, hardly seem so clear cut. With her striped leggings, denim shorts, and lightweight trainers. Kazyra was sporting a distinctly Western look that seemed to belong to a different world as she stood in the snow outside her grandmother's ramshackle wooden house (Fig. 5.2). It would be tempting to conclude from the images that the photographer was deliberately trying to invoke stereotypes of a backward Belarus. The pictures were lapped up in the US media.[165]

---

[163] Tatstsiana Kazyra, interviewed by Melanie Arndt, 21 July 2016. Skype.

[164] Viktor Kuz'min, 'Poshli s Kozyro', *NTV*, 24 November 2008, www.ntv.ru/novosti/145 035/ (4 May 2024).

[165] Paul Payne, 'Aid Worker Says Girl Who Sparked International Feud Worried about Sanctions', *The Press Democrat*, 25 November 2008, www.pressdemocrat.com/news/2 188197-181/aid-worker-says-girl-who (4 May 2024).

Tatstsiana Kazyra's life was by no means straightforward again now. She faced animosity and vilification because people blamed her for the suspension of trips abroad for Chernobyl children. She was subjected to abuse on the telephone, online, and out on the street – the diatribes included claims that she was 'brainless' and that the whole affair could have come straight out of *Santa Barbara* (a bad soap opera).[166] Kazyra even drew attention to how she had been cursed by a 'gypsy'-sounding voice on the telephone – a reflection of the fact that superstition, often accompanied by racist overtones, was widespread in the former Soviet Union. Looking back, she blamed the curse for an illness that began exactly one year after the sinister call. She needed multiple operations. Seeking financial assistance to pay for medication, her grandmother turned to a representative of the Ministry of Internal Affairs who had pledged to help Tatstsiana in the United States – but to no avail. Instead, the cost of the medication was covered by the Zapatas, with whom Kazyra had been in touch again for quite some time.[167] It was also not unknown for her to be praised for the courage and patriotism she had – it was thought – shown in deciding to return to her homeland after all. In such cases, the blame for all the fuss was laid squarely at the feet of her host family – everything had, after all, unfolded in a 'very American manner'.[168]

Almost exactly after a year after Kazyra's return, Ruth Williams, president of the Chernobyl Children's Project, visited her and her grandmother in Belarus together with another member of the organization. In a post on the *Little Chernobyl* blog, they too gave the impression that everything was going swimmingly: Tatstsiana was 'going to college and getting good grades', and the accompanying photographs showed her posing with grandmother, cats, and cousins.[169] Shortly afterward, Kazyra finished school and started working as a saleswoman. In 2013 she gave birth to a daughter and went on maternity leave for three years. The Zapatas were still there for her when money was needed for medicine, clothes, or a car.[170]

---

[166] Cf., e.g., the comment sections in Siry, 'Tat'iana Kozyro'; BelaPan, 'Tat'iana Kozyro vernulas' na rodinu iz SShA', *Tut.by*, 23 November 2008, https://bit.ly/3YvYhWx (4 May 2024).

[167] Siry, 'Tat'iana Kozyro'.

[168] Cf., e.g., the comment sections in ibid.; BelaPan, 'Tat'iana Kozyro vernulas''.

[169] Ruth Williams, 'Tanya Kazyra', *Little Chernobyl*, 19 November 2009, https://littlechernobyl.blogspot.com/2009/11/tanya-kazrya.html (4 May 2024).

[170] Zapata, McClelland, interviewed by Arndt; Kozyro, interviewed by Arndt.

Together with the Children of Chernobyl United States Alliance, the Chernobyl Children's Project attempted to engage with representatives of the Belarusian state to resolve the suspension of trips for Belarusian children. Oleg Kravchenko, the new chargé d'affaires at the Belarusian embassy, was invited to present the Belarusian position at the alliance's annual conference in Raleigh, North Carolina, in October 2008. He did so diplomatically, but did not change his stance.[171]

Two lawyers working on a pro bono basis had drafted a host family contract that Ruth Williams presented to various members of Congress in Washington, D.C., after the Raleigh conference in an effort to win them over. She did the same with Belarusian embassy officials. These efforts, however, failed to bring about further progress. The future of recuperation visits remained in limbo because the United States was not prepared to enter into the agreement the Belarusians were demanding.[172] After consultation with the consular section of the US embassy in Minsk, the Children of Chernobyl United States Alliance advised against inviting Chernobyl children privately on the grounds that the Department of State was working on an agreement and private efforts were likely to jeopardize this.[173]

In the eyes of Julie McClelland, the US government would never have acceded to the demands of the Belarusian president. This ex-serviceman's daughter was in no doubt that it 'would not have been the reaction of our government to do something like that'. It was a matter of political and moral principle: 'We are not going to concede to a dictator.'[174]

5.5.3 'Gud-Bai, Amerika'

In September 2008, a journalist at the Russian newspaper *Novye Izvestiia* came up with the headline 'Gud-Bai, Amerika' – thereby anticipating what Lukashenka was to confirm in law barely a month later.[175] On 13 October, the Belarusian president issued decree 555, which put a temporary stop to trips abroad for Chernobyl children. Children could now only travel to countries that had concluded an international agreement with Belarus in which the inviting state guaranteed their safety and

---

[171] CofCUSA, *Alliance News*, 2009, https://bit.ly/3YAXBPK (2 May 2024), 1–3.
[172] Payne, 'Aid Worker'.
[173] CofCUSA, *Private Invitations*, 26 January 2010, https://cofcusa.wordpress.com/2010/01/26/private-invitations/ (4 May 2024).
[174] Zapata, McClelland, interviewed by Arndt.
[175] Kagan, 'Gud-bai, Amerika'.

their timely return. The upper age limit was lowered from eighteen to fourteen, and no more than three trips to the same country were now permitted.[176] Italy was an exception: because it had already concluded an agreement with Belarus prior to 1 October 2008 (in response to the Viktoriia Moroz incident), there was still no limit on the number of times children could go there before they turned eighteen.[177]

The decree sent a clear message to initiatives that were not interested in cooperating with the state. Grushevoi thought the days of civil society efforts to support the Chernobyl children were numbered[178] – but it was not quite that simple. The governments of several countries soon decided to follow the Italian example: Ireland, Germany, Great Britain, Liechtenstein, Spain, and the Netherlands concluded agreements with Belarus in order not to jeopardize the children's visits. Even so, in 2009 around 25 per cent fewer children went abroad than in the previous year. Eleven countries – Bulgaria, Denmark, Greece, Japan, Canada, Lithuania, Luxembourg, Norway, Poland, the Czech Republic, and the United States – stopped hosting Belarusian Chernobyl children for recuperation.[179]

Despite the cautions from the Children of Chernobyl United States Alliance, a number of host families invited 'their' children privately, thereby circumventing the travel ban, which applied only to group trips for Chernobyl children. Private trips, however, involved a considerable amount of additional work and extra costs for the families because they had to organize everything themselves.[180]

5.5.4 Children from Ukraine, Adults in Belarus

As it became increasingly apparent that an agreement between the United States and Belarus was not on the horizon, those US organizations that

---

[176] Aleksandr Lukashenko, *Ukaz no. 555: O vneseni i dopolneni v Ukaz Prezidenta Respubliki Belarus ot 18 fevralia 2004 g. No. 98*, 13 October 2008, https://bit.ly/4ggep Db (4 May 2024). Lukashenka changed the age limit again in 2010, allowing children up to the age of eighteen to make such trips once more. Lukashenko, *Ukaz no. 59. O vnesenii izmenenii v Ukaz Prezidenta Respubliki Belarus ot 18 fevralia 2004 g. No. 98*, 28 January 2010, https://bit.ly/4fsdwGy (4 May 2024).

[177] In principle, the decree made such provisions for any state that had already entered into an agreement with Belarus prior to 1 October 2008; in practice, this meant Italy alone. Cf. Lukashenko, *Ukaz no. 555*.

[178] 'Predsedatel' fonda "Detiam Chernobylia" Gennadii Grushevoi: Unichtozhatsia poslednie vozmozhnosti detei ozdorovliat'sia za granitsei', *Belorusskie Novosti*.

[179] For an overview, see Arndt, *Tschernobylkinder*, 442–3.

[180] Dillon, interviewed by Arndt.

wanted to keep inviting groups of Chernobyl children switched their attention to Ukraine. A year after the Kazyra fiasco, for instance, the Chernobyl Children's Project in California invited Ukrainian Chernobyl children rather than Belarusian ones. Ruth Williams remained optimistic, however, and continued to hope that an understanding with the Belarusian government would be reached.[181] This never transpired. Belarusian Chernobyl children could now go to the United States only if they needed special medical attention that they could not get in their own country.[182] In January 2010, the Children of Chernobyl United States Alliance officially announced on its blog that it was shifting its attention to Ukrainian Chernobyl children.[183]

Switching to Ukraine had been largely straightforward, Kate Van Dyck of Children of Chernobyl – Christ Church Alexandria, Virginia, concluded after her programme's first summer without Belarusian Chernobyl children. The Ukrainian children were just as much in need of help, so it was 'not really a change of heart'; 'We found our new partners to be honest, sincere ... For the most part, the host families did not even notice any differences', she wrote.[184] She hoped that other programmes in the United States would follow their example and 'keep their hosting programs alive by inviting some of these children who are waiting and hoping for their turn to be given a safe haven in a US respite program'. It was, she concluded, 'really not the time for us to give up!'[185]

At the same time – as a side effect of the new situation, as it were – what Lukashenka had been calling for came to pass: foreign initiatives became active inside Belarus. They organized holidays at facilities within the country for 'their' Chernobyl children and others as well.[186] For Laurann Schlapper, who ran Project Restoration International in Kansas City, what mattered was to make sure that children who were no longer able to

---

[181] Paul Payne, 'With Belarus Out, Ukrainian Kids Visit', *The Press Democrat*, 15 August 2009, https://bit.ly/4jodMzB (4 May 2024).

[182] Programmes such as the Cultural and Educational Enrichment Project in Ellensburg, Washington State, were able to continue because they were not officially Chernobyl children initiatives. See Cec Calhoun, 'Educational Programs: Cultural and Educational Enrichment Project (CEEP)' in CofCUSA, *Alliance News*, no. 2 (2010).

[183] CofCUSA, *Ukraine Respite Video*, 26 January 2010, https://cofcusa.wordpress.com/20 10/01/29/ukraine-respite-vide/ (6 November 2019).

[184] Kate van Dyck, 'Change is Difficult? Not Really!' in CofCUSA, *Alliance News*, no. 2 (2010).

[185] Ibid.

[186] Elizabeth Tennison, 'Hope for Chernobyl's Child' in CofCUSA, *Alliance News*, no. 2 (2010).

go abroad knew 'that we and their host families still cared for them, loved them and hadn't forgotten them'.[187] Accordingly, Project Restoration International had turned its attention to organizing summer respite at the Nadezhda rehabilitation and education centre, north of Minsk, which had originated as a German–Belarusian joint venture in 1994.

This new way of getting involved entailed an even greater readiness to engage with Belarusian society, and its social practices and codes, than before. This was not without friction and, at least initially, could lead to disagreements and difficult paths to consensus. The division of areas of responsibility was often the main challenge. Learning processes were also involved, as Laurann Schlapper pointed out: 'It was impossible to plant ourselves in another culture from the perspective of our culture, nor should we dictate from the US what WE wanted at a camp in their country. Some issues that our board felt uncomfortable about are the norm there. Camping is simply done differently in Belarus!'[188] Unspoken expectations and the associated disappointments were a greater hurdle. 'Quite frankly, the entire experience was very challenging', as Schlapper put it. The Belarusian chaperones at Nadezhda had, she wrote, been disappointed when they realized she had come on her own and only brought 'a relatively small supply' of 'camp stuff' with her. The American side, meanwhile, would have appreciated a firmer basis on which to plan. Schlapper reiterated advice that she herself had been given before her first trip to Belarus: 'Be sure to bring your FLEX POWDER.' Belarusians who had previously been chaperones in the United States were also an important source of assistance, she said.

The Americans – they were usually women – who took part in initiatives such as the Nadezhda holiday camp introduced the same educational elements that they had built into their activities in the United States; volunteering, attitudes to people with disabilities, and intercultural skills figured prominently here.[189] Some US organizations arranged for Belarusian counsellors who could provide professional support for the Chernobyl children to be present during their recuperation. As one account put it, some children said that recuperating in Belarus was 'almost as good as America!!'[190] Other US organizations redoubled their efforts to make life better for children in residential facilities through measures such

---

[187] Laurann Schlapper, 'Project Restoration Camp Nadezhda August 11–23, 2010' in CofCUSA, *Alliance News*, no. 2 (2010).

[188] Some members of the board, for instance, had had 'concerns about the ... "teenage boys and girls" and the various challenges when dealing with that age group'. Ibid.

[189] Ibid.

[190] Jill Tyson, 'CCP of Greater Chattanooga' in CofCUSA, *Alliance News*, no. 2 (2010).

as renovating buildings and creating new playgrounds.[191] One initiative helped purchase a dacha so that children from an orphanage could visit it all year round and grow vegetables in the garden there.[192]

Aid work inside Belarus brought with it a shift in perspective that echoed – as was clearly the political plan – the new official line on Chernobyl in Belarus: 'Chernobyl' was to disappear.[193] Writing in a newsletter of the Children of Chernobyl United States Alliance, Jill Tyson of the Chernobyl Children's Program of Greater Chattanooga explained how her organization had been given to understand that it was time to let go of the term 'Chernobyl':

> As many older groups have probably noticed, the term 'Chernobyl' seems rather dated. It has been suggested to us that if we hope to continue ministry there, that we change our stated focus from summer programs for child respite, since we've all gotten past hope that we can bring groups of children in the near future. We are now widening the aims to all Belarusian children and families.[194]

5.6 'FOR ME, CHERNOBYL MEANS . . .'

'Chernobyl is inescapable; there's no getting rid of the radioactive matter in the ground now, no pretending it's not part of our lives. But that doesn't mean it's particularly topical.'[195] Thus spoke Irina Narkevich, herself a Chernobyl child who also went on to chaperone groups, on what Chernobyl meant to her twenty-five years after the accident. Her observation that it did not attract much attention was in part a comment on public discourse in Belarus, but in part also a reflection of the fact that the accident is hardly present in the everyday lives of the now grown-up Chernobyl children any more. This is not to say that there is no awareness of the health issues facing those affected or the risks associated with nuclear power. The Chernobyl children all see the accident as a terrible catastrophe, but they have largely pushed Chernobyl out of their everyday lives, irrespective of whether they are still living in contaminated areas or not. The accident and its legacy have become 'normal' for them – not in the sense of Charles Perrow when he talks about the inevitability of

---

[191] Patricia Lundeberg, 'Children of Chernobyl Foundation, San Diego' in CofCUSA, *Alliance News*, no. 2 (2010).

[192] Eric Hicks, 'North Alabama Belarus Mission' in CofCUSA, *Alliance News*, no. 2 (2010).

[193] On the 'disappearance' of Chernobyl from public discourse, see Stepanov, 'Tschernobyl ist niemals passiert?'; Kuchinskaya, *Politics of Invisibility*.

[194] Tyson, 'CCP of Greater Chattanooga'.

[195] Narkevich, interviewed by Arndt.

accidents in the use of complex high-risk technologies,[196] but in the sense that the ongoing disaster has become a routine part of everyday life.

5.6.1 The Disaster as Normality

For the Chernobyl children, the shadow of Chernobyl will always be there.[197] It has become a normal part of their everyday lives. It hardly ever figures directly, only indirectly in the form of references to illnesses or its impact on how people go about things. A striking feature of my interviews was the fact that more women than men said they tried to err on the side of safety in everyday life (e.g. when choosing food).

For Nadzeya Husakouskaya, who holds a doctorate in gender studies, Chernobyl had long been very distant. 'It was something that belonged to my childhood, to the Soviet Union, to the 1980s', she reflected. Only when she was living in London, she recounted, did she find a connection with it, through reading Aleksievich's *Chernobyl Prayer*. To her surprise, she realized that Chernobyl is one of the 'central concepts' that define how she, as a Belarusian, sees things: 'It was a profound sense of belonging – not to the country as such, but to the earth in the material sense of the word, to the place, to a history that concerns everyone.'[198]

Vladimir from Mahileu captured the feelings of many Chernobyl children when he spoke of 'just getting on with life'.[199] Sergei Kamornikov from Homel likewise refused to let Chernobyl dominate his existence: 'You have to live with it, ignore it, that's all.' Kamornikov's pointed indifference here, however, sits somewhat uneasily with his description of the accident as 'a terrible catastrophe' that 'changed the lives of millions of people'.[200] Similar contradictions will be found in the comments of many Chernobyl children.

For most of those whose views I sought, 'Chernobyl child' was not a negatively connoted label. Svetlana Bodrunova from the Homel oblast, now a professor of communication studies in St Petersburg, was very

---

[196] Charles Perrow, *Normal Accidents: Living with High-risk Technologies* (Princeton, NJ: Princeton University Press, 1999).

[197] Greg Bankoff, *Cultures of Disaster: Society and Natural Hazards in the Philippines* (London: RoutledgeCurzon, 2003); Bankoff, 'Cultures of Disaster, Cultures of Coping: Hazard as a Frequent Life Experience in the Philippines' in Christof Mauch, Christian Pfister (eds.), *Natural Disasters, Cultural Responses: Case Studies toward a Global Environmental History* (Lanham, MD: Lexington Books, 2009), 265–84.

[198] Husakouskaya, interviewed by Arndt.

[199] Vladimir, interviewed by Melanie Arndt, 6 September 2010. Mahileu.

[200] Sergei Kamornikov, interviewed by Melanie Arndt, March 2012. Online.

clear: 'That I was a Chernobyl child did not bring anything other than benefits.' This assessment, though, followed a catalogue of health issues that she put down to radiation exposure. The fact that her overall position is a favourable one is closely connected to the multiple trips she made abroad: she had been to Germany, France, and Italy. Chernobyl, she said, had ultimately made her 'cosmopolitan'.[201] Many other Chernobyl children see themselves in similar terms. Such ways of finding meaning can reflect subconscious self-defence mechanisms while also affecting how the disaster is remembered. Confrontation with traumatic experience is constrained by a synthesis of control and integration in a process where contradictions in rational terms can be perfectly straightforward.[202] The defence mechanism at work in a remark by Artem from Mahileu is obvious: 'You don't speak about it, because you don't want to remember it strongly.'[203] Another Chernobyl child explained that not talking about Chernobyl meant that it receded into the background, making it akin to a 'background noise that you simply forget about as the years go by'.[204]

There are situations, however, in which that background noise grows louder again, such as when those affected see signs warning of radiation, as Ekaterina from the Mahileu area explained.[205] She said that she made a point of eating food from 'clean' sources and 'of course' did not pick mushrooms and berries in affected areas. This would tend to put her in a minority. Young men, above all, claimed not to be bothered about such things.[206]

For many years, the continued existence of Prypiat as a virtual town online was also part of the normality of the disaster for evacuees.[207] It was the manifestation of a real-life community: 'I still see myself as belonging to Prypiat, and I will probably continue to do so', Alena Oginets told me. Her circle of friends largely consisted of other people from Prypiat: 'And I only got to know them in Kiev, paradoxically.' Becoming godparents for one another's children, she explained, meant that many of them had effectively become united as a family. What they had suffered together,

[201] Bodrunova, interviewed by Arndt, 2012.
[202] Hans Becker, 'Bewusste und unbewusste Abwehr- und Anpassungsmechanismen gegen das Erinnern an kollektive Traumen und Taten' in Günter H. Seidler, Wolfgang U. Eckart (eds.), *Verletzte Seelen. Möglichkeiten und Perspektiven einer historischen Traumaforschung* (Gießen: Psychosozial-Verlag, 2005), 303–14, here 304.
[203] Artem, interviewed by Melanie Arndt, 6 September 2010. Mahileu.
[204] Anna, interviewed by Melanie Arndt, 6 September 2010. Mahileu.
[205] Ekaterina, interviewed by Melanie Arndt, 6 September 2010. Mahileu.
[206] Aleksandr, interviewed by Arndt.
[207] See Section 2.3.1.

she said, created a bond between them. Oginets counts her father's death, which she blames on the accident – 'Otherwise, he'd have been a healthy person, a perfectly healthy person, positive, good' – as part of this suffering. She described how her mother remained single after his death, 'broken in mind and body'. Even if Chernobyl no longer played a role in her own everyday life, Oginets explained, she could not forget about this aspect. 'There is no turning a page here, for those dearest to you live on for as long as your memory of them lives on. And that memory – it's real, it has no expiry date, it's always with you.'[208] Just how strong the sense of community is can also be seen from the acerbic debate about the *lzhe-chernobyltsy*, the 'fake Chernobylers', who – rightly or wrongly – are suspected of having obtained formal recognition as victims by underhand means.[209] Forums where 'fake Chernobylers' are exposed and shamed have existed online ever since communication over the Internet became established in Eastern Europe.[210]

5.6.1.1 *Views on Nuclear Power*

In spite of their experiences of the disaster, most of the Chernobyl children whose views I obtained supported the use of nuclear power. Where Belarus was concerned, they justified this primarily on the grounds that the country has no oil or natural gas reserves and no coal deposits of any significance – that there is no alternative, in other words. Their views on the issue, however, also reflected an unshaken faith in the power of technology; it is possible that the legacy of the Soviet Union's technological euphoria was still at work here. Nuclear power stations just had to be built so as to be completely safe, some said.[211] Narkevich argued that the construction and operation of nuclear power plants should be subject to public oversight. She said that she could see herself playing a role in such a regulatory body; she was also more afraid of nuclear weapons than of nuclear power stations.[212]

Committed opponents of nuclear power plants, such as Alena Oginets or Olga O., were in a minority. Their opposition stemmed not so much from concerns about the technological safety of nuclear power as from

---

[208] Oginets, interviewed by Arndt.
[209] Ibid.
[210] Cf., e.g., the public Facebook page of the activist and writer Liubov Sirota (www.faceb ook.com/lyubov.sirota?locale=de_DE).
[211] On the shift in public opinion that was encouraged by targeted state campaigns and resulted in support for nuclear power, see Stepanov, 'Tschernobyl ist niemals passiert?'.
[212] Ibid.

human fallibility. Oginets was clear about the fact that the experience of Chernobyl was part of why she felt the way she did: 'Solar plants, anything else, as long as it's not nuclear power. Nuclear power is terrible.' She also, however, expressed a general scepticism about work ethics in Ukraine that figured just as prominently in her concerns. In her eyes, there was an inherent safety risk in the fact that the facilities are constructed and operated by human beings; hence, 'even if Japanese or Germans were to build it [a nuclear power station], the human factor is still there'.[213]

For Oginets, living in Ukraine involved the constant fear that she might not learn about an incident promptly: 'Try finding anything out here!' Since Chernobyl, she said, she no longer felt that official sources could be trusted to provide reliable and truthful information. Her lack of trust stems from conversations that she overheard at the Novovoronezh nuclear power plant while waiting to have her body and clothes checked for radiation – the volunteers who brought the *Chernobyltsy* food and fresh clothes there talked about how they were regularly checked with dosimeters because leaks were a common occurrence.

Olga O. also expressed her views in no uncertain terms: 'On the basis of my experience, the idea that anyone could think nuclear power is worth risking the lives and health of the whole nation for again is completely incomprehensible for me.'[214] When asked about the risks and benefits of nuclear power, many individuals mentioned a kind of inertia that stops people reflecting properly on the use of this technology. Iulia from Mahileu, for instance, who took part in demonstrations against nuclear power and advocating for greater use of wind power (a form of protest she had first encountered abroad), said that nuclear energy would never be absolutely safe but nobody in Belarus was interested in this. Her own position on the construction of a Belarusian nuclear power plant, which was not yet a certainty at the time of the interview, was not straightforward: she was really completely against it, she said, but also felt that the country simply did not have any other resources. Her compromise was that nuclear power could be endorsed if plants were staffed by 'capable specialists'.[215]

On the whole, the attitudes of the Chernobyl children I spoke to and surveyed were apolitical. For many of them, political activism was out of the question for the simple reason that they were afraid of the repercussions.

---

[213] Oginets, interviewed by Arndt.
[214] Ol'ga O., interviewed by Arndt.
[215] Juliia, interviewed by Melanie Arndt, 6 September 2010. Mahileu.

'You can end up in prison for that here', Iulia explained.[216] Only a small number, including Aleksandr and Tatiana Chulitskaia from Mahileu, and Nadzeya Husakouskaya from Minsk, had ever got involved in this way. Oginets captured the reticence when it came to politics with *zhivi da raduisia* – a Russian proverb meaning 'live and be happy'.[217]

5.6.2 Chernobyl Children Who Stayed Abroad

5.6.2.1 *Tatsiana Khvitsko: The Role Model*

As a baby, Tatsiana Khvitsko met with the same fate as many other children with disabilities: her parents handed her over to a home when she was born. In all likelihood, they did so because they felt they had no other option and the social pressure was too great. Four years later, they revoked their decision and took their daughter in again. They did not tell her that they were her biological parents until she was thirteen, when Tatsiana noticed her resemblance to her mother and confronted them. After returning to her family, Tatsiana got her first prostheses when she was four, but they were made of heavy wood and had to be attached to her leg stumps with uncomfortable leather straps.

Her parents later sent her to a boarding school for children with disabilities – in order to shield her, as Khvitsko herself emphasized: 'People saw me differently and would stare at me or point. My parents wanted to protect me so they sent me to this school.'[218] It was there that Project Restoration, a Chernobyl children initiative from Kansas City, found her in 1996; an invitation to the United States followed. 'That's when my life changed because I was loved by so many strangers', Khvitsko recalled in a magazine interview in 2017.[219] From then on, she returned to the United States every summer. She felt that she belonged there in spite of her disability. She was encouraged, in contrast to what she was used to in Belarus, to show her hands and leg stumps openly rather than concealing them. Tatsiana wore skirts for the first time in the United States. Three host families untiringly provided her with new prostheses that helped her to build confidence as she grew up.

For the organization that had invited her, Khvitsko's story was a total success. At first, though, she – like many others – stood out because her

---

[216] Ibid.
[217] Oginets interviewed by Arndt.
[218] Quoted in McDermott, 'Tatsiana Khvitsko'.
[219] Quoted in ibid.

behaviour had been somewhat challenging: Laurann Schlapper, the head of Project Restoration, recalled that 'during her first summer, she yanked the hair of a little girl in her first host family and loved to kick their dog!' Looking back, Khvitsko explained her frustration as a result of insecurity and rejection of her own body. These feelings increasingly began to recede, not least because her religious host families never gave up on her, despite her feistiness.[220] As she grew older, and later as a young woman, she herself supported the programme that had first brought her to the United States, describing herself as a 'graduate of Project Restoration'.[221] The initiative had been witness to an extraordinary transformation: 'Who would have ever imagined what lay ahead for that little girl from Belarus.'[222]

At the age of eighteen, Khvitsko decided to go to the United States for good. She studied corporate communication at the MidAmerica Nazarene University in Olathe, Kansas, with the help of scholarships and support from her three host families. Alongside her studies, she became involved in sports groups for girls from socially disadvantaged backgrounds, such as Girls on the Run: 'How many times are you going to be coached by a girl with no legs? I encourage the girls to push themselves and I challenge them to do something different because I'm different. I want to empower young women.'[223] Many girls took her message to heart. One ten-year-old with large glasses and spindly legs told a journalist: 'She just inspires me to keep going even when you feel like you want to give up. So, whenever I feel like I want to give up, I will think of her.'[224]

Khvitsko herself, though, prefers not to be seen as an inspiration: 'I'm not an inspiration; I want to be thought of as giving encouragement. I want people to remember my story, how it reflects on life and how much passion you can have for life.'[225] After her studies, Khvitsko started working as a PR specialist for a company that makes medical textile products, mainly for customers with physical disabilities. She writes openly about her life as an amputee on the company's blog. In 2017, she

---

[220] Ibid.

[221] Quoted in CofCUSA, 'From Kansas City Area' in CofCUSA, *Alliance News*, no. 4 (2010), 1.

[222] Ibid.

[223] Quoted in McDermott, 'Tatsiana Khvitsko'; see also Sara Bean, 'Runner Brings a Message of Inspiration to Gardner', *The Kansas City Star*, 11 May 2015, https://bit.ly/3VNMoMy (4 May 2024).

[224] Quoted in Bean, 'Runner Brings a Message of Inspiration'.

[225] Quoted in McDermott, 'Tatsiana Khvitsko'.

married an American she had met at the gym; their child was born in 2018. Khvitsko is now part of US society.

Khvitsko believes that her disability was caused by the Chernobyl accident, but she is just as clear about the positives that came out of it: 'Because of Chernobyl I'm wearing prosthetics, because of Chernobyl I'm disabled, but also because of Chernobyl I'm here in the US. I can't complain that I'm missing a limb or two, I mean, I have so much.'[226]

5.6.2.2 *Olga O.: Keeping a Low Profile*

'For one child, it was all worth it', Nadya Neal Hinson concluded when Olga O. got her first wig. It had, Neal Hinson wrote, been 'like Christmas morning' when hair-replacement specialists in Seattle gave the girl her long blonde hair.[227] Olga O. herself recalled how she had not known what to feel at first because having a wig was so new to her: 'Everyone was friendly and happy for me, but as an eight-year-old, I didn't know what I was meant to feel. I think I felt like a princess when I saw myself with the hair.'[228] In her Belarusian home town of Druzhny, not far from Minsk, her schoolmates knew that she did not have any hair. She wore headscarves that her mother sewed for her; once they got used to this, her class accepted her, but outside the classroom, she was bullied. This got worse when she returned from the United States with a wig. Even teachers branded her as conceited. Some children pulled at the artificial hair to see whether the wig would stay on – and, if not, to see what was underneath it. Others mocked her. As a result, Olga O. became increasingly wary of contact with other children. Even in her own family, Olga O. was not spared discrimination: her younger cousins taunted her. Their parents did not approve of this, but for Olga O. it demonstrated the post-Soviet attitude to people who were simply a little bit different from everyone else: 'It just shows how post-Soviet parents did not find it necessary to talk to their children about such issues and how to act around other kids with health problems.'[229]

Olga came from Druzhny, an *atomgrad* that was founded as recently as 1985 to serve as a home for workers at a planned nuclear heat and power plant; the name, which means 'welcoming', had apparently been chosen by children in a competition. Construction work on the plant began in

---

[226] Barcroft TV clip accompanying Rosenbaum, 'Chernobyl'.
[227] HA, Center for Civil Society International, Box 48–4.
[228] Ol'ga O., interviewed by Arndt.
[229] Ol'ga O. to Arndt, email, 7 November 2017.

1986 but was halted following the Chernobyl accident. Numerous specialists from Druzhny were deployed as 'liquidators' in the disaster area; the town later became home to resettled people as well. Although the disaster did not involve Olga's family directly and her parents were not deployed as 'liquidators', she put her own illness, as well as the various health problems that affected her parents, down to the accident. She recalled that elevated levels of caesium-137 were detected in her blood at a medical examination in the United States.[230]

From 1995 to 2003, Olga O. went to Seattle every summer to recuperate. She said that her visits had always made her 'happier and more optimistic': 'They gave hope of something better. There were always new clothes, tasty treats to take back home, toys, things that my parents would never have been able to afford.'[231] Not least, she said, she had hoped that her illness might be cured after all. Her host family organized medical appointments and arranged for various treatments. Some of them seemed to work for a bit, and her hair began to grow back, but it soon fell out again; some patches remained completely bald.[232] She ended up wearing a wig all time.

When she was fifteen, her host family invited her to come to the United States to study. She took up their offer: 'It was a chance to make my life and my family's future better.'[233] Getting a visa, however, was no easy matter, particularly for a young woman. Olga O. is of the view that her application would not have been approved if it had not been for her illness: when she told the consular immigration officer about it and her recuperation trips to the United States, she 'saw an immediate change' in his eyes.[234] Olga O. moved to Seattle in 2004, at the age of sixteen; she finished high school and took an associate of arts degree there, before moving to the east coast to study psychology. After three years, she returned to Washington State to be closer to her host family again. She was afraid of the Belarusian bureaucracy, so it was nine years before she went back, by which time she had a green card that would ensure her ability to return to the United States.[235]

In the same year that she got her own green card, her father was a winner in the green card lottery, so her parents were able to accompany

[230] Ol'ga O., interviewed by Arndt.
[231] Ibid.
[232] Ol'ga O. to Arndt, 7 November 2017.
[233] Ol'ga O., interviewed by Arndt.
[234] Ol'ga O. to Arndt, 23 October 2017.
[235] Ibid.

her when she went back to the United States at the end of her visit to Belarus. Olga O. has since become a US citizen and married an American.[236] She visits Belarus every two years.[237] Looking back, Olga O. sees her artificial hair as a mixed blessing.[238] She thinks it likely that the wig prevented her from developing a self-confident relationship with her illness. She lives in constant fear that her alopecia might be exposed.[239]

5.6.3 Chernobyl Children Who Went Home

The majority of Chernobyl children returned to their home countries, primarily because of their attachment to their families and *rodina*, or motherland. Some gave themselves the option of moving abroad later – as in the case of Tatiana Sinitskaia, who had made eight visits to California and picked up English very quickly there. She later studied in Minsk to train as an English teacher. Teachers do not earn much in Belarus, so she also works as a cosmetics sales representative. She could still see herself living in the United States now.[240]

Tatstsiana is one of many. Almost a quarter of those surveyed by Gatalskaia and Zaitseva in their Chernobyl children study could see themselves living and working abroad. Self-fulfilment and improving one's material circumstances were the main reasons given for this, the study reported, but the environmental situation at home and better health-care provisions abroad also played a role. There were also personal reasons, such as wanting to distance oneself from one's parents if relations with them were strained.[241] Many Chernobyl children also considered university study after their trips, following encouragement from their host families.[242]

Empowerment and mobilizing civil society were central aims of the Belarusian 'For the Children of Chernobyl' fund. The results of a survey early in 1994 were a welcome sign that its efforts were starting to bear fruit. Of those questioned, more than 50 per cent said they wanted to get involved in existing initiatives to address Chernobyl-related issues, and

---

[236] Ibid.
[237] Ol'ga O. to Arndt, 7 November 2017.
[238] Ol'ga O., interviewed by Arndt.
[239] Ol'ga O. to Arndt, 23 October 2017.
[240] Tatiana Bibikova (previously Sinitskaia), interviewed by Melanie Arndt, 17 September 2017. Online.
[241] Gatal'skaia, Zaitseva, *Sotsial'no-psikhologicheskii analiz*, 54.
[242] Ibid.

30 per cent expressed a desire to take an active role in these efforts or set up new initiatives themselves.[243] Taking stock, Irina Grushevaia said: 'I'm not saying that all the children have undergone a revolutionary transformation; but having seen a different life has changed the way they view their own lives forever. They have started to compare what could be with what is.'[244] Her husband's assessment of humanitarian work was less positive. He feared that it might give rise to passivity: 'People today want good medicine to heal the children who are sick, but they don't understand how those children who are still well can be protected. . . . And the matter is increasingly off the table.' Even among activists, he registered a growing sense of indifference, and observed that silence about the accident was returning to Belarusian society.[245] The Chernobyl children who went abroad, though, could help to break this silence indirectly – if, that is, Chernobyl was discussed with them there.

CONCLUSION

In the period covered by this study, 'Chernobyl child' was a very broad category that encompassed far more than just children who had been directly affected by the disaster. The power to determine who counted as a Chernobyl child and was sent to recuperate as such lay not with the children themselves or their families, but with the organizations at home and abroad that had devoted themselves to protecting them. Even so, parents were, at least to some extent, able to influence whether their child was treated as a Chernobyl child by being particularly persistent or using personal connections. The category also increasingly became a form of self-description: as they grew older, the Chernobyl children made it their own in various ways. They often gave the label positive connotations, only describing it as discriminatory in rare cases. They made symbolic capital out of their (often repeated) trips abroad, the cultural and linguistic skills associated with them, and the networks that they built in other countries; in the best case, they were able to rely on these contacts and connections for years. At least on the surface, the negative consequences of the accident slipped into the background as a result.

For orphans in particular, and the more so if they had disabilities, being categorized as a Chernobyl child could extend their horizons in ways that

---

[243] Author's own records, '"Die Welt nach Tschernobyl 8 Jahre danach"'.
[244] Tamkovich, *Filasofiia dabryni*, 40.
[245] Author's own records, '"Die Kinder von Tschernobyl" heute'.

would otherwise have been impossible in the Soviet Union and the social upheaval of the 1990s. The staffing situation and conditions in many homes were desperate. Being adopted or fostered in their own country opened up opportunities for Chernobyl children that would have been unthinkable in the past. Later, foreign interest and associated humanitarian work at facilities in Belarus, Ukraine, and the Russian Federation began making a difference too. Life became better there, and people with disabilities, who had been excluded from society for so long, began appearing in public more and more often.

Sending children away to recuperate was a political matter from the outset, as we have seen in the discussion of the culture shock theory. The arguments for and against recuperation abroad exposed the fractures in a society caught up in transition, as well as the persistence of Cold War clichés. For their part, the Chernobyl children interpreted any culture shock positively, as a broadening of their horizons, or applied the concept to their own society and the upheaval there. Things were not always so easy for the chaperones and interpreters that went abroad with the Chernobyl children – in some cases, they were confronted with realities that called their own past and values into question.

The political tensions culminated in the controversies surrounding Viktoriia Moroz and Tatstsiana Kazyra, the two 'Chernobyl girls' who refused to return home to Belarus. The reactions to these affairs illustrate the differences between East and West when it came to (?) ideas of order in state and society in the mid-2000s. Perspectives carried over from the Cold War played a role here too, often hindering smooth communication. When President Lukashenka stopped Belarusian children being sent abroad to countries that had not concluded an intergovernmental agreement with Belarus, US organizations quickly turned their attention to Ukraine – but also started lending their backing to initiatives inside Belarus. The authoritarian president, therefore, got what he wanted: instead of corrupting Chernobyl children abroad, as he claimed, foreign organizations invested in recuperation infrastructure and residential care for children within Belarus.

The host families were a cornerstone of efforts to support the Chernobyl children. Out of nowhere, thousands of families came forward to take children into their homes during the 1990s. A great many of them knew nothing about the world behind the Iron Curtain that had now been lifted. The organizations that recruited them drew on centuries-old stereotypes of a backward, impoverished East, often with a tendency to overdramatize, resulting in expectations that had to be reconciled with the realities of the

Chernobyl children. There were processes of accommodation and compromise on every level: between host families and children; between host families and chaperones; between local organizations and organizations in the children's home countries; and between organizations, the Chernobyl children's parents, and, last but not least, state institutions. If host families wanted to invite 'their' Chernobyl child again, they often had to start networking to fund the trips. Initiatives also engaged increasingly in networking with one another in order to share resources (above all, knowledge). The visits had an impact on the host families and the Chernobyl children's own families as well; tensions could result if the different words involved were not immediately compatible. Looking back, though, the Chernobyl children, at least, felt that such clashes had been enriching for them. Many Chernobyl children have since become citizens of other countries, living in the lands that once took them in to help them recover from the effects of radiation exposure.

Concluding Remarks

The Chernobyl children and their networks of concern are one of the reasons why a nuclear accident that was initially branded 'typically Soviet' in many parts of the world came into focus as a transnational disaster. These children brought the reality of Chernobyl into the everyday lives of hundreds of thousands of people in Europe and North America. They were living proof of the characteristic distress and fears associated with the accident. Far removed from the scene of the disaster, abstract headlines were replaced by the stories of actual individuals – albeit not always ones that confirmed the ideas and expectations of the children's hosts.

The Chernobyl children also witnessed – and stood for – a collapsing political and social system, the end of the bipolar world order, and life in the Anthropocene. They embody what it means to be at home on a damaged planet.[1] The features of natural 'innocence, vulnerability, and malleability'[2] ascribed to children all over the world meant that young people growing up in the shadow of Chernobyl met with empathy, a readiness to help, and even self-sacrifice on the part of people who were (or had been) living on the other side of the ideological divide. The fact that those harmed by the accident also turned out to be on the losing side in the Cold War provided an opportunity to demonstrate humility in victory, further confirming the ethical credentials of those who came to their aid.

As well as hundreds of thousands of Chernobyl children, thousands of accompanying adults and activists also criss-crossed between East and

[1] I am alluding here to Tsing et al., *Arts of Living*.
[2] Winkler, 'Kindheitsgeschichte'.

West during the 'transformation decade'.[3] From the mid-1990s, they were joined by a growing number of host parents who used their new personal contacts to see for themselves what kind of life 'their' Chernobyl children had in the former Soviet republics. Belarusian and Ukrainian organizations helped to promote such exchange. They ran congresses and set up various programmes aimed both at facilitating dialogue and knowledge circulation and at building networks. Chernobyl set in motion developments in East–West relations that went far beyond the provision of material assistance and humanitarian aid. It was not unusual for Western involvement to be bound up with specific political aims, including the democratization of post-Soviet societies and opposition to the use of nuclear power. Although it is often thought to be apolitical, humanitarian aid is in reality more 'part confidence trick and part self-delusion';[4] the case of the Chernobyl children was no exception to this. In practice, certain values and expectations lay behind the interest that was taken in them; the movement developed its own dynamic and its own language. New terms such as 'Chernobyl children' or 'Chernobyl AIDS' entered the lexicon, and old terms such as 'clean' or 'zone' acquired new meanings in the changed reality that followed the disaster.

THE 'COLLAPSE OF STATE AUTHORITY' AND GORBACHEV'S 'HUMAN FACTOR'

By exposing latent crises,[5] the accident at Chernobyl accelerated the 'collapse of state authority' in the Soviet republics that were affected – and far beyond their borders as well.[6] In the first three years following the accident, the Soviet leadership fell back on well-established strategies in its efforts to 'control' the situation and create a sense of order. It concealed the true extent of contamination and – with only a few exceptions, which were often not made public either – turned down offers of assistance at home and abroad. Despite the warning signs, the disaster caught the system off guard. It was incompatible with the technological and ideological self-image of the Soviet Union – which could only mean that there *was* no disaster.[7] There was nothing unique about how the socialist Soviet

---

[3] Dieter Segert, *Transformationen in Osteuropa im 20. Jahrhundert* (Bonn: Bundeszentrale für Politische Bildung 2014), 151.
[4] Barnett, *Empire*, 34.
[5] García-Acosta, 'Historical Disaster Research', 65.
[6] Altrichter, *Russland 1989*, 123, 213.
[7] Guski, 'Stimme der Opfer', 74.

state had played down the possibility that the worst could happen. It was simply doing exactly what the global nuclear industry and its supporters had also been doing. The Soviet leadership looked on with satisfaction as industry representatives played down the implications of the explosion and thereby lent credibility to the flawed Soviet response. The legacy of the minimization strategy was (and is) most readily apparent in the debates surrounding the victims of radiation exposure and the medical and environmental consequences of the disaster. There is no shortage of studies by now, but advocates of nuclear power continue to cite gaps in the research and operate with data that suits their purposes. Because of this, there is no clear prognosis for the individuals whose bodies remind them of the disaster on a daily basis.

It is wrong to suggest that the Soviet state did not take any steps to safeguard its children or the rest of the population. Instead, a more nuanced perspective on the response to the disaster is needed – one that takes its unprecedented nature into account without turning it into a way of relativizing or excusing each and every mistake, unacceptable delay, or failure to act. In an ideal world, Prypiat would have been evacuated as soon as the accident became known – but when the town and the surrounding area *were* finally abandoned, the process took place efficiently in a matter of hours. Emergency procedures that were fit for purpose did, therefore, exist; they simply had not been drawn up with a disaster of this magnitude in mind. Many measures were thwarted by shortcomings in the Soviet apparatus. Prypiat could have been evacuated earlier – had the political will been there and had the responsible parties been prepared to take action accordingly.

Particularly on a regional level, a number of Party and state functionaries tried to protect at least some parts of the population, above all children. Kyiv schoolchildren, for example, were evacuated against the wishes of Moscow in May 1986. On the whole, however, the state did not do enough, encumbered as it was by the ideologically ingrained lack of flexibility and slow response mechanisms that were built into the bureaucratic system through which it exercised its power. Free speech and free markets were not present to provide a corrective. Only when clear signs of social and economic crisis manifested themselves were processes for reallocating resources, effecting structural change, and addressing inconsistencies set in motion.[8] This is an obvious feature of the response to

---

[8] For a contemporary analysis that still makes for worthwhile reading, cf. Egor Gajdar, Konstantin Kogalovskij, 'Tendenzen der Wirtschaftskrise in der UdSSR' in Klaus Segbers

Chernobyl in the Soviet state. The Fukushima disaster, however, shows that even liberal democracies can be overwhelmed by the complexities of nuclear accidents, apparently to the point that authoritarian reactions can materialize in such systems too. The encroachments on freedom of expression and access to information, the treatment of the general public – particularly children – and clean-up workers, the underhand dealings of TEPCO, and, not least, the widespread sense of helplessness in Japan after Fukushima should all give cause to reconsider the idea of Chernobyl as an inherently Soviet accident.[9]

By the time the Soviet leadership grasped how bad things really were for the Chernobyl children, it was too late. Even though the Soviet Union was starting to fall apart, they still managed to draw up a comprehensive plan for systematically improving the long-term situation of the affected children. But it was not possible to implement this centralized Chernobyl children programme before the Soviet Union collapsed. Everything was in short supply: financial resources, the necessary political will on the part of the union-level leadership and the governments of the republics, appropriately trained personnel, and medical facilities and equipment.

The Soviet system was not even in a position to make sure 'its' children were safe. When the public realized this, whatever hopes they still attached to the socialist promise of a better future began to dissolve. It was clear to them that the Soviet Union had lost the battle when it came to which system was best for humanity in the modern world. The loss of legitimacy on the part of state and Party organs went hand in hand with social action. Disappointment, fear, and anger found a way out – 'in the name of our children' was effectively shorthand for 'the future' – in unprecedented protests. Several functionaries in the Communist Party switched their allegiance and backed the new and diverse socio-environmental and national movements instead. It was in this context that the first independent Chernobyl children organizations came into being at the end of the 1980s. This amounted to the 'mobilization of the

---

(ed.), *Perestrojka. Zwischenbilanz* (Frankfurt a. M.: Suhrkamp, 1990), 230–65, here 230–1.

[9] On Fukushima, see, e.g., Steffi Richter, *Japan nach 'Fukushima': Ein System in der Krise* (Leipzig: Leipziger Universitäts-Verlag, 2012); Lisette Gebhardt, Steffi Richter (eds.), *Lesebuch 'Fukushima': Übersetzungen, Kommentare, Essays* (Berlin: EB-Verlag, 2013); and also Jens Kersten, Markus Vogt, Frank Uekoetter (eds.), *Europe after Fukushima: German Perspectives on the Future of Nuclear Power* (Munich: Rachel Carson Center, 2012); Eiji Oguma, *Tell the Prime Minister*, Japan, 2015: UPLINK.

human factor' that Gorbachev had called for,[10] except that the ideological straitjacket of state socialism was gone. The majority of independent groups no longer saw any point in trying to reform the Soviet system and turned their backs on their own state instead. The case of the Belarusian 'For the Children of Chernobyl' fund is instructive: the idea that the state was its enemy incarnate was a constant presence throughout the fund's existence and the various fallings-out that accompanied it. The restrictive and repressive measures implemented under President Lukashenka were hardly conducive to a reassessment of this position. In this case, the collective presence of the Chernobyl children had contributed to the disintegration of the Soviet Union, and led to control over various aspects of society returning to the hands of the state in an authoritarian climate.

POLYCENTRIC GLOBAL NETWORKS OF CONCERN

If there was one thing on which the Party and state apparatus and the general public did agree during the transformation decade, it was the idea that sending children away was an effective way to protect them from the consequences of the accident. In the initial phase, from immediately after the first evacuations to the appearance of independent Chernobyl children organizations in 1989, the trips were arranged and funded by the state. The destinations were always within the borders of the Soviet empire. When perestroika morphed into *katastroika*, it became clear that something had to change. The true scale of the accident was being interrogated and discussed in public, and more and more people were demanding better provisions for children from contaminated areas. They called for better-equipped facilities where the children could recuperate and their health would improve. At the same time, the state's ability to provide such assistance was rapidly disappearing.

After Ukraine and Belarus declared sovereignty and, ultimately, independence, Moscow started relinquishing responsibility for dealing with the consequences of the disaster. The republics, however, did not have sufficient means of their own to implement their Chernobyl children plans and continue sending children away for recuperation. Even state institutions now found themselves pinning their hopes on international assistance. In pursuing it, they departed, radically at times, from Soviet values. Particularly in the early days, they also showed a remarkable

---

[10] Quoted in Altrichter, *Russland 1989*, 20–1.

willingness to compromise (e.g. when dealing with religious donors) in order to secure the commitments and support they needed.

In the final two years of the Soviet Union's existence, the Soviet Chernobyl children became 'children of the whole planet'. The idea of the globe in this contemporary description is, of course, selective and reflects a Western bias: the children were, in the vast majority of cases, invited by and sent to countries in the Global North. The nature and extent of the attention that the Chernobyl children received in over thirty countries – from Australia to the United States – was nonetheless unprecedented.

Non-state initiatives collaborated with foreign partners from the outset. Considering the global polycentric networks of concern that grew up around the Chernobyl children allows the processes of decentralization that were gathering pace in Soviet history at the end of the 1980s to be understood in much broader terms than has previously been the case. The processes of separation now present themselves as a move away not only from the centre in Moscow but also from the national context as a whole. Foreign interest and transnational networking had an integral part to play in coming to terms with the disaster; this applied to the environmental tragedy just as much as it did to the social, cultural, and economic crisis in (post-)Soviet society. Far more was involved here than simply providing the children with 'radiation holidays'. Foreign organizations and host families supported initiatives and families in the East for years in both material and financial terms. Involvement in these trips generally resulted in the accumulation of social and cultural capital: cross-cultural experiences, networks, foreign languages. Many people were able to convert the symbolic capital into economic capital in the context of their careers. Humanitarian work could open up hitherto unimaginable horizons for orphans and seriously ill Chernobyl children in particular, either in their own country or in their host country (which could also become their new homeland). It is often taken as a given that children belong to a particular nation with its own specific territory;[11] the Chernobyl children, typified as they are by adoption, emigration, and repeated extended visits abroad, call this view into question.

For adults, getting involved could open the way to a post-Soviet career, or at least make it easier to find one. This was true as much for former representatives of state and Party organizations as it was for activists in the civil society humanitarian movement. In the midst of the social and

---

[11] See Winkler, 'Kindheitsgeschichte'.

economic turmoil, this provided a source of stability for identities and biographies, or even a way to reinvent them for the better, as well as allowing social divisions to be overcome or, in some cases, simply set aside entirely. During perestroika, even state institutions could become places to learn about and experiment with civil society activism.

Only towards the end of the transformation decade did the Belarusian and, to some extent, Ukrainian states change their positive view of the children's trips. In the early years, humanitarian assistance had, in the vast majority of cases, been welcomed and facilitated with a minimum of bureaucracy. From 1996, however, state intervention and moves to centralize control became more and more common. Those sceptical about sending children abroad increasingly claimed that the trips subjected the children to a culture shock that was bound to be damaging for the young generation and would alienate them from their own country. This fear of losing authority over the minds and bodies of 'its' children provided the Belarusian state with a pretext – if not the motivation – for cementing authoritarian practices and reinforcing its own position. State institutions, including the security service, now obstructed the work of independent organizations in Belarus – or prohibited it entirely. Those organizations consequently turned their back on the state even more emphatically. The lack of engagement on both sides reflected a missed opportunity to develop new forms of collaboration.

HUMANITARIANISM IN THE TRANSFORMATION DECADE

Starting in the 1950s at the latest, human influence on the environment began to increase as a result of a previously unprecedented acceleration on several levels: fossil-fuel consumption, population growth, urbanization, and the employment of new (and high risk) technologies such as nuclear power.[12] Around the end of the millennium, scientists drew these developments together under the concept of the Anthropocene.[13] These processes of change were also affecting the environment in a way that led to the development of a 'moral impetus of the Anthropocene'.[14] This new

---

[12] See McNeill, Engelke, *Great Acceleration*.

[13] See Melanie Arndt, 'Umweltgeschichte: Version 3.0', *Docupedia-Zeitgeschichte*, 10 November 2015, https://docupedia.de/zg/Arndt_umweltgeschichte_v3_de_2015 (4 May 2024).

[14] William Cronon, 'The Portage: Time, Memory, and Storytelling in the Making of an American Place'. Paper at conference *Arts of Living in a Damaged Planet*, University of Santa Cruz, CA, 9 May 2014.

moral imperative, which brought into focus human responsibility for the fundamental shifts that were occurring, not only involved protection of the environment in an immediate sense but also underpinned the self-image of the international peace and anti-nuclear movement. As this book has shown, it also shaped humanitarian aid for the Chernobyl children: many of the pioneers in this work came from the New Social Movements, or were at least associated with them.

The twentieth century was the nuclear age and the environmental age – but also the age of volunteering.[15] As Michael Barnett has shown, in the course of the century, compassion developed into 'a virtue, so much so that it has become a status symbol, and individuals, organizations, and states compete to be recognized for their generosity'.[16] This trend extended into commercial activity; Paul Newman's not-for-profit enterprise, from which the Chernobyl children also benefited when they stayed at his Hole in the Wall Camps, is one example among many.

The Chernobyl children movement was distinctively underpinned by 'the desire to demonstrate and create a global spirit'.[17] This desire was shared by twentieth-century humanitarian movements in general, of course, but the end of the Cold War created a particularly wide window of opportunity for turning the One World vision into practical action.[18] Barnett's theory that humanitarianism flourishes when 'nations worry about losing a sense of mission and people seek to restore their humanity'[19] is clearly true of the transformation decade. Faced with an ailing world power in ecological ruins, the West embraced solidarity with the collapsing Soviet Union when the Soviets raised the Iron Curtain (which had long since started to rust anyway), led the Chernobyl children onto the stage, and asked for help. There were no boasts of victory on the Western side in the context of the humanitarian effort; this did not, however, preclude more subtle demonstrations of perceived superiority.

There is a sense in which sending Chernobyl children away for respite was anachronistic – an attempt to meet the challenges of the nuclear age by turning to a bygone era. Most of those involved in the schemes benefited materially and intangibly from supporting the Chernobyl children. Their work shows how much personal growth on an individual level

---

[15] See Hilton, McKay (eds.), *Ages of Voluntarism*.
[16] Barnett, *Empire*, 220.
[17] Ibid., 20.
[18] Joachim Radkau, *Ära der Ökologie: Eine Weltgeschichte* (Munich: C. H. Beck, 2011), 139.
[19] Barnett, *Empire*, 227.

and the development of civil society structures have to gain from transnational encounters and confrontations with very different lifeworlds. But none of this can change the health risks associated with higher-than-normal exposure to radiation.

Recent studies on international humanitarian aid in philosophy and the social sciences point out that it is practically impossible to resolve the 'tension between inequality and solidarity, between a relation of domination and a relation of assistance'.[20] Paternalistic quandaries of this kind were everywhere in the Chernobyl children movement; they were particularly fraught in moral terms here precisely because children were involved. For all the professions to the contrary and all the efforts to pursue a partnership of equals, the power dynamic between Chernobyl children and those who helped them remained an asymmetric one – equality was, to borrow a formulation of Didier Fassin, always an illusion.[21] The sociomaterial context in which relationships evolved led perforce to a mixture of concern and control, to an 'operational inevitability of "circumstantial inequality"'.[22] As Barnett pointedly puts it: 'Some can choose altruism, others have no choice but to play the role of the vulnerable but always grateful pauper.'[23] This disparity became ingrained because power relations were often considered in a one-sided manner: activists were well attuned to the power that the state had over them and those they sought to protect, but they lost sight of the power that they themselves had over those same individuals.[24] One reason this happened so easily was that, at least on the surface, many Chernobyl children and their families acquiesced in the hierarchies that became established. As time went by, a large number of Belarusian and Ukrainian NGOs did develop a degree of self-assurance in their dealings with those who offered help, as well as finding strategies with which to steer it to fit their needs. This was not, however, enough to remove the imbalance in favour of the Western partners – a state of affairs that became increasingly unacceptable for leading organizations such as the

---

[20] Didier Fassin, *Humanitarian Reason: A Moral History of the Present Times* (Berkeley: University of California Press 2012), 3; Barnett, *Empire*.

[21] Fassin, *Humanitarian Reason*, 189.

[22] Kate Manzo, 'Imaging Humanitarianism: NGO Identity and the Iconography of Childhood', *Antipode* 40, no. 4 (2008), 632–57, here 640. See further David Chandler, 'The New International Paternalism: International Regimes' in Michael N. Barnett (ed.), *Paternalism beyond Borders* (Cambridge: Cambridge University Press, 2017), 132–58, here 153.

[23] Barnett, *Empire*, 34.

[24] See ibid., 232–3.

'For the Children of Chernobyl' fund. As their self-confidence grew, they ended alliances that they had entered into for pragmatic reasons in the initial phase of efforts to support the children.

As 'socially positive action on the part of an individual', charitable action always has an emphatically subjective component. Its aim is to deliver help to those who appear to need it in the eyes of the benefactor, and it is rooted in the benefactor's – not necessarily fixed – emotional and spiritual needs.[25] Motivations for socially positive action can include solidarity, eagerness to help, or concern for the common good – but also less altruistic ones such as self-interest, boosting self-esteem, and personal fulfilment. These are not necessarily mutually exclusive alternatives; instead, they constitute charity's two sides. Efforts to support the Chernobyl children are no exception to this.

The fundamental inequality between aid organizations and the Chernobyl children was also reflected in how the Chernobyl children were represented. It was, as a rule, those offering help who asserted control over the form and content of such representations. They often presented the children's fate in such a way that their own actions or objectives stood out in a positive light and the Chernobyl children were left with little room to express their perspectives.[26] Even original output, such as the drawings by the children from Briansk or Aleinikova's cancer hospital, was generally produced at the prompting of adults. It was also primarily composed with adults in mind; it is to be assumed that the children were often trying to meet their expectations. This was particularly likely to be the case, and had particularly important implications, if a trip abroad was to be expected in return for 'successful' work. The extent to which those helping the children were invested in this imbalance can also been seen in their own comments and publications; it allowed them to keep their own insecurities at bay and made it easier for them to put their ideas into practice.

Representations of the children were dominated by elements of naturalization, sacralization, emotionalization, and idealization. Almost all activists were alike in the zeal with which they framed the role of the children – as natural bringers of peace, messengers from a better future, or 'people's diplomats' – in highly charged ethical terms. This placed the children, and everyone else involved in the humanitarian response to Chernobyl, under considerable pressure to meet expectations that they

[25] See Lingelbach, *Spenden*, 14, 20; Barnett, *Empire*, 223.
[26] See Barnett, *Empire*, 34.

were never realistically going to live up to. At the same time, such representations were reductive when it came to the true complexity of reality after the disaster and the humanitarian workers' own work; this can in turn be read as a – conscious or subconscious – reaction to the expectations of society in their home countries. Plain, inspiring words and clear images were needed to get the attention of potential donors and host families and secure their commitment. Those helping the children produced flyers, gave talks, and were interviewed; they came across as credible authorities and, in the process, generated, ordered, and transferred knowledge (whether it was actually accurate or not is another matter). In many instances, this did not foster a substantive appreciation of the environmental, political, and social situation in the home countries of the Chernobyl children. This in turn meant that, in most cases, neither the local nor the foreign organizations covered here considered the possibility of alternative courses of action such as facilitating rehabilitation in the affected areas themselves or supporting state initiatives.

The idealizing representations all presented the Chernobyl children in the same light, but several different concepts of childhood were still vying with one another in the background. Collective ideas of childhood were dominant in the Soviet Union and its successor states, whereas adults in Western Europe and the United States saw children as independent individuals to a much greater degree. This distinction has rightly been criticized as an oversimplification,[27] but the general tendency holds true for the Chernobyl children's interactions with host families and local and foreign organizations.

Seen as a whole, these relationships never completely managed to leave the perspectives, practices, and linguistic habits of the Cold War behind. Sometimes – as is typical of humanitarianism – engagement and exchange actually reinforced the stereotypes.[28] Surprisingly, the presence of these stereotypes could become particularly apparent after a close East–West relationship – for instance, between a Chernobyl child and their host family – had developed. The sense of responsibility that resulted from this brought with it a considerable risk of disappointment should the child respond to the host parents in a way that was not (or was no longer) in line with their expectations; in such cases, the parents often fell back on traditional clichés to make sense of what had happened.

[27] For the criticism, cf. Winkler, 'Kindheitsgeschichte'.
[28] Barnett, *Empire*, 34–7.

Most of the foreign partner organizations had difficulty in adapting their mindsets and strategies to post-Soviet realities. Their perceptions of the societies in transformation were generally selective and lacking in nuance, and were still marked by the legacy of the language and ideas of the Cold War. Facts that did not fit this mould – that alternative youth movements existed in the East, or that the borders were generally open, for instance – were largely ignored. Even in Belarus, the situation during the period under consideration was far more complex than the humanitarian initiatives with their ingrained mindsets assumed. Organizations in the former Soviet Union did not reflect the multifaceted reality of the situation either when they continued to perpetuate old black-and-white ways of thinking. The Belarusian 'For the Children of Chernobyl' fund even insisted that its Western partner organizations adopt its own mentality, specifically where collaboration with state institutions was concerned. The fund and other civil society initiatives justified their position by – legitimately – citing the state repression to which they were repeatedly subjected. Competition for resources and contacts among humanitarian organizations – which could become a battle for survival – also played a significant role here, of course. On occasion, it was now no longer the state but civil society organizations that displayed paternalism. That changed as Lukashenka widened his powers and sought to bring humanitarian work relating to Chernobyl back under tighter state control.

Serguei Oushakine likened the process whereby trauma is 'normalized' to 'pain becoming identical with the air'. He used this image to characterize the state of people who have learnt to live with their trauma: those who 'neither suppress nor repress its negative consequences but instead find a place for them in their destinies and narratives about themselves'.[29] By their own account, many of the Chernobyl children introduced in the present study managed to achieve this. Their experiences abroad often made it easier to give the disaster a positive meaning where their own lives were concerned; in the best case, the fear that they might have suffered long-term health damage became little more than background noise. This assessment is subject to the caveat of the inevitable selectivity of the historical record.[30] I had only limited control over my choice of interviewees, for I established contact with many of them with the help of the

---

[29] Sergey Ushakin (ed.), *Travma. Punkty* (Moscow: Novoe literaturnoe obozrenie, 2009), 7–8.

[30] See Arnold Esch, 'Überlieferungs-Chance und Überlieferungs-Zufall als methodisches Problem des Historikers', *Historische Zeitschrift* 40, no. 240 (1985), 529–70, here 540.

organizations that had arranged their trips as Chernobyl children: there was, in other words, an element of prior selection. It is also likely that the people who opened up to me were primarily those who felt they had something to say that mattered. I attempted to pre-empt potential distortions by using snowball sampling to arrange further interviews and by taking into account written sources as well as numerous conversations and personal observations in Belarus, Ukraine, Russia, Germany, and the United States, but the potential for imbalances remains.

PLUTOPIA DISMANTLED?

In and beyond the affected countries, the humanitarian response to Chernobyl created a climate in which people who had not previously been politically active or socially minded became involved in various causes of common concern, such as protecting the environment.[31] The Chernobyl children were often appropriated by anti-nuclear movements to help them make their case. The majority of Western host families, however, did not politicize their personal support for the children in any meaningful sense, developing instead a greater sensitivity for the nuances of the one world that they had stepped into.

In the post-Soviet states, Chernobyl was far from the end of the Soviet 'love affair between technology and power'.[32] The intense initial criticism of the use of nuclear power did not continue for very long. Even most Chernobyl children saw nuclear power as a safe, clean, and necessary energy source in the period covered by this study. Shortly after the accident at Fukushima Daiichi, Belarus signed the preliminary construction contract for its first – and so far only – nuclear power plant; there was no significant public opposition.[33] As of today, Plutopia remains to be dismantled.

*

Sunshine, blue sky. Shaky footage of a small group of young males clambering over a barbed-wire fence that no longer seems particularly forbidding. Their excitement is palpable as they hurry through verdant,

---

[31] The title of this section alludes to Shane, *Dismantling Utopia* and Brown, *Plutopia*. On the 'politicization of caring', see Krista Harper, *Wild Capitalism: Environmental Activists and Post-socialist Political Ecology in Hungary* (Boulder, CO: East European Monographs, 2006).

[32] Gestwa, *Die Stalinschen Großbauten*, 556.

[33] See Stepanov, 'Tschernobyl ist niemals passiert?'.

knee-high grass. In a forest clearing, they encounter a hunched-over old woman with a brightly coloured headscarf and a sizeable mushroom basket on her back. They ask her the way. They're looking for Prypiat. They want to visit the abandoned town that they know only from photographs and the *STALKER* computer game. On the way there, they check out abandoned houses and film themselves drinking water from a pond. It tastes 'great', one of them announces to the camera, with a 'light uranium aftertaste'.

These scenes appear in Holly Morris and Anne Bogart's documentary *Babushkas of Chernobyl*.[34] By the first decade of the new millennium, the 'zone' had become a destination for leisure travellers in search of added adrenalin. Even stag and hen parties were held there, largely by foreigners.[35] This 'death zone' tourism received a further boost from HBO's *Chernobyl* mini-series. 'Stalkers' were present when Russian troops occupied the exclusion zone and power plant on 24 February 2022, the opening day of the war on Ukraine.[36] In fact, the first footage of the Russian invasion in the north of the country was captured by a Chernobyl tour company: their booth at the exclusion zone checkpoint had a small camera which now happened to be filming the advancing Russian military columns.[37] These days, some Chernobyl tourism providers have replaced trips into the evacuation zone with visits to towns that have been particularly impacted by the war.[38]

In Chapter 1, I took issue with an excessively broad use of the term 'Chernobyl children'. In a sense, though, the disaster tourists are also children of Chernobyl. In their way, they too are confronting the disaster – perhaps, indeed, so that they do not have to engage with it more deeply. Many of the Chernobyl children presented in this study, however, would utterly reject any idea of such an equivalence.

---

[34] Holly Morris, Anne Bogart, *The Babushkas of Chernobyl*, USA, 2015: Powderkeg Studios.

[35] Florence Wilkinson, 'How the Abandoned Nuclear Wasteland of Chernobyl Became a Bachelorette Party Town', *VICE*, 7 June 2017, https://bit.ly/4fkRUM9 (1 May 2024) 'Bachelorette Party Town'.

[36] Organizatsiia Ob"edinennykh Natsii, 'Reportazh c Chernobyl'skoi AES: godovshchina avarii, voina, pomoshch MAGATE', *United Nations*, 26 April 2022, https://news.un.org/ru/story/2022/04/1422642 (4 May 2024).

[37] Miriam Berger, 'A Chernobyl Tour Group Secretly Helped Track Russia's Invasion', *The Washington Post*, 21 August 2022, www.washingtonpost.com/world/2022/08/21/ukraine-spy-tour-group-russians/ (4 May 2024).

[38] Chat with Alana from Chernobyl Story Tours, 12 July 2023.

More than two decades after her evacuation on 27 April 1986, Alena Oginets went back to Prypiat. For many years, she had organized trips for the Ukrainian Chernobyl Association (Soiuz Chernobyl Ukrainy) so that former residents could pay their respects to the dead in cemeteries that were now otherwise off-limits. But she never joined them. She could not bring herself to do so. She was afraid of the experiences reported by other people who visited the accident area: 'People have terrible dreams afterward, it's just awful.' But one day, some time after she stopped working for the Chernobyl Association – she had asked them to let her go when she was pregnant, so that they would not have to pay for her maternity leave – she stepped onto one of the buses heading for the 'zone'.[39] 'What I remembered', Oginets said of what she found there, 'and what it is like now are utterly worlds apart. The town is so different in my memories. The streets are so narrow, so tiny, and they seemed so … you would be able to get a bike down there where the prospekt was, but that's about it.' Only the little forest around her apartment block, where the squirrels that she could feed from her balcony used to live, looked the same. She was followed by wild boar now, though, not squirrels. She even went into her former flat. All she found there was the bathtub, turned on its side. It had clearly been too big to get through the door. Apart from that, there was nothing left.

Oginets has no wish to return to the town of her childhood again. Her son would like to go, but she told him it was better not to because there was nothing to do there. Having seen the images on television is more than enough anyway, she says. And, she cautions, 'who knows what the consequences of such a visit might be'. Similarly, her husband – also from Prypiat – has never gone back, despite having worked in the 'zone' for several years after being discharged from the army. He was never compensated for it. His employment record book merely contains an entry for 'construction work'. The location is not specified. Oginets points out that they did not see anything unusual in this at the time: 'We were foolish, the way young people are; we didn't have a clue.'[40]

---

[39] Oginets, interviewed by Arndt.
[40] Ibid.

Bibliography

PRIMARY SOURCES

'10 let posle Chernobylia. Situatsiia, problemy, deistviia' in Belorusskii blagotvoritel'nyi fond 'Detiam Chernobylia' (ed.), *III Mizhnarodny Kangres 'Svet paslia Charnobylia': Osnovnye nauchnye doklady* (Minsk, 1996), 7–42.

'A Bill for the Relief of Alexandre Malofienko, Olga Matsko, and their son, Vladimir Malofienko', *Congressional Bills 106th Congress, 2nd session*, 8 September 1999, www.congress.gov/bill/106th-congress/senate-bill/199/text (1 May 2024).

Abukov, Aleksei K. *Turizm na novom etape: Sotsial'nye aspekty razvitiia turizma v SSSR* (Moscow: Profizdat, 1983).

Adamushko, I. V. et al. (eds.). *Chernobyl': 26 aprelia 1986 – dekabr' 1991: Dokumenty i materialy* (Minsk: NARB, 2006).

Akbar, Jay. 'Woman Born with Only Four Fingers and No Legs after Exposure to Chernobyl Radiation while in the Womb Defies the Odds to Become a Bodybuilder', *Daily Mail Online*, 14 October 2015, https://bit.ly/3Ar72ZU (4 May 2024).

Alexievich, Svetlana. *Chernobyl Prayer: A Chronicle of the Future*, tr. Anna Gunin and Arch Tait (London: Penguin Books, 2016).

Anndo7. 'Telemarafon Detskogo fonda', *LiveJournal*, 7 January 2010, http://anndo7.livejournal.com/14506.html (2 May 2024).

Arndt, Melanie, Margarethe Steinhausen (eds.). *'Wir mussten völlig neu anfangen': Opfer der Tschernobylkatastrophe berichten* (Bielefeld: Luther-Verlag, 2011).

Associated Press. 'Protests, Telethon Mark Chernobyl Anniversary', *The Los Angeles Times*, 27 April 1990, www.latimes.com/archives/la-xpm-1990-04-27-mn-183-story.html (2 May 2024).

Babaev, N. S., V. F. Demin, I. I. Kuz'min, V. I. Stepanchikov. 'Problemy bezopasnosti na atomnykh elektrostantsiiakh', *Priroda*, no. 6 (1980), 30–43.

Balke, Eva. '"Das" Tschernobylkind gibt es nicht', *Bundesarbeitsgemeinschaft 'Den Kindern von Tschernobyl'*, 16 April 2008, https://bit.ly/3Cbca57 (4 May 2024).

Baranov, A. A. 'Maternal and Child Health Problems in the USSR', *Archives of Disease in Childhood 66*, no. 4 (1991), 542–5.

Baranov, A. E., G. D. Selidovkin, A. Butturini, R. P. Gale. 'Hematopoietic Recovery after 10-Gy Acute Total Body Radiation', *Blood 83*, no. 2 (1994), 596–9.

Baranovs'ka, Nataliia P. *Ukraina – Chornobyl' – Svit: Chornobyl's'ka problema u mizhnarodnomu vymiri 1986–1999* (Kyiv: Nika-Tsentr, 1999).

Barringer, Felicity. 'From Children of Chernobyl, Stories of Flight and of Fears', *New York Times*, 5 June 1986, https://bit.ly/3NSrGoR (1 May 2024).

Barringer, Felicity. 'Reactor Diplomacy', *New York Times*, 31 July 1988, www.nytimes.com/1988/07/31/books/reactor-diplomacy.html (1 May 2024).

Bean, Sara. 'Runner Brings a Message of Inspiration to Gardner', *The Kansas City Star*, 11 May 2015, https://bit.ly/4fhkSx9 (4 May 2024).

BelaPAN. 'Belarus' priostanavlivaet programmy ozdorovleniia belorusskikh detei v SShA', *Naviny.by*, 14 August 2008, https://naviny.by/rubrics/society/2008/08/14/ic_news_116_295692 (4 May 2024).

BelaPAN. 'Tat'iana Kozyro vernulas' na rodinu iz SShA', *Tut.by*, 23 November 2008, https://bit.ly/3YvYhWx (4 May 2024).

Belorusskii blagotvoritel'nyi fond 'Detiam Chernobylia'. *Chernobyl', Posledstviia dlia okruzhaiushchei sredy, zdorov'ia i prav cheloveka*, 1996.

Belorusskii Detskii Fond. *Istoriia fonda*, www.bcf.by/ru/fund/history/history.html (2 May 2024).

Belorusskii Detskii Fond. *Raduga nadezhdy*, www.bcf.by/ru/programs/rainbow.html (2 May 2024).

Belous, Svetlana. 'Ukradennaia mechta Viki Moroz: Kak slozhilas' zhizn' siroty, kotoruiu rasluchili s ital'ianskoi sem'ei', *Tut.by*, 5 June 2014, http://web.archive.org/web/20161013231133/http://news.tut.by/society/401904.html (4 May 2024).

Biermann, Renate, Gerd Biermann, Heiner Biermann. *Die Angst unserer Kinder im Atomzeitalter* (Frankfurt a. M.: Fischer, 1988).

'Blagotvoritel'nyi fond "Detiam Chernobylia" prekrashchaet deiatel'nost'', *Naviny.by*, 21 December 2012, https://naviny.by/rubrics/society/2012/12/21/ic_news_116_407752 (2 May 2024).

Bubich, Ol'ga. '"Chernobyl'tsy" za rubezhom: Pravo na druguiu zhizn', *Belarusskii Zhurnal*, 23 April 2016, http://journalby.com/news/chernobylcy-za-rubezhom-pravo-na-druguyu-zhizn-393 (4 May 2024).

Calhoun, Cec. 'Educational Programs: Cultural and Educational Enrichment Project (CEEP)' in CofCUSA, *Alliance News*, no. 2 (2010).

Carter, Michelle. *From under the Russian Snow* (Fairfield, CA: Bedazzled Ink Publishing, 2017).

Carter, Michelle, Michael J. Christensen. *Children of Chernobyl: Raising Hope from the Ashes* (Minneapolis: Augsburg Fortress, 1993).

Chalenko, Aleksandr. 'Nashi na Kube: Domoi ne khotim!', *Segodnia*, 3 September 2009, www.segodnya.ua/lifestyle/fun/nashi-na-kube-domoj-ne-khotim-162124.html (2 May 2024).

Chali, Petr. 'Vinovnikom gibeli chernobyl'skich detei budet priznan bel'giiskii shofer'', *Narodnaia Volia*, 9 July 1997, 1.

Cheban, Aleksandr. 'Chernobyl' 30 let spustia: Evakuatsiia glazami detei', *LiveJournal*, 27 April 2016, https://alexcheban.livejournal.com/290807.html (1 May 2024).

'Chernobyl and Its Legacy: A Special Report', *EPRI Journal* 12, no. 4 (June) (1987), 5–21.

'"Chernobyl" Is the Highest-Rated TV Series Ever: HBO's Show Rates Higher on IMDb than Other Historical Dramas and Even "Breaking Bad"', *The Economist*, 4 June 2019, https://bit.ly/3AzudRK (1 May 2024).

'Chernobyl (serial 2022)'. Wink Universal Distribution (Dir.: Aleksei Muradov): www.kinopoisk.ru/series/1169262/ (1 May 2024).

'Chernobyl Victim's Immigration Status Left Hanging by Congress', *CNN*, 22 October 1998, http://edition.cnn.com/WORLD/europe/9810/22/chernobyl .boy/index.html (1 May 2024).

Children of Chernobyl Foundation San Diego. Host Family Guidelines, https://bit .ly/3YCjQEX (4 May 2024).

'Chornobyl'ska trahediia v dokumentakh ta materialakh' (Kyiv: Sfera, 2001), special issue of *Z arkhiviv VUChK, GPU, NKVD, KGB* 16, no. 1 (2001).

CitiHope. *Belarus 1990: CitiHope's Journey Begins*, www.youtube.com/watch? v=X3ZorA_V-hI (uploaded 28 September 2015) (4 May 2024).

CitiHope, *Chernobyl's Children in Crisis: CitiHope Responds*, 2015, www.you tube.com/watch?v=qZxZECyXZj4 (4 May 2024).

CitiHope. *History*, https://web.archive.org/web/20160317011743/http://citi hope.org/history/ (uploaded 17 March 2016) (4 May 2024).

'Clifford McClain', *The Press Democrat*, 15 November 2013, www.legacy.com/ obituaries/pressdemocrat/obituary.aspx?pid=168002879 (4 May 2024).

CofCUSA. *Alliance News* 3, no. 1, 2009, https://bit.ly/3YAXBPK (4 May 2024).

CofCUSA. 'From Kansas City Area' in CofCUSA, *Alliance News*, no. 4 (2010), 1.

CofCUSA. *Private Invitations*, 26 January 2010, https://cofcusa.wordpress.com/ 2010/01/26/private-invitations/ (4 May 2024).

CofCUSA. *Ukraine Respite Video*, 26 January 2010, https://cofcusa.wordpress .com/2010/01/29/ukraine-respite-vide/ (4 May 2024).

Deti zapada. Sbornik dlia detei i iuniushchestva (Moscow: n. p., 1926).

Diamond, Stewart. 'How Chernobyl Alters the Nuclear Equation', *New York Times*, 25 May 1986, https://bit.ly/4eiayDP (1 May 2024).

Dolzhenko, Gennadii P. *Istoriia turizma v dorevoliutsionnoi Rossii i SSSR* (Rostov na Donu: Izdatel'stvo Rostovskogo Universiteta, 1988).

Dowdy, Zachary R. 'City Rally Remembers Chernobyl', *Boston Herald*, 27 April 1991.

DSP ChAES. *NBK. Dostupno pro novyi bezpechnyi konfainment*, https:// chnpp.gov.ua/nbk/index.html (5 May 2024).

Dudin, Mikhail Aleksandrovich. *Derevo dlia aista: Stichotvoreniia 1968–1978* (Moscow: Molodaia Gvardiia, 1980).

Dyck, Kate van. 'Change Is Difficult? Not Really!' in CofCUSA, *Alliance News*, no. 2 (2010).

Edwards, Mike, Steve Raymer, Pierre Mion. 'Chernobyl: One Year After'. *National Geographic* (1987), 632–53.

Epstein, Edward Jay. *Dossier: The Secret History of Armand Hammer* (New York: Random House, 1996).

Feriado-Tour. *Kuba – Ukraina: 1990–2011 roki: Vsestoronnaia meditsinskaia programma pomoshchi detiam, postradavshim ot chernobyl'skoi avarii*, https://feriado-tour.com.ua/treatment-in-cuba/childsprograms/chernobil-ruso/ (2 May 2024).

Feriado-Tour. *Rezul'taty lecheniia detei na Kube*, https://bit.ly/4fqGbvB (2 May 2024).

'Fond "Detiam Chernobyl'ia" vstal vlastiam peperek gorla', *Belorusskaia Delovaia Gazeta*, 20 May 2002, https://bdg.by/news/news.htm%3F27337,3 (2 May 2024).

Gale, Robert Peter, Thomas Hauser. *Final Warning: The Legacy of Chernobyl* (New York: Warner Books, 1988).

Gale, Robert Peter, Eric Lax. *Radiation: What It Is, What You Need to Know* (New York: Alfred A. Knopf, 2013).

Gale, Robert Peter, Andrej I. Vorob'ev. 'First Use of Myeloid Colony-stimulating Factors in Humans', *Bone Marrow Transplantation* 48, no. 1358 (2013).

Geerhart, Bill, Ken Sitz. *Atomic Platters: Cold War Music from the Golden Age of Homeland Security*, Bear Family Records 2005, 4 Sound Discs.

Glionna, John M. 'Girl from Chernobyl Refuses to Go Back', *The Los Angeles Times*, 8 September 2008, https://bit.ly/4fspdwI (4 May 2024).

Gosatomnadzor. *Istoriia sozdaniia i perspektivy razvitiia*, https://bit.ly/4gG5cUn (2 May 2024).

Gosatomnadzor. *Rukovodstvo i struktura Gosatomnadzora*, https://bit.ly/3VK4ayW (2 May 2024).

Grossman, Karl. *Power Crazy: Is LILCO Turning Shoreham into America's Chernobyl?* (New York: Grove Press, 1986).

Grushevoi, Gennadii. 'Dlia ozdorovleniia detei v Belarusi "poka net podkhodiashchikh uslovii"', *Telegraf.by*, 14 November 2008.

Grushevoi, Gennadii. 'Esli by Chernobyl' sluchilsia segodnia, my by tozhe ne uznali pravdu', *Khartyia '97*, 27 April 2006.

Grushevoi, Gennadii. 'Eto gore ne dolzhno lishit' Vas muzhestva i reshimosti prodolzhat' dal'she nelegkoe, no takoe nuzhnoe delo gummanoi pomoshchi liudiam', *Narodnaia Volia*, 9 July 1997, 1.

Gruschewaja, Irina, Alexander Tamkowitsch. *Der Tschernobyl-Weg: Von der Katastrophe zum Garten der Hoffnung* (Berlin: RMF, 2017).

Gruschewoi, Gennadi. 'Monolog über kartesianische Philosophie' in Swetlana Alexijewitsch, 'Stimmen aus Tschernobyl', *Aus Politik und Zeitgeschichte* 13 (2006), 3–11.

Gubarev, V. S., Odinets M. 'Stantsiia i vokrug nee. Nashi spetsial'nye korrespondenty peredaiut iz raiona Chernobyl'skoi atomnoi elektrostantsii', *Pravda*, 6 May 1986, 6.

Hammer, Armand, Neil Lyndon. *Hammer: Witness to History* (Sevenoaks: HarperCollins Distribution Services, 1988).

Hammer, Armand, Neil Lyndon, Gerda Bean. *Mein Leben*, 5th ed. (Bern: Scherz, 1989).

Hevesi, Dennis. 'Nora Bredes, Who Fought Long Island Nuclear Plant, Dies at 60', *New York Times*, 22 August 2011, https://bit.ly/4gDQa10 (1 May 2024).

Hicks, Eric. 'North Alabama Belarus Mission' in CofCUSA, *Alliance News*, no. 2 (2010).

Hirschkorn, Phil. 'Chernobyl Boy's Family Fears Deportation Could Mean Death'. *CNN*, 20 October 1998, http://edition.cnn.com/WORLD/europe/9810/20/chernobyl.boy/ (1 May 2024).

Hole in the Wall Gang. *Founder & History*, www.holeinthewallgang.org/about/Founder-and-History/ (2 May 2024).

Hushtyn, Adar'ia. 'Professor Bandazhevskii: Segodniashnie deti podverzheny radiatsii bol'she, chem ikh roditeli', *Tut.by*, 26 April 2019, https://bit.ly/4iyR4OI (1 May 2024).

Illesh, Andrei V. 'Obstanovka pod kontrolem', *Izvestiia*, 7 May 1986, 6.

Institut teorii i istorii sotsialisma TsK KPSS (ed.). *XXVIII s''ezd Kommunisticheskoi partii Sovetskogo Soiuza: 2 – 13 iiulia 1990 goda*; Stenograficheskii otchet (Moscow: Izdatel'stvo politicheskoi literatury, 1991).

International Movie Database. *Chernobyl: TV Mini-Series 2019*, www.imdb.com/title/tt7366338/ (1 May 2024).

'Introduction', *Radical America*, 20, no. 2/3 (1986), 2.

Izrael, I. '"Chernobyl": Proshloe i prognoz na budushchee', *Pravda*, 20 March 1989, 4.

Johnson, Julie. 'Constance McClain', *The Press Democrat*, 25 February 2013, https://bit.ly/49Ko3Zi (4 May 2024).

Johnson, Shane. *Nuclear Power 2010: Program Overview* (Office of Nuclear Energy, Science and Technology, 2002), https://bit.ly/4gctJ3y (1 May 2024).

Jones, Natasha. 'A Reprieve from the Shadow of Chernobyl', *Langley Times*, 16 July 1994, 1.

Kagan, Vladislav. 'Gud-bai, Amerika', *Novye Izvestiia*, 19 September 2008.

Kamenev, Maxim. 'How Russia Took Over Chernobyl', *openDemocracy*, 22 June 2022, https://bit.ly/3ZVlJhM (1 May 2024).

Kenigsberg, Iakov. 'Chernobyl' i belorusskoe obshchestvo'. *Svobodnaia mysl'*, no. 6 (2008), 145–156.

'Kinder von Tschernobyl': Geheimdienst besetzt Stiftungsbüro in Minsk', *Frankfurter Rundschau*, 22 March 1997, 5.

Klatt, Karola. '"Dort arbeiten, wo der Staat versagt": Wie sich zivilgesellschaftliche Gruppen in einer Diktatur behaupten: Ein Gespräch mit der belarussischen Bürgerrechtlerin Irina Gruschewaja', *Institut für Auslandsbeziehungen*.

Korelina, Ol'ga. '"Prezident pozhimal mne ruku i blagodaril". Perevodchiki vstrech na vysshem urovne rasskazali o svoei rabote', *Tut.by*, 19 February 2015, https://bit.ly/3ZOSEmK (4 May 2024).

Korotkova, Elena. 'Kastro potratil $350 mln na lechenie detei iz Chernobylia: Fidel' skazal: "Pomogat" budem stol'ko, skol'ko budet nuzhno', *Moskovskii komsomolets*, 27 November 2016, https://bit.ly/3Bwf4Bp (2 May 2024).

Korovenkova, Tatiana. 'Apelliatsionnyi sud Genui prigovoril k vos'mi mesiatsam tiur'my suprugov Dzhusto-Bornakhin i vsekh, prichastnykh k ukryvatel'stvu belorusskoi siroty Viki Moroz', *BelaPAN*, 22 March 2019.

Korovenkova, Tatiana. 'Taniu Kozyro vernuli: A drugikh ne vypustiat', *Naviny. by*, 27 November 2008, https://bit.ly/49M5syY (4 May 2024).

Kozhemiako, V. 'O glasnosti: V polnii golos', *Pravda*, 26 April 1986, 2.

Kravchenko, Petr. *Belarus' na perelome: Diplomaticheskii proryv v mir* (Minsk: BIP-S Plius, 2009).

Kushnir, Lina. 'Valentyna Shevchenko: "Provesti demonstratsiiu 1 travnia 1986-ho nakazaly z Moskvy', *Istorychna pravda*, 25 April 2011, www.istpravda.co m.ua/articles/2011/04/25/36971/ (1 May 2024).

Kuz'min, Viktor. 'Poshli s Kozyro', *NTV*, 24 November 2008, www.ntv.ru/nov osti/145035/ (4 May 2024).

Legasov, Valeri, Lev Feoktistov, Igor Kuzmin. 'Nuclear Power Engineering and International Security', *Soviet Life*, no. 2 (353) (1986), 14–5.

Leitman, A. 'Rok-n-roll na vozdusiakh', *Molodoi Dal'nevostochnik*, 1 June 1991, http://sovr.narod.ru/articles/91015.html (4 May 2024).

Lezard, Nicholas. 'Chernobyl Prayer by Svetlana Alexievich Review: Witnesses Speak', *The Guardian*, 27 April 2016, https://bit.ly/3ZO3ZVo (1 May 2024).

Likhanov, Al'bert. 'Obernut'sia k detstvu: Net zaboty vazhnee', *Pravda*, 13 May 1987, 3, 6.

'Liudi stroiat khramy', *Sobor.by*, http://sobor.by/kamen.php (2 May 2024).

Longman, Jere. 'Ukraine's Poisoned Past', *New York Times*, 12 June 2012, https:// bit.ly/3VT1f6K (1 May 2024).

Lorenzetto, Stefano. 'L'industria dei bimbi di Chernobyl', *Il Giornale*, 30 September 2006, www.ilgiornale.it/news/l-industria-dei-bimbi-chernobyl.html (4 May 2024).

Lukashenko, Aleksandr. *Ukaz no. 555: O vneseni i dopolneni v Ukaz Prezidenta Respubliki Belarus ot 18 fevralia 2004 g.* No. 98, 13 October 2008, https://bit .ly/4ggepDb (4 May 2024).

Lukashenko, Aleksandr. *Ukaz no. 59: O vnesenii izmenenii v Ukaz Prezidenta Respubliki Belarus ot 18 fevralia 2004 g.* No. 98, 28 January 2010, https://bit .ly/4fsdwGy (4 May 2024).

Lukashenko, Aleksandr. *Vystuplenie na tseremonii podpisaniia resheniia, priniatogo respublikanskim referendumom*, 17 November 2004, https://bit.ly/ 3DnScVh (2 May 2024).

'Lukashenko rasporiadilsia proverit' v kakikh usloviiakh nakhodilas' Vika Moroz v belorusskom internate', *Newsru.com*, 7 October 2006, www.newsru.com/ world/07oct2006/tobecntnd.html (4 May 2024).

Luk'ianov, A. 'Postanovlenie "Ob edinoi programme po likvidatsii posledstvii avarii na Chernobyl'skoi AES i situatsii, sviazannoi s etoi avarii"', 1452–1, 25 April 1990, in Valerii Stanislavovich Levonevskii, *Zakonodatel'stvo*, http:// pravo.levonevsky.org/baza/soviet/sssr0933.htm (2 May 2024).

Lundeberg, Patricia. 'Children of Chernobyl Foundation, San Diego' in CofCUSA, *Alliance News*, no. 2 (2010).

Macy, Joanna. *Despair and Personal Power in the Nuclear Age* (Philadelphia: New Society Publishers, 1993).

Macy, Joanna. *Widening Circles: A Memoir* (Gabriola Island, BC: New Society Publishers, 2007).

Macy, Joanna, Molly Young Brown. *Coming Back to Life: Practices to Reconnect Our Lives, Our World* (Philadelphia : New Society Publishers, 1999).

Marx, Trish, Dorita Beh-Eger, Cindy Karp. *I Heal: The Children of Chernobyl in Cuba* (Minneapolis: Lerner Publications Company 1996).

McCally, Michael. 'Hospital Number Six: A First-hand Report', *Bulletin of the Atomic Scientists* 42, no. 7 (1986), 10–12.

McDermott, Cindy. 'Tatsiana Khvitsko: "The Joy of Running Makes Me Unstoppable!"'. *Her Life Magazine*, 2017, https://bit.ly/3ZYuuHT (4 May 2024).

Medvedev, Grigorij. *No Breathing Room: The Aftermath of Chernobyl* (New York: Basic Books, 1993).

Medvedev, Zhores A. *The Legacy of Chernobyl* (New York: W. W. Norton, 1992).

Miller, Alan C. 'Donor to Democrats Sentenced', *The Los Angeles Times*, 21 December 1999, www.latimes.com/archives/la-xpm-1999-dec-21-mn-4616 9-story.html (2 May 2024).

Mindlin, Leanid. *Sad nadzei*. Belarus, 2016: Belsat.

Molotsky, Irvin, Warren Weaver, Jr. 'Washington Talk: Briefing; Letters from Asselstine', *New York Times*, 26 October 1986, https://bit.ly/3ZVwbpG (1 May 2024).

Moore, Paul. *My CitiHope Story: Invitation*, www.youtube.com/watch?v=6ifoB L-dK5E (4 May 2024).

Morris, Holly, Anne Bogart. *The Babushkas of Chernobyl*, USA, 2015: Powderkeg Studios.

Morrissey, Janet. 'Charity That Begins with Spaghetti Sauce', *New York Times*, 2 November 2016, www.nytimes.com/2016/11/06/giving/charity-that-begins-with-spaghetti-sauce.html (2 May 2024).

Nazarenko, Mikhail. *Mne dorogi eti mesta* (Pripiat'-Fil'm, 1986), www.youtube .com/watch?v=-M_tYuGfXdY (1 May 2024).

Nazarenko, Mikhail. *Nezabyvaemoe* (Pripiat'-Fil'm, 1988), www.youtube.com/ watch?v=fg6q7uCsv3E (1 May 2024).

Neal, Nadya. *On Silent Clouds of Butterfly Wings: Growing Up in the Nuclear Age*, 3rd ed. (Chattaroy, WA: Booksurge Publishing, 2009).

Neal, Nadya. *The Butterfly's Effect: A Story about Children of the Atomic Age*, 2nd rev. ed. (Chattaroy, WA: Alchemy Farm Press, 2005).

Neal-Oldenettel, Nancy. 'Belarusian Culture, History & Children: Host Family Guidelines', *Children of Chernobyl Foundation*, https://bit.ly/4gmFoyA (4 May 2024).

Nesterenko, V. B. *Einwirkung der Radiation auf die Gesundheit von Kindern in Belarus 12 Jahre nach dem Tschernobylunfall* (n. p., 1998).

Nesterenko, V. B. et al. *Sravnenie effektivnosti ozdorovleniia detei chernobyl'skikh regionov Belarusi v respublikanskikh sanatoriiakh i za rubezhom* (Minsk: Belrad, [2007]).

Net-film.ru. *Telemarafon Detskogo Fonda (1990)*, www.net-film.ru/film-20931/ (2 May 2024).

Newman, Paul, Aaron E. Hotchner. *Shameless Exploitation in Pursuit of the Common Good* (New York: Nan A. Talese 2003).

Nichiporuk, T. '"Ia otvetstven za to: chtoby pobezhdalo Dobro"', *Belaruskaia Maladzezhnaia*, [1996], 3.

Norton, Laura. 'Despite Incentives from Her Government, Girl in Petaluma Intents on Staying Here', *The Press Democrat*, 23 August 2008, www.pressde mocrat.com/news/2203140-181/despite-incentives-from-her-government (4 May 2024).

Oguma, Eiji. *Tell the Prime Minister*, Japan, 2015: UPLINK.

Organizatsiia Ob"edinennykh Natsii. 'Reportazh c Chernobyl'skoi AES: godovshchina avarii, voina, pomoshch MAGATE', *United Nations*, 26 April 2022, https://news.un.org/ru/story/2022/04/1422642 (4 May 2024).

Ostrovtsova, Inga. 'Sad nadezhdy: Fil'm o beloruse, kotoryi spas 600 tys. detei', *Belsat*, 22 April 2016, https://bit.ly/3P2qzn3 (2 May 2024).

'Ot Soveta Ministrov SSSR', *Pravda*, 29 April 1986, 1.

'Ot Soveta Ministrov SSSR', *Pravda*, 30 April 1986, 2.

'Ozdorovlenie "chernobyl'skikh detei". Kubinskii fenomen', *Cuba.com.ua*, www.cuba.com.ua/articulos/cuba-tourist-4.html (2 May 2024).

Page, Anthony. *Chernobyl: The Final Warning*. USA, 1991: Carolco Pictures.

Pankratova, Elena. 'Grustnyi pessimizm' in Aliaksandr Tamkovich (ed.), *Filasofiia dabryni. Ad katastrofy – da sada nadzei* (Minsk: self-pub., 2016), 303–9.

Paskevich, Sergei. *Chernobyl'-2. Sekretnyi dvoinik goroda Chernobyl'*, http://ch ornobyl.in.ua/chernobyl-2.html (1 May 2024).

Payne, Paul. 'Aid Worker Says Girl Who Sparked International Feud Worried about Sanctions', *The Press Democrat*, 25 November 2008, www.pressdemo crat.com/news/2188197-181/aid-worker-says-girl-who (4 May 2024).

Payne, Paul. 'With Belarus Out, Ukrainian Kids Visit', *The Press Democrat*, 15 August 2009, www.pressdemocrat.com/news/2263334-181/with-berlarus-out-ukrainian-kids (4 May 2024).

Petrochenkova, Valentina. 'Nashim detiam v Tarara lushche chem doma', *ZN.ua*, 16 May 1997, https://zn.ua/HEALTH/nashim_detyam_v_tarara__luchshe,_ch em_doma.html (2 May 2024).

Petrochenkova, Valentina. *Ostanovis', mgnovenie!* (Kyiv: self-pub., 2010).

Podgoronikov, Mikhail. 'My ot prirody ne sposobny k demokratii? Sostoianie genofonda v SSSR vnushaet trevogu: No mozhno li govorit' o "debilizatsii" naseleniia?', *Literaturnaia Gazeta*, 11 December 1991, 7.

Postman, Neil. *The Disappearance of Childhood* (New York: Delacorte, 1982).

'Predsedatel' fonda "Detiam Chernobylia" Gennadii Grushevoi. Unichtozhatsia poslednie vozmozhnosti detei ozdorovliat'sia za granitsei', *Belrusskie Novosti*.

Prezident Respubliki Belarus. Polozhenie o Departamente po gumanitarnoy pomoshchi pri Prezidente Respubliki Belarus N. 404, 24 July 1997.

Prezident Respubliki Belarus. Zakon Respubliki Belarus' ob obshchestvennykh ob"edineniiakh. No. 3254-XII, 4 October 1994, http://pravo.by/document/?g uid=3871&po=v19403254 (2 May 2024).

Pripyat.com. http://web.archive.org/web/20220313094141/http://pripyat.com/ (1 May 2024).

Prypiat': Fotorozpovid pro odne z naymolodshykh mist Ukrainy (Kyiv: n. p., 1986).

PSR/IPPNW Schweiz, Kinder von Tschernobyl Belarus / Frankreich, Weimarer Verein 'Hilfe für die Kinder von Tschernobyl': JANUN e.V. (eds.). *Können Pektinkuren die Kinder von Tschernobyl schützen? Programm Internationales Pektinhearing*, Hanover, 16 February 2007.

Pustovaia, Daina. *Chelovek, kotoryi ostalsia za kadrom*, https://bit.ly/4g3iGtw (1 May 2024).

Rada BNR. www.radabnr.org/ (2 March 2024).

Radina, Natal'ia. 'Izderzhki total'nogo kontrolia: Blagotvoritel'nuiu pomoshch liudi okazyvaiut liudiam, no ne totalitarnomu gosudarstvu', *Nasha Svaboda*, 18 May 2000, https://bit.ly/3VGkDnK (2 May 2024).

Ravo, Nick. 'For Victims of Chernobyl, a Respite at Camp', *New York Times*, 10 September 1990, https://bit.ly/4iIStSU (2 May 2024).

Reagan, Ronald. 'Evil Empire Speech', 8 March 1983, *Voices of Democracy: The US Oratory Project*, https://voicesofdemocracy.umd.edu/reagan-evil-empire-s peech-text/ (1 May 2024).

'Reaktorkatastrophe. Retter nicht in Sicht', *Der Spiegel*, 7 May 1990.

Renck, Johan. *Chernobyl*, USA, 2019: Sister Pictures, The Mighty Mint. HBO.

RIA Novosti. 'Ministr Rad'kov o dal'neishei sud'be Viki Moroz', *Tut.by*, 2 October 2006, https://bit.ly/3VKkWoQ (4 May 2024).

Riemann, Achim, V. B. Nesterenko. *Häufig gestellte Fragen zum Thema Pektin* (2007).

Rimer, Sara. 'For City's Ukrainians, Ordeals of Waiting for Word on Disaster', *New York Times*, 1 May 1986, https://bit.ly/4fjntWI (1 May 2024).

RKMBR, *Nasha istoriia*, https://aksakovschina.by/o-больнице/наша-история (2 May 2024).

Rolff, Hans-Günter, Peter Zimmermann. *Kindheit im Wandel: Eine Einführung in die Sozialisation im Kindesalter* (Weinheim: Beltz 1985).

Rosenbaum, Sophia. 'Chernobyl Woman Born without Legs Is Competitive Bodybuilder', *New York Post*, 14 October 2015, https://bit.ly/49O8gvJ (4 May 2024).

Rossiiskii Detskii Fond. *Deti Chernobylia*', www.detfond.org/about/stranicy-ist orii/deti-Chernobylia/ (6 November 2019).

Rudolph, Richard, Scott Ridley. 'Chernobyl's Challenge to Anti-Nuclear Activism', *Radical America* 20, no. 2/3 (1986), 8–11.

Rylsky, Maxim. 'A Town Born of the Atom', *Soviet Life*, no. 2 (353) (1986), 13.

Schlapper, Laurann. 'Project Restoration Camp Nadezhda August 11–23, 2010' in CofCUSA, *Alliance News*, no. 2 (2010).

Schonbrun, Zach. 'Paul Newman Who? Salad Dressing Company Adjusts To Reach Millenials', *New York Times*, 13 November 2016, https://bit.ly/3ZVgA pX (2 May 2024).

Schuchardt, Erika, Lev Z. Kopelev. *Die Stimmen der Kinder von Tschernobyl: Geschichte einer stillen Revolution*, 2nd ed. (Freiburg: Herder, 1996).

Shakhanovich, Tat'iana. 'Vrach i mnogodetnaia mama Ol'ga Aleinikova: Poslerodovaia depressiia – u tekh, kto vyros v teplichnykh usloviiakh', *Komsomol'skaia Pravda*, 5 July 2016.

Shcherbak. Iouri *Chernobyl'*. *Dokumental'noe povestvovanie* (Moscow: Sovetskii pisatel', 1991).

Sinowjew, Alexander. *Katastroika: Gorbatschows Potemkinsche Dörfer* (Frankfurt a. M., Berlin: Ullstein, 1988).

Siry, Ales'. 'Tat'iana Kozyro, geroinia mezhdunarodnogo skandala v 2008 godu, rodila rebenka i zhivet v Borisove', *Tut.by*, 5 February 2014, https://bit.ly/3P9 5TKl (4 May 2024).

Smolnikowa, Valentina. *Nachruf*, 10 February 2014, https://bit.ly/3P8zEL3 (2 May 2024).

Solntsev, E. 'Blagotvoritel'nost'' in *Bol'shaia Sovetskaia Entsiklopediia* 466–71.

'Sovetskii detskii fond imeni V.I. Lenina sozdan!', *Smena* 23, no. 1453 (Dekabr') (1987), 1–6, http://smena-online.ru/stories/sovetskii-detskii-fond-imeni-vi-leni na-sozdan (1 May 2024).

'Soviet Statements on Nuclear Plant Accident', *New York Times*, 1 May 1986, 10, https://bit.ly/3ZUA268 (1 May 2024).

Spanhel, Dieter, Stefanos Hotamanidis (eds.). *Die Zukunft der Kindheit: Die Verantwortung der Erwachsenen für das Kind in einer unheilen Welt* (Weinheim: Deutscher Studien-Verlag, 1988).

Springer, Gudrun. 'Das Erbe von 1945: Kinder, quer durch Europa verschickt,' *Der Standard*, 25 April 2015, www.derstandard.at/story/2000014693464/kin der-quer-durch-europa-verschickt (1 May 2024).

SSE ChNPP. *NSC Construction Is Formally Completed*, 3 September 2019, https:// bit.ly/4gt4w4O (1 May 2024).

Survila, Ivonka. *Daroha: Stouptsy – Kapenhahen – Paryzh – Madryd – Atava – Mensk* (n. p.: Radye Svaboda, 2008).

Tamkovich, Aliaksandr (ed.). *Filasofiia dabryni: Ad katastrofy – da sada nadzei* (Minsk: self-pub., 2016).

Taran, Valentina. 'Viktoriia Moroz: goriachii spor vokrug problemy', *Trud*, 12 October 2006.

Taubman, Philip. 'Moscow Rock Concert Aids Chernobyl Victims', *New York Times*, 31 May 1986, https://bit.ly/3ZYkezp (4 May 2024).

Tennison, Elizabeth. 'Hope for Chernobyl's Child' in CofCUSA, *Alliance News*, no. 2 (2010).

Tershakovets, Tamara. 'Mria Airplane Takes Off for CCRF's Thanksgiving Airlift', *The Ukrainian Weekly*, 1 December 1991, 3.

'The Nuclear Power Industry in the Ukraine', *Soviet Life* no. 2 (353) (1986), 8.

The Only One Foundation. *About Us*, https://bit.ly/3VPl3Z3 (4 May 2024).

Theisen, Alois. *Die Kinder von Tschernobyl*, ZDF-Frontal, 30 April 1996.

Tkachuk, Anatoly N. *Ich war im Sarkophag von Tschernobyl: Der Bericht des Überlebenden* (Vienna: Styria Premium, 2011).

'Tol'ko vsem mirom. Televizionnyi marafon', *Argumenty i Fakty*, January 1990, https://annd07.livejournal.com/14506.html (2 May 2024).

Tomkovich, Aleksandr. 'Naezd' na dobrotu, *Svobodnye Novosti*, 25 April 1997, 1–2.

Tomkovich, Aleksandr. 'Negromkaia data: Oni spasli nashikh detei … ', *Svobodnye Novosti Plius*, 1 June 2015, www.sn-plus.com/ru/page/society/582 5 (2 May 2024).

Tomkovich, Aleksandr. 'Vzroslye igry – detiam', 30 October 2008, https://bit.ly/4frtQXY (4 May 2024).

Torricelli, Giuseppina, Carla Baroncelli. *Piccoli ospiti e parenti del cuore: Non chiamiamoli i bambini di Chernobyl* (Modena: Edizioni del Loggione, 2016).

Tschernobyl-Initiative der Propstei Schöppenstedt e. V. *Die Spur der schwarzen Wolke: Die Katastrophe von Tschernobyl mit den Augen der betroffenen Kinder und Eindrücke einer deutsch-weissrussischen Reisegruppe* (Klitzschen: Elbe-Dnjepr-Verlag, 2000).

Tyson, Jill. 'CCP of Greater Chattanooga' in CofCUSA, *Alliance News*, no. 2 (2010).

Uhlmann, Irene, Irene, Klemm, Günther Liebig (eds.). *Kleine Enzyklopädie Gesundheit*, 3rd ed. (Leipzig: Bibliographisches Institut, 1957).

'Ukrains'ka hromads'ka blahodiina orhanizatsiia mizhnarodnyi Chornobyl's'kyi fond', *Who-is-who.ua*, https://who-is-who.ua/main/page/kiev/156/116 (2 May 2024).

'Ukrainskii molodezhnyi Chernobyl'skii Fond: Bozhko Aleksandr Fedorovich', *Who-is-who.ua*, https://who-is-who.ua/main/bookmaket/donetsk/4/96.html (2 May 2024).

UN General Assembly. *International Co-operation to Address and Mitigate the Consequences of the Accident at the Chernobyl Nuclear Power Plant*, 45/190, 21 December 1990, https://digitallibrary.un.org/record/105595/files/A_RES_45_190-EN.pdf (2 May 2024).

USAID. *Changing Societal Perceptions of People with Disabilities in Belarus: Art and Media Put the Spotlight on Talent and Potential*. www.usaid.gov/node/226636 (4 May 2024).

Verkhovnii Sovet SSSR. *Pervyi s"ezd narodnykh deputatov SSSR: 25 maia–9 iiunia 1989 g.* Stenograficheskii otchet, 6 vols. (Moscow: 1989).

Vakulenko, N. 'Dokladnaia zapiska', 12 March 1981, *Z arkhiviv VUChK, GPU, NKVD, KGB* 16, no. 1 (2001), 37–9.

Vakulenko, N. 'O nedostatochnoi nadezhnosti', 16 October 1981, *Z arkhiviv VUChK, GPU, NKVD, KGB* 16, no. 1 (2001), 41–3.

Vakulenko, N. 'Spetsial'noe soobshchenie', 20 April 1981, *Z arkhiviv VUChK, GPU, NKVD, KGB* 16, no. 1 (2001), 40.

Vakulenko, N. 'Spravka', 26 August 1980, *Z arkhiviv VUChK, GPU, NKVD, KGB* 16, no. 1 (2001), 35–6.

Vorob'ev, Pavel Andreevich, Andrei Ivanovich Vorob'ev. 'Do i posle Chernobyl'ia', *Nezavisimaia Gazeta*, 28 April 2006, www.ng.ru/health/2006-04-28/8_chernobyl.html (1 May 2024).

Wald, Matthew L. 'Retiring US Official Assails Nuclear Plant Safety', *New York Times*, 7 June 1987, https://bit.ly/3VHkPTB (1 May 2024).

Walker, Linda, Julie Gater, Mike Allison, Mags Whiting, Ken Whiting, John Gater, Harold Bowden. 'Guidelines for Host Families', *Chernobyl Children's Project (UK)*, https://bit.ly/3BPYxIt (4 May 2024).

Ward, Barbara E. *Spaceship Earth* (Boston: Columbia University Press, 1966).

Ward, Barbara E., René Jules Dubos. *Only One Earth: The Care and Maintenance of a Small Planet* (New York: Norton, 1972).

Wechlin, Daniel. 'Die sterbende Stadt der Liquidatoren', *Neue Zürcher Zeitung*, 22 April 2016, www.nzz.ch/international/europa/die-sterbende-stadt-der-liqui datoren-ld.15314 (1 May 2024).

Weinberg, Alvin M. 'A Nuclear Power Advocate Reflects on Chernobyl', *Bulletin of the Atomic Scientists* 43, no. 1 (1986), 57–60.

Weinberg, Steve. 'Armand Hammer's Unique Diplomacy', *Bulletin of the Atomic Scientists* 42, no. 7 (1986), 50–2.

Weinberg, Steve. *Armand Hammer: The Untold Story* (Boston: Little, Brown and Co., 1989).

WHO. *TV Address by Mikhail Gorbachev*, 14 May 1986, https://bit.ly/4gi7lG2 (1 May 2024).

Wilkinson, Florence. 'How the Abandoned Nuclear Wasteland of Chernobyl Became a Bachelorette Party Town', *VICE*, 7 June 2017, https://bit.ly/4fkRU M9 (1 May 2024).

Williams, Ruth. 'Tanya Kazyra', *Little Chernobyl*, 19 November 2009, https://li ttlechernobyl.blogspot.com/2009/11/tanya-kazrya.html (4 May 2024).

World Information Service on Energy. *Telethon Chernobyl*, 6 April 1990, www .wiseinternational.org/nuclear-monitor/330/telethon-chernobyl (2 May 2024).

World Nuclear Association. *RBMK Reactors*, February 2022, www.world-nucle ar.org/information-library/appendices/rbmk-reactors (1 May 2024).

XXVIII s''ezd KPSS. 'Obsuzhdenie i priniatie rezoliutsii XXVIII s''ezda KPSS "O politicheskoi otsenke katastrofy na Chernobyl'skoi AES i khoda rabot po likvidatsii ee posledstvii"' in Institut teorii i istorii sotsializma TsK KPSS (ed.), *XXVIII s''ezd Kommunisticheskoi partii Sovetskogo Soiuza. 2–13 iiulia 1990 goda*; Stenograficheskii otchet (Moscow: Izdatel'stvo politicheskoi literatury, 1991), 585–98.

Zhitniuk, Anzhela. *Serial 'Chernobyl'. Novaia versiia ot NTV*, 7 August 2019, https://kinofilmpro.ru/chernobyil-ot-ntv.html (1 May 2024).

Zhirina, Ludmila S. *Chernobyl through the Eyes of Children: A Compilation of Children's Drawings and Tales* (n. p., 2013).

SECONDARY LITERATURE

Adam, Thomas, Simone Lässig, Gabriele Lingelbach (eds.). *Stifter, Spender und Mäzene: USA und Deutschland im historischen Vergleich* (Stuttgart: Steiner, 2009).

Altrichter, Helmut. *Russland 1989: Der Untergang des sowjetischen Imperiums* (Munich: Beck, 2009).

Alvesson, Mats, André Spicer. 'Critical Leadership Studies: The Case for Critical Performativity', *Human Relations* 65, no. 3 (2012), 367–90.

Ariès, Philippe. *Geschichte der Kindheit* (Munich: Hanser 1975).

Arndt, Melanie. 'Einleitung. Ökologie und Zivilgesellschaft' in Melanie Arndt (ed.), *Politik und Gesellschaft nach Tschernobyl: (Ost-)Europäische Perspektiven* (Berlin: Ch. Links, 2016), 10–24.

Arndt, Melanie. 'Environmental History', *Docupedia-Zeitgeschichte*, 23 August 2016, https://docupedia.de/zg/zgArndt_environmental_history_v3_en_2016 (1 May 2024).

Arndt, Melanie. 'Ever Been to a Brothel?', *Plotki* (2001), 27–30.

Arndt, Melanie. 'From Nuclear to "Human Security"? Prerequisites and Motives for the German Chernobyl Commitment in Belarus', *Historical Social Research* 35, no. 4 (2010), 289–308.

Arndt, Melanie. 'Grün nach der Katastrophe? Die Entwicklung der Umweltbewegungen in Litauen und Belarus nach Tschernobyl' in Martin Sabrow (ed.), *ZeitRäume 2009* (Göttingen: Wallstein, 2010), 8–21.

Arndt, Melanie (ed.). 'Memories, Commemorations, and Representations of Chernobyl', special issue of *Anthropology of East Europe Review* 30 (2012).

Arndt, Melanie. 'Merle Hilbk: Tschernobyl Baby: Wie wir lernten, das Atom zu lieben – Anatoly N. Tkachuk: Ich war im Sarkophag von Tschernobyl: Der Bericht des Überlebenden – Alla A. Yaroshinsrkaya: Chernobyl. Crime without Punishment', *Osteuropa* 62, no. 6–8 (2012), 542–4.

Arndt, Melanie (ed.). *Politik und Gesellschaft nach Tschernobyl: (Ost-) Europäische Perspektiven* (Berlin: Ch. Links, 2016).

Arndt, Melanie. 'The Babushkas of Chernobyl: Directed by Holly Morris and Anne Bogart', *Environmental History* 23, no. 2 (2018), 396–400.

Arndt, Melanie. *Tschernobyl: Auswirkungen des Reaktorunfalls auf die Bundesrepublik Deutschland und die DDR*, 3rd rev. ed. (Erfurt: Landeszentrale für politische Bildung, 2012).

Arndt, Melanie. 'Tschernobyl in Deutschland' in Bernd Greiner, Tim B. Müller, Klaas Voss (eds.), *Erbe des Kalten Krieges* (Hamburg: Hamburger Edition, 2013), 364–82.

Arndt, Melanie. *Tschernobylkinder: Die transnationale Geschichte einer nuklearen Katastrophe* (Göttingen: Vandenhoeck & Ruprecht, 2020).

Arndt, Melanie. 'Umweltgeschichte. Version 3.0', *Docupedia-Zeitgeschichte*, 10 November 2015, https://docupedia.de/zg/Arndt_umweltgeschichte_v3_d e_2015 (4 May 2024).

Augustine, Dolores L. *Taking on Technocracy: Nuclear Power in Germany, 1945 to the Present* (New York: Berghahn Books, 2018).

Awotona, Adenrele (ed.). *Rebuilding Sustainable Communities for Children and their Families after Disasters: A Global Survey* (Newcastle-upon-Tyne: Cambridge Scholars Publishing, 2010).

Babiracki, Patryk, Kenyon Zimmer (eds.). *Cold War Crossings: International Travel and Exchange across the Soviet Bloc, 1940s–1960s* (College Station: Texas A & M University Press, 2014).

Bankoff, Greg. 'Cultures of Disaster, Cultures of Coping: Hazard as a Frequent Life Experience in the Philippines' in Christof Mauch, Christian Pfister (eds.), *Natural Disasters, Cultural Responses: Case Studies toward a Global Environmental History* (Lanham, MD: Lexington Books, 2009), 265–84.

Bankoff, Greg. *Cultures of Disaster: Society and Natural Hazards in the Philippines* (London: RoutledgeCurzon, 2003).

Bankoff, Greg, Georg Frerks, Thea Hilhorst (eds.). *Mapping Vulnerability: Disasters, Development, and People* (London: Earthscan Publications, 2004).

Baranovs'ka, Natalia P. (ed.). *Chornobyl's'ka trahediia: Dokumenty i materialy* (Kyiv: Naukova Dumka, 1996).

Barnett, Michael N. *Empire of Humanity: A History of Humanitarianism* (Ithaca, NY: Cornell University Press, 2011).

Barnett, Michael N. (ed.). *Paternalism beyond Borders* (Cambridge, MA: Cambridge University Press, 2017).

Bauer, Susanne, Tanja Penter (eds.). *Tracing the Atom: Nuclear Legacies in Russia and Central Asia* (New York: Routledge, 2022).

Becker, Hans. 'Bewusste und unbewusste Abwehr- und Anpassungsmechanismen gegen das Erinnern an kollektive Traumen und Taten' in Günter H. Seidler, Wolfgang U. Eckart (eds.), *Verletzte Seelen: Möglichkeiten und Perspektiven einer historischen Traumaforschung* (Gießen: Psychosozial-Verlag, 2005), 303–14.

Beger, Kathleen. *Erziehung und 'Unerziehung' in der Sowjetunion: Das Pionierlager Artek und die Archangelsker Arbeitskolonie im Vergleich* (Göttingen: Vandenhoeck & Ruprecht, 2019).

Bekus, Nelly. *Struggle over Identity: The Official and the Alternative 'Belarusianness'* (Budapest: Central European University Press, 2010).

Berger, Miriam. 'A Chernobyl Tour Group Secretly Helped Track Russia's Invasion', *The Washington Post*, 21 August 2022, www.washingtonpost.com/world/2022/08/21/ukraine-spy-tour-group-russians/ (4 May 2024).

Berridge, Virgina, Alex Mold. 'Professionalisation, New Social Movements and Voluntary Action in the 1960s and 1970s' in McKay Hilton, James McKay (eds.), *The Ages of Voluntarism: How We Got to the Big Society* (Oxford: Oxford University Press, 2011), 114–34.

Biess, Frank. 'Die Sensibilisierung des Subjekts: Angst und "neue Subjektivität" in den 1970er Jahren', *Werkstatt Geschichte* 49 (2008), 51–71.

Biess, Frank. '"Everybody Has a Chance": Nuclear Angst, Civil Defence, and the History of Emotions in Postwar West Germany', *German History* 27, no. 2 (2009), 215–43.

Birkland, Thomas A. 'Natural Disasters as Focusing Events: Policy Communities and Political Response', *International Journal of Mass Emergencies and Disasters* 14, no. 2 (1996), 221–43.

Blumay, Carl, Henry Edwards. *The Dark Side of Power: The Real Armand Hammer* (New York: Simon & Schuster, 1991).

Bodrunova, Svetlana. 'Chernobyl in the Eyes: Mythology as a Basis of Individual Memories and Social Imaginaries of a "Chernobyl Child"', *Anthropology of East Europe Review* 30 (2012), 13–24.

Bohn, Thomas M., Victor Šadurskij. *Ein weißer Fleck in Europa: Die Imagination der Belarus als Kontaktzone zwischen Ost und West* (Berlin: De Gruyter, 2011).

Bonnerjea, Lucy. 'Disasters, Family Tracing and Children's Rights: Some Questions about the Best Interests of Separated Children', *Disasters* 18, no. 3 (1994), 277–83.

Bösch, Frank (ed.). *Geteilte Geschichte: Ost- und Westdeutschland 1970–2000* (Göttingen: Vandenhoeck & Ruprecht, 2015).

Boym, Svetlana. *Common Places: Mythologies of Everyday Life in Russia* (Cambridge, MA: Harvard University Press, 1994).

Bradford, Martin. 'Musical Cultural Exchanges in the Age of Détente: Cultural Fixation, Trust, and the Permeability of Culture', *Journal of Contemporary History* 51, no. 2 (2016), 364–84.

Bronfenbrenner, Urie, John C. Condry. *Two Worlds of Childhood* (London: Allen & Unwin, 1971).

Brown, Archie. *Seven Years that Changed the World: Perestroika in Perspective* (Oxford: Oxford University Press, 2008).

Brown, Archie. *The Gorbachev Factor* (Oxford: Oxford University Press, 1997).

Brown, Kate. *Manual for Survival: A Chernobyl Guide to the Future* (New York: Norton, 2019).

Brown, Kate. *Plutopia: Nuclear Families, Atomic Cities, and the Great Soviet and American Plutonium Disasters* (Oxford: Oxford University Press, 2013).

Bühler-Niederberger, Doris. 'Einleitung: Der Blick auf das Kind – gilt der Gesellschaft' in Doris Bühler-Niederberger (ed.), *Macht der Unschuld: Das Kind als Chiffre* (Wiesbaden: Verlag für Sozialwissenschaften, 2015), 9–22.

Bühler-Niederberger, Doris (ed.). *Macht der Unschuld: Das Kind als Chiffre* (Wiesbaden: Verlag für Sozialwissenschaften, 2015).

Caldwell, Melissa L. *Living Faithfully in an Unjust World: Compassionate Care in Russia* (Oakland, CA: University of California Press, 2017).

Cantelon, Philip L., Richard G. Hewlett, Robert C. Williams (eds.). *The American Atom: A Documentary History of Nuclear Policies from the Discovery of Fission to the Present*, 2nd ed. (Philadelphia: University of Pennsylvania Press, 1991).

Carr, Lowell J. 'Disaster and the Sequence-Pattern Concept of Social Change', *The American Journal of Sociology* 38, no. 2 (1932), 207–18.

Chandler, David. 'The New International Paternalism: International Regimes' in Michael N. Barnett (ed.), *Paternalism beyond Borders* (Cambridge, MA: Cambridge University Press, 2017), 132–58.

Chaumont, Jean-Michel. *Die Konkurrenz der Opfer: Genozid, Identität und Anerkennung* (Lüneburg: zu Klampen, 2001).

Cole, Alyson M. *The Cult of True Victimhood: From the War on Welfare to the War on Terror* (Stanford, CA.: Stanford University Press, 2007).

Collinson, David. 'Critical Leadership Studies' in Alan Bryman (ed.), *The SAGE Handbook of Leadership* (Los Angeles: Sage, 2014), 179–92.

Conze, Eckart, Martin Klimke, Jeremy Varon. *Nuclear Threats, Nuclear Fear and the Cold War of the 1980s* (New York: Cambridge University Press, 2016).

Cronon, William. 'The Portage: Time, Memory, and Storytelling in the Making of an American Place'. Paper at conference *Arts of Living in a Damaged Planet*, University of Santa Cruz, CA., 9 May 2014.

Dalhouski, Aliaksandr. *Tschernobyl in Belarus: Ökologische Krise und sozialer Kompromiss (1986–1996)* (Wiesbaden: Harrassowitz, 2015).

Davies, John L., Edward Kaufman. 'Second Track /Citizens' Diplomacy: An Overview' in John L. Davies, Edward Kaufman (eds.), *Second Track Diplomacy for Ethnic and Nationalist Conflicts: Applied Techniques for Conflict Transformation* (Lanham, MD: Rowman & Littlefield, 2002), 1–12.

Dawson, Jane I. *Eco-nationalism: Anti-nuclear Activism and National Identity in Russia, Lithuania, and Ukraine* (Durham: Duke University Press, 1996).

Depkat, Volker. 'Autobiographie und die soziale Konstruktion von Wirklichkeit', *Geschichte und Gesellschaft* 29 (2003), 441–76.

Doering-Manteuffel, Anselm, Lutz Raphael, *Nach dem Boom: Perspektiven auf die Zeitgeschichte seit 1970* (Göttingen: Vandenhoeck & Ruprecht, 2008).

Doose, Katja. 'Green Nationalism? The Transformation of Environmentalism in Soviet Armenia, 1969–1991', *Ab Imperio* 1 (2019), 181–205.

Doose, Katja. *Tektonik der Perestroika: Das Erdbeben und die Neuordnung Armeniens, 1985–1998* (Vienna: Böhlau 2019).

Dunn, Elizabeth C. 'The Chaos of Humanitarian Aid: Adhocracy in the Republic of Georgia', *Humanity* 3, no. 1 (2012), 1–23.

Edele, Mark. *Soviet Veterans of the Second World War: A Popular Movement in an Authoritarian Society 1941–1991* (Oxford: Oxford University Press 2008).

Edelman, Lee. *No Future: Queer Theory and the Death Drive* (Durham: Duke University Press, 2004).

Eißler, Friedmann. *Vereinigungskirche (Moon-Bewegung)*, April 2016, www.ez w-berlin.de/html/3_3065.php (2 May 2024).

Eitler, Pascal. '"Alternative" Religion: Subjektivierungspraktiken und Politisierungsstrategien im "New Age" (Westdeutschland 1970–1990)' in Sven Reichardt, Detlef Siegfried (eds.), *Das alternative Milieu: Antibürgerlicher Lebensstil und linke Politik in der Bundesrepublik Deutschland und Europa 1968–1983* (Göttingen: Wallstein, 2010), 335–52.

Enarson, Elaine P., Betty H. Morrow (eds.). *The Gendered Terrain of Disaster: Through Women's Eyes* (Westport, CT: Bloomsbury, 1998).

Enarson, Elaine P., Alice Fothergill, Lori Peek. 'Gender and Disaster: Foundations and New Directions for Research and Practice' in Havidán Rodríguez, William W. Donner, Joseph E. Trainor (eds.), *Handbook of Disaster Research* (Cham: Springer, 2018).

Epple, Angelika. 'Globale Mikrogeschichte: Auf dem Weg zu einer Geschichte der Relationen' in Ewald Hiebl, Ernst Langthaler (eds.), *Im Kleinen das Große suchen. Mikrogeschichte in Theorie und Praxis* (Innsbruck: Studienverlag, 2012), 37–47.

Esch, Arnold. 'Überlieferungs-Chance und Überlieferungs-Zufall als methodisches Problem des Historikers', *Historische Zeitschrift* 40, no. 240 (1985), 529–70.

Fábián, Katalin, Elzbieta Korolczuk (eds.), *Rebellious Parents: Parental Movements in Central-Eastern Europe and Russia* (Bloomington: Indiana University Press, 2017).

Fassin, Didier. *Humanitarian Reason: A Moral History of the Present Times* (Berkeley: University of California Press 2012).

Fassin, Didier, Richard Rechtman. *The Empire of Trauma: An Inquiry into the Condition of Victimhood* (Princeton: Princeton University Press, 2009).

Fieseler, Beate. *Arme Sieger. Die Invaliden des 'Großen Vaterländischen Krieges' der Sowjetunion 1941–1991* (Cologne: Böhlau, 2008).

Fletcher, Joyce K. 'The Paradox of Postheroic Leadership: An Essay on Gender, Power, and Transformational Change', *The Leadership Quarterly* 15, no. 5 (2004), 647–61.

Fordham, Maureen. 'Gendering Vulnerability Analysis: Towards a More Nuanced Approach' in Greg Bankoff, Georg Frerks, Thea Hilhorst (eds.),

Mapping Vulnerability: Disasters, Development, and People (London: Earthscan Publications, 2004), 174–82.

Friedman, Alexander, Rainer Hudemann. *Diskriminiert, vernichtet, vergessen: Behinderte in der Sowjetunion, unter nationalsozialistischer Besatzung und im Ostblock 1917–1991* (Stuttgart: Franz Steiner Verlag, 2016).

Gaydar, Egor, Konstantin Kogalovskij, 'Tendenzen der Wirtschaftskrise in der UdSSR' in Klaus Segbers (ed.), *Perestrojka. Zwischenbilanz* (Frankfurt a. M.: Suhrkamp, 1990), 230–65.

García-Acosta, Virginia. 'Historical Disaster Research' in Susanna Hoffman, Anthony Oliver-Smith (eds.), *Catastrophe & Culture: The Anthropology of Disaster* (Santa Fe, NM: School of American Research Press, 2002), 49–66.

Gatal'skaia, Galina V., Nina M. Zaitseva, 'Sotsial'no-psikhologicheskii analiz opyta ozdorovleniia belorusskikh detei za rubezhom v postchernobyl'skii period' *Psikhologicheskii zhurnal*, no. 1 (2008), 44–54.

Gebhardt, Lisette, Steffi Richter (eds.). *Lesebuch 'Fukushima': Übersetzungen, Kommentare, Essays* (Berlin: EB-Verlag, 2013).

Geist, Edward. 'Political Fallout: The Failure of Emergency Management at Chernobyl'', *Slavic Review* 74, no. 1 (2015), 104–26.

Gestwa, Klaus. *Die Stalinschen Großbauten des Kommunismus: Sowjetische Technik- und Umweltgeschichte, 1948–1967* (Munich: Oldenbourg, 2010).

Gestwa, Klaus. 'Katastrojka und Super-GAU. Die Nuklearmoderne in Zeiten von Tschernobyl und Fukushima' in Katharina Kucher, Gregor Thum, Sören Urbansky (eds.), *Stille Revolutionen: Die Neuformierung der Welt seit 1989* (Frankfurt a. M., New York: Campus, 2013), 57–68.

Gestwa, Klaus. 'Ökologischer Notstand und sozialer Protest: Ein umwelthistorischer Blick auf die Reformunfähigkeit und den Zerfall der Sowjetunion', *Archiv für Sozialgeschichte* 43 (2003), 349–83.

Gestwa, Klaus, Stefan Rohdewald. 'Verflechtungsstudien: Naturwissenschaft und Technik im Kalten Krieg', *Osteuropa* 59, no. 10 (2009), 5–14.

Gibbs, Joseph. *Gorbachev's Glasnost: The Soviet Media in the First Phase of Perestroika* (College Station: Texas A & M University Press, 1999).

Gill, Rebecca. 'Networks of Concern, Boundaries of Compassion: British Relief in the South African War', *The Journal of Imperial and Commonwealth History* 40, no. 5 (2012), 827–44.

Goltermann, Svenja. 'Der Markt der Leiden, das Menschenrecht auf Entschädigung und die Kategorie des Opfers: Ein Problemaufriss', *Historische Anthropologie* 23, no. 2 (2015), 70–92.

Goltermann, Svenja. *Opfer: Die Wahrnehmung von Krieg und Gewalt in der Moderne* (Frankfurt a. M.: S. Fischer, 2017).

Gorsuch, Anne E. *Youth in Revolutionary Russia: Enthusiasts, Bohemians, Delinquents* (Bloomington: Indiana University Press, 2000).

Greiner, Bernd. 'Angst im Kalten Krieg: Bilanz und Ausblick' in Bernd Greiner et al. (eds.), *Angst im Kalten Krieg* (Hamburg: Hamburger Edition, 2009), 7–31.

Greiner, Bernd, Tim B. Müller, Klaas Voss (eds.). *Erbe des Kalten Krieges* (Hamburg: Hamburger Edition, 2013).

Guski, Andreas. 'Die Stimme der Opfer: Vom Umgang mit Katastrophen in Russland', *Osteuropa* 58, nos. 4–5 (2008), 69–80.

Hall, Peter D. 'Philanthropie, Wohlfahrtsstaat und die Transformation der öffentlichen Institutionen in den USA, 1945–2000' in Thomas Adam, Simone Lässig, Gabriele Lingelbach (eds.), *Stifter, Spender und Mäzene: USA und Deutschland im historischen Vergleich* (Stuttgart: Steiner, 2009), 69–98.

Harper, Krista. *Wild Capitalism: Environmental Activists and Post-socialist Political Ecology in Hungary* (Boulder, CO: East European Monographs, 2006).

Harwin, Judith. *Children of the Russian State, 1917–95* (Aldershot: Avebury, 1996).

Hecht, Gabrielle. *Being Nuclear: Africans and the Global Uranium Trade* (Cambridge, MA: MIT Press, 2012).

Heijmans, Annelies. 'From Vulnerability to Empowerment' in Greg Bankoff, Georg Frerks, Thea Hilhorst (eds.), *Mapping Vulnerability: Disasters, Development, and People* (London: Earthscan Publications, 2004), 115–27.

Henningsen, Monika. *Der Freizeit- und Fremdenverkehr in der (ehemaligen) Sowjetunion unter besonderer Berücksichtigung des Baltischen Raums* (Frankfurt a. M.: Lang, 1994).

Higginbotham, Adam. *Midnight in Chernobyl: The Untold Story of the World's Greatest Nuclear Disaster* (New York: Simon & Schuster, 2019).

Hilton, Matthew, James McKay (eds.). *The Ages of Voluntarism: How We Got to the Big Society* (Oxford: Oxford University Press, 2011).

Hilton, Matthew, Nick Crowson, Jean-François Mouhot, James McKay (eds.). *A Historical Guide to NGOs in Britain: Charities, Civil Society and the Voluntary Sector since 1945* (Houndmills: Palgrave Macmillan, 2012).

Hoffman, Susanna M. 'The Regenesis of Traditional Gender Patterns in the Wake of Disaster' in Anthony Oliver-Smith, Susanna Hoffman (eds.), *The Angry Earth: Disaster in Anthropological Perspective* (New York: Routledge, 2008), 173–91.

Hoffmann, Stefan-Ludwig (ed.). *Human Rights in the Twentieth Century* (Cambridge, MA: Cambridge University Press, 2010).

Iarskaia-Smirnova, Elena R., Pavel V. Romanov. *Sovetskaia sotsial'naia politika: Stseny I deistvuiushchie litsa, 1940–1985* (Moscow: TsSPGI, 2008).

Ilchman, Warren F., Stanley N. Katz, Edward L. Queen II (eds.). *Philanthropy in the World's Traditions* (Bloomington: Indiana University Press, 1998).

Iriye, Akira, Petra Goedde, William I. Hitchcock (eds.). *The Human Rights Revolution: An International History* (Oxford: Oxford University Press, 2012).

Ivanova, Evgenija. 'Vom Tod zum Leben: Tschernobyl-Politik durch die Gender-Brille' in Melanie Arndt (ed.), *Politik und Gesellschaft nach Tschernobyl: (Ost-) Europäische Perspektiven* (Berlin: Ch. Links, 2016), 130–50.

Jacobi, Robert. *Die Goodwill-Gesellschaft. Die unsichtbare Welt der Stifter, Spender und Mäzene* (Hamburg: Murmann, 2009).

Jacobs, Robert A. *The Dragon's Tail: Americans Face the Atomic Age* (Amherst: University of Massachusetts Press, 2010).

Jacobs, Robert A., Mick Broderick. The Global Hibakusha Project, https://bit.ly/4f9jdd6 (1 May 2024).

Jain, S. Lochlann. *Malignant: How Cancer Becomes Us* (Berkeley: University of California Press, 2013).

Jordan, Katrin. *Ausgestrahlt: Die mediale Debatte um 'Tschernobyl' in der Bundesrepublik und in Frankreich 1986/1987* (Göttingen: Wallstein-Verlag, 2018).

Josephson, Paul R. 'Atomic-Powered Communism: Nuclear Culture in the Postwar USSR', *Slavic Review* 55, no. 2 (1996), 297–324.

Josephson, Paul R. *Industrialized Nature: Brute Force Technology and the Transformation of the Natural World* (Washington, DC: Island Press 2002).

Josephson, Paul R. *Lenin's Laureate: Zhores Alferov's Life in Communist Science* (Cambridge, MA: MIT Press, 2010).

Josephson, Paul R. '"Projects of the Century" in Soviet History: Large-scale Technologies from Lenin to Gorbachev', *Technology and Culture* 36, no. 2 (1995), 519–59.

Josephson, Paul R. *Red Atom: Russia's Nuclear Power Program from Stalin to Today* (Pittsburgh : University of Pittsburgh Press, 2005).

Josephson, Paul R. *Resources under Regimes: Technology, Environment, and the State* (Cambridge, MA: Harvard University Press, 2006).

Josephson, Paul R. *Totalitarian Science and Technology*, 2nd ed. (Amherst, NY: Humanities Press, 2005).

Josephson, Paul R. *Would Trotsky Wear a Bluetooth: Technological Utopianism under Socialism 1917–1989* (Baltimore: Johns Hopkins University Press, 2010).

Kalmbach, Karena. 'Frankreich nach Tschernobyl: Eine Rezeptionsgeschichte zwischen "Nichtereignis" und "Apokalypse"' in Melanie Arndt (ed.), *Politik und Gesellschaft nach Tschernobyl: (Ost-)Europäische Perspektiven* (Berlin: Ch. Links, 2016), 237–55.

Kalmbach, Karena. 'Revisiting the Nuclear Age: State of the Art Research in Nuclear History', *Neue Politische Literatur* 62, no. 1 (2017), 49–69.

Kalmbach, Karena. *The Meanings of a Disaster: Chernobyl and Its Afterlives in Britain and France* (New York: Berghahn, 2021).

Karbalevich, Valerii. 'Put' Lukashenko k vlasti' in Dmitrii E. Furman (ed.), *Belorussiia i Rossiia: Obshchestva i gosudarstva* (Moscow: Prava Cheloveka, 1998), 226–57.

Kasperski, Tatiana. *Les politiques de la radioactivité: Tchernobyl et la mémoire nationale en Biélorussie contemporaine* (Paris: Éditions Petra, 2020).

Kasperski, Tatjana. 'Nation versus Gedächtnis: Die Nationalisierung kollektiver Vorstellungen über Tschernobyl als Faktor zum Vergessen der Katastrophe' in Melanie Arndt (ed.), *Politik und Gesellschaft nach Tschernobyl: (Ost-) Europäische Perspektiven* (Berlin: Ch. Links, 2016), 152–81.

Kelly, Catriona. *Children's World: Growing Up in Russia, 1890–1991* (New Haven, CT: Yale University Press, 2007).

Kersten, Jens, Markus Vogt, Frank Uekoetter (eds.). *Europe after Fukushima: German Perspectives on the Future of Nuclear Power* (Munich: Rachel Carson Center, 2012).

Kharash, Adol'f. 'Gumanitarnaia ekspertiza v ekstremal'nykh situatsiiakh: Ideologiia, metodologiia, protsedura' in A. I. Dontsov, I. M. Zhukov (eds.),

Vvedenie v prakticheskuiu sotsial'nuiu psikhologiiu: Uchebnoe posobie dlia vysshikh uchebnykh zavedenii (Moscow: Nauka, 1994), 60–88.

Kind-Kovács, Friederike. 'The Great War, the Child's Body and the American Red Cross', *European Review of History/Revue européenne d'histoire* 23, nos. 1–2 (2016), 33–62.

Kirk, John M. *Healthcare without Borders: Understanding Cuban Medical Internationalism* (Gainesville, FA: University Press of California, 2015).

Kirschenbaum, Lisa A. *Small Comrades: Revolutionizing Childhood in Soviet Russia, 1917–1932* (New York: RoutledgeFalmer, 2001).

Klein, Ansgar, Hans-Josef Legrand, Thomas Leif. *Neue soziale Bewegungen: Impulse, Bilanzen und Perspektiven* (Opladen: Westdeutscher Verlag, 1999).

Kołodziejczyk, Dorota, Cristina Şandru. *Postcolonial Perspectives on Postcommunism in Central and Eastern Europe* (London: Routledge, 2016).

Kotkin, Stephen. *Armageddon Averted: The Soviet Collapse, 1970–2000* (Oxford: Oxford University Press, 2001).

Krige, John. *American Hegemony and the Postwar Reconstruction of Science in Europe* (Cambridge, MA: MIT Press, 2006).

Krige, John. 'Hybrid Knowledge: The Transnational Co-production of the Gas Centrifuge for Uranium Enrichment in the 1960s', *The British Journal for the History of Science* 45, no. 3 (2012), 337–57.

Kristof, Nicholas D. 'The Power of One' in Scott Slovic, Paul Slovic (eds.), *Numbers and Nerves: Information, Emotion, and Meaning in a World of Data* (Corvallis, OR: Oregon State University Press, 2015), 85–8.

Kuchinskaya, Olga. *The Politics of Invisibility: Public Knowledge about Radiation Health Effects after Chernobyl* (Cambridge, MA: MIT Press, 2014).

Kuletz, Valerie. *The Tainted Desert: Environmental Ruin in the American West* (New York: Routledge, 1998).

Laqua, Daniel, Charlotte Alston (eds.). 'Ideas, Practices and Histories of Humanitarianism', special issue of *Journal of Modern European History* 2 (2014).

Laufs, Paul. *Reaktorsicherheit für Leistungskernkraftwerke: Die Entwicklung im politischen und technischen Umfeld der Bundesrepublik Deutschland* (Berlin: Springer 2013).

Lindenmeyr, Adele. 'From Repression to Revival: Philanthropy in Twentieth-century Russia' in Warren F. Ilchman, Stanley N. Katz, Edward L. Queen II (eds.), *Philanthropy in the World's Traditions* (Bloomington: Indiana University Press, 1998), 309–31.

Lindner, Jörg. *Den svenska Tysklands-hjälpen 1945–1954* (Umeå: Universitet Umeå, Almqvist & Wiksell International, 1988).

Lingelbach, Gabriele. *Spenden und Sammeln: Der westdeutsche Spendenmarkt bis in die 1980er Jahre* (Göttingen: Wallstein, 2009).

Lorenz, Astrid. 'Politischer Wandel in Belarus: Tendenzen, Probleme, Perspektiven', *WeltTrends* 29 (2001/2002), 59–77.

Manzo, Kate. 'Imaging Humanitarianism: NGO Identity and the Iconography of Childhood', *Antipode* 40, no. 4 (2008), 632–57.

Marples, David R. *Belarus: From Soviet Rule to Nuclear Disaster* (Houndmills: Macmillan Press, 1996).

Marples, David R. *Chernobyl and Nuclear Power in the USSR* (New York: St. Martin's Press, 1986).

Marples, David R. *The Lukashenka Phenomenon: Elections, Propaganda, and the Foundations of Political Authority in Belarus* (Trondheim: Norwegian University of Science and Technology, 2007).

Marples, David R. *The Social Impact of the Chernobyl Disaster* (Basingstoke: Macmillan, 1988).

Marples, David R. *Ukraine under Perestroika: Ecology, Economics and the Workers' Revolt* (London: Palgrave Macmillan, 1991).

Masco, Joseph. *The Nuclear Borderlands: The Manhattan Project in Post-Cold War New Mexico* (Princeton, NJ: Princeton University Press, 2006).

Matlock, Jack F. *Reagan and Gorbachev: How the Cold War Ended* (New York: Random House, 2004).

McCagg, William O., Lewis Siegelbaum (eds.). *The Disabled in the Soviet Union: Past and Present, Theory and Practice* (Pittsburgh: University of Pittsburgh Press, 2009).

McCannon, John. 'Technological and Scientific Utopias in Soviet Children's Literature, 1921–1932', *The Journal of Popular Culture* 34, no. 4 (2001) 4, 153–69.

McCleary, Rachel M. *Global Compassion: Private Voluntary Organizations and US Foreign Policy since 1939* (Oxford: Oxford University Press, 2009).

McNeill, John Robert, Peter Engelke, *The Great Acceleration: An Environmental History of the Anthropocene Since 1945* (Cambridge, MA: Harvard University Press, 2016).

Melosi, Martin V. *Atomic Age America* (Boston: Pearson, 2013).

Mihalisko, Kathleen. 'Belarus: Retreat to Authoritarianism' in Karen Dawisha, Bruce Parrott (eds.), *Democratic Changes and Authoritarian Reactions in Russia, Ukraine, Belarus, and Moldova* (New York: Cambridge University Press, 1997), 223–81.

Mikkonen, Simo, Pekka Suutari (eds.). *Music, Art and Diplomacy: East–West Cultural Interactions and the Cold War* (London: Routledge, 2016).

Morris, David B. *Illness and Culture in the Postmodern Age* (Berkeley: University of California Press 1998).

Moyn, Samuel. *The Last Utopia: Human Rights in History* (Cambridge, MA: Belknap Press of Harvard University Press, 2010).

Natchyk. 'Referendum' in Vital' Silitski et al. (eds.), *Nainoushaia historyia belaruskaha parliamentaryzmu* (Minsk: Analitychny hrudok, 2005).

Nixon, Rob. *Slow Violence and the Environmentalism of the Poor* (Cambridge, MA: Harvard University Press, 2011).

Noack, Christian. 'Tourismus in Russland und der UdSSR als Gegenstand historischer Forschung. Ein Werkstattbericht', *Archiv für Sozialgeschichte* 45 (2005), 477–98.

Noack, Christian. 'Von "wilden" und anderen Touristen. Zur Geschichte des Massentourismus in der UdSSR', *Werkstatt Geschichte* 36 (2004), 24–41.

Nyseter, Tor, Axel Valen-Sendstad, *Forsoningens bru: Vest-Berlins flyktningebarn kommer til Norge* (Bonn: n. p., 1986).

O'Sullivan, Kevin, Matthew Hilton, Juliano Fiori. 'Humanitarianisms in Context', *European Review of History/Revue européenne d'histoire* 23, no. 1–2 (2016), 1–15.

Oushakine, Serguei A. 'Realism with Gaze-Appeal: Lenin, Children, and Photomontage', *Jahrbücher für Geschichte Osteuropas* 67, no. 1 (2019), 11–64.

Paige, R. Michael. (ed.). *Education for the Intercultural Experience* (Yarmouth, ME: Intercultural Press Inc, 1993).

Paulmann, Johannes (ed.). *Dilemmas of Humanitarian Aid in the Twentieth Century* (Oxford: Oxford University Press, 2016).

Paulmann, Johannes. 'The Dilemmas of Humanitarian Aid: Historical Perspectives' in Johannes Paulmann (ed.), *Dilemmas of Humanitarian Aid in the Twentieth Century* (Oxford: Oxford University Press, 2016), 1–31.

Peacock, Margaret. *Innocent Weapons: The Soviet and American Politics of Childhood in the Cold War* (Chapel Hill: The University of North Carolina Press, 2014).

Pedersen, Paul. *The Five Stages of Culture Shock: Critical Incidents around the World* (Westport, CT: Greenwood Press, 1995).

Penter, Tanja (ed.). 'Vernichtungskrieg, Besatzung und juristische Aufarbeitung: Opferperspektiven', special issue of *Jahrbücher für Geschichte Osteuropas* 68, nos. 3–4 (2020).

Perrow, Charles. *Normal Accidents: Living with High-risk Technologies* (Princeton, NJ: Princeton University Press, 1999).

Petryna, Adriana. 'Biological Citizenship: The Science and Politics of Chernobyl-exposed Populations', *Osiris* 19 (2004), 250–65.

Petryna, Adriana. *Life Exposed: Biological Citizens after Chernobyl* (Princeton, NJ: Princeton University Press, 2002).

Phillips, Sarah D. '"There Are No Invalids in the USSR!": A Missing Soviet Chapter in the New Disability History', *Disability Studies Quarterly* 29, no. 3 (2009).

Pilkington, Hilary. *Russia's Youth and Its Culture: A Nation's Constructors and Constructed* (London: Routledge, 1994).

Piotukh, Volha. *Biopolitics, Governmentality and Humanitarianism: 'Caring' for the Population in Afghanistan and Belarus* (Hoboken: Routledge, 2015).

Plokhy, Serhii. *Chernobyl: History of a Tragedy* (London: Basic Books, 2019).

Pritchard, Sara B. 'An Envirotechnical Disaster: Nature, Technology, and Politics at Fukushima', *Environmental History* 17, no. 2 (2012), 219–43.

Pritchard, Sara B., Carl A. Zimring (eds.). *Technology and the Environment in History* (Baltimore: John Hopkins University Press, 2020).

Proctor, Robert N. 'Agnotology: A Missing Term to Describe the Cultural Production of Ignorance (and Its Study)' in Robert Proctor, Londa L. Schiebinger (eds.), *Agnotology: The Making and Unmaking of Ignorance* (Stanford, CA: Stanford University Press, 2008) 1–33.

Raab, Nigel A. *All Shook Up: The Shifting Soviet Response to Catastrophes, 1917–1991* (Montreal: McGill-Queen's University Press, 2017).

Radkau, Joachim. *Ära der Ökologie: Eine Weltgeschichte* (Munich: C. H. Beck, 2011).

Radkau, Joachim. *Nature and Power: A Global History of the Environment* (New York: Cambridge University Press, 2008).

Radkau, Joachim. *The Age of Ecology: A Global History*, tr. Patrick Camiller (Cambridge: Polity Press, 2014).

Raleigh, Donald J. *Soviet Baby Boomers: An Oral History of Russia's Cold War Generation* (Oxford: Oxford University Press, 2012).

Rasell, Michael, Elena Iarskaia-Smirnova, *Disability in Eastern Europe and the Former Soviet Union. History, Policy, and Everyday Life* (London: Routledge, 2014).

Reichhardt, Sven, Detlef Siegfried (eds.). *Das alternative Milieu: Antibürgerlicher Lebensstil und linke Politik in der Bundesrepublik Deutschland und Europa 1968–1983* (Göttingen: Wallstein, 2010).

Renner, Andreas. 'Globale Ikone des Kalten Krieges? Der Atompilz und die sowjetische Nuklearkultur', *Osteuropa* 66, nos. 6–7 (2016), 215–36.

Richers, Julia. 'Die erste Kosmonautin: Valentina Tereshkova und der transkontinentale Geschlechterkampf im Kalten Krieg' in Martina Ineichen et al. (eds.), *Gender in Trans-it: Transkulturelle und transnationale Perspektiven* (Zürich: Chronos, 2009), 235–45.

Richmond, Yale. *Cultural Exchange and the Cold War: Raising the Iron Curtain.* (University Park: Pennsylvania State University Press, 2003).

Richter, Steffi. *Japan nach 'Fukushima': Ein System in der Krise* (Leipzig: Leipziger Universitäts-Verlag, 2012).

Ries, Nancy. *Russian Talk: Culture and Conversation during Perestroika* (Ithaca, NY: Cornell University Press, 1997).

Risch, Julia. *Russen und Amis im Gespräch: Die sowjetisch-amerikanische Telebrücke (1982–1989): Ein vergessener Beitrag zur Beendigung des Kalten Krieges* (Berlin: SAXA-Verlag, 2012).

Roisko, Pekka. *Gralshüter eines untergehenden Systems: Zensur in Massenmedien in der UdSSR 1981–1991* (Cologne: Böhlau 2013).

Romanov, P., E. Iarskaia-Smirnova, S. Vaitfild, S. Kelli (eds.). *Sotsiologicheskoe issledovanie problem invalidnosti i reabilitatsii invalidov v Rossiiskoi Federatsii: Analiz osnovnykh rezul'tatov issledovaniia* (Moscow: n. p., 2008).

Rouda, Uladzimir. *Palitychnaia sistema respubliki Belarus'* (Vilnius: EHU, 2011).

Rucht, Dieter. *Modernisierung und neue soziale Bewegungen: Deutschland, Frankreich und USA im Vergleich* (Frankfurt a. M.: Campus, 1994).

Rüthers, Monica. 'Picturing Soviet Childhood: Photo Albums of Pioneer Camps', *Jahrbücher für Geschichte Osteuropas* 67, no. 1 (2019), 65–95.

Sahm, Astrid. *Die weissrussische Nationalbewegung nach der Katastrophe von Tschernobyl. 1986–1991* (Münster: Lit, 1994).

Sahm, Astrid. *Transformation im Schatten von Tschernobyl: Umwelt- und Energiepolitik im gesellschaftlichen Wandel von Belarus und der Ukraine* (Münster: Lit, 1999).

Said, Edward. *Orientalismus*, 5th ed. (Frankfurt a. M.: S. Fischer, 2010).

Sal'nikova, Alla A. *Rossiiskoe detstvo v XX veke: Istoriia, teoriia i praktika issledovaniia* (Kazan': Kazanskii Gosudarstvennyi Universitet, 2007).

Schattenberg, Susanne. *Stalins Ingenieure: Lebenswelten zwischen Technik und Terror in den 1930er Jahren*, re-print 2014 (Berlin: Oldenbourg, 2002).

Schlott, Wolfgang. 'Abgekoppelt, auf anderen Gleisen. Wohin strebt Rußlands Jugend?' in Wolfgang Schlott (ed.), *Die enterbte Generation: Russische Jugend nach der Perestroika* (Leipzig: Reclam Leipzig, 1994), 10–19.

Schlott, Wolfgang (ed.). *Die enterbte Generation: Russische Jugend nach der Perestroika* (Leipzig: Reclam Leipzig, 1994).

Schmid, Sonja D. *Producing Power: The Pre-Chernobyl History of the Soviet Nuclear Industry* (Cambridge, MA: MIT Press, 2015).

Schregel, Susanne. 'Konjunktur der Angst: "Politik der Subjektivität" und "neue Friedensbewegung"' in Bernd Greiner, Christian T. Müller, Dierk Walter (eds.), *Angst im Kalten Krieg* (Hamburg: Hamburger Edition, 2009), 495–520.

Segert, Dieter. *Transformationen in Osteuropa im 20: Jahrhundert* (Bonn: Bundeszentrale für Politische Bildung 2014).

Shane, Scott. *Dismantling Utopia: How Information Ended the Soviet Union* (Chicago: Ivan R. Dee, 1994).

Shaw, Claire L. *Deaf in the USSR: Marginality, Community, and Soviet Identity, 1917–1991* (Ithaca: Cornell University Press, 2017).

Shaw, Claire L. '"We Have No Need to Lock Ourselves Away": Space, Marginality, and the Negotiation of Deaf Identity in Late Soviet Moscow', *Slavic Review* 74, no. 1 (2015), 57–78.

Sidorenko, V. A. 'Istoriia RBMK' in V. A. Sidorenko (ed.), *Istoriia atomnoi energetiki Sovetskogo Soiuza i Rossii*, vol. 3 (Moscow: IzdAT, 2003).

Sidorenko, V. A. 'Uroki avarii na Chernobyl'skoi AES' in V. A. Sidorenko (ed.), *Istoriia atomnoi energetiki*, vol. 4 (Moscow: IzdAT, 2002).

Slim, Hugo (ed.). 'Children and Childhood in Emergency Policy and Practice 1919–1994', special issue of *Disasters* 18, no. 3 (1994).

Smircich, Linda, Gareth Morgan. 'Leadership: The Management of Meaning', *Journal of Applied Behavioural Studies* 18, no. 3 (1982), 257–73.

Stepanov, Andrej. 'Tschernobyl ist niemals passiert? Praktiken der Legitimierung des technopolitischen Regimes in Belarus' in Melanie Arndt (ed.), *Politik und Gesellschaft nach Tschernobyl: (Ost-)Europäische Perspektiven* (Berlin: Ch. Links, 2016), 256–82.

Stephens, Sharon (ed.). *Children and the Politics of Culture* (Princeton, NJ: Princeton University Press, 1995).

Stephens, Sharon. 'Children and the Politics of Culture in "Late Capitalism"' in Sharon Stephens (ed.), *Children and the Politics of Culture* (Princeton, NJ: Princeton University Press, 1995), 3–48.

Sutherland, Neil, Christopher Land, Steffen Bohm. 'Anti-leaders(hip) in Social Movement Organizations: The Case of Autonomous Grassroots Groups', *Organization* 21, no. 6 (2014), 759–81.

Szulecki, Kacper. 'Von Czarnobyl zu Żarnobyl: Die Auswirkungen Tschernobyls auf die grüne Opposition in Polen' in Melanie Arndt (ed.), *Politik und Gesellschaft nach Tschernobyl: (Ost-)Europäische Perspektiven* (Berlin: Ch. Links, 2016), 26–52.

Tsing, Anna L., Heather Anne Swanson, Elaine Gan, Nils Bubandt (eds.). *Arts of Living on a Damaged Planet: Ghosts of the Anthropocene* (Minneapolis: University of Minnesota Press, 2017).

Tsuchiya, Yuka. '"I Was Not Afraid of the Atom Bomb": Young Japanese Tuna Fishermen and Thermonuclear Tests in the Pacific, 1953–1962'. Paper presented at the *8th Biennial Society for the History of Children and Youth Conference*, Vancouver, 2015.

Ushakin, Sergey (ed.). *Travma. Punkty* (Moscow: Novoe literaturnoe obozrenie, 2009).

Vorona, V., E. Golovacha, Iu. Saenko. *Sotsial'ni naslidky Chornobyl's'koi katastrofy. Rezul'taty sotsiolohichnykh doslidzhen' 1986–1995 rr.* (Charkiv: Folio, 1996).

Weart, Spencer R. *The Rise of Nuclear Fear* (Cambridge, MA: Harvard University Press, 2012).

Weiner, Amir. *KGB: Ruthless Sword, Imperfect Shield* (forthcoming).

Weiner, Douglas R. *A Little Corner of Freedom: Russian Nature Protection from Stalin to Gorbachёv* (Berkeley: University of California Press, 1999).

Wellock, Thomas R. 'The Children of Chernobyl: Engineers and the Campaign for Safety in Soviet-Designed Reactors in Central and Eastern Europe', *History and Technology* 29, no. 1 (2013), 3–23.

Wellock, Thomas. *'Too Cheap to Meter': A History of the Phrase*, 3 June 2016, https://bit.ly/4gv3B3P (1 May 2024).

Wellock, Thomas. *Safe Enough? A History of Nuclear Power and Accident Risk* (Oakland: University of California Press, 2021).

Wendland, Anna V. 'Tschernobyl: (K)eine visuelle Geschichte: Nukleare Bilderwelten in der Sowjetunion und ihren Nachfolgestaaten' in Melanie Arndt (ed.), *Politik und Gesellschaft nach Tschernobyl: (Ost-)Europäische Perspektiven* (Berlin: Ch. Links, 2016), 182–210.

Wenzlhuemer, Roland. 'Globalization, Communication and the Concept of Space in Global History', *Historical Social Research* 35, no. 1 (2010), 19–47.

Werth, Karsten. *Ersatzkrieg im Weltraum: Das US-Raumfahrtprogramm in der Öffentlichkeit der 1960er Jahre* (Frankfurt a. M.: Campus, 2006).

White, Anne. 'Charity, Self-help and Politics in Russia, 1985–91', *Europe-Asia Studies* 45, no. 5 (1993), 787–811.

White, Richard. *The Organic Machine*, 6th ed. (New York: Hill and Wang, 2001).

Wilson, Andrew. *Belarus: The Last Dictatorship in Europe* (New Haven: Yale University Press, 2011).

Winkler, Martina (ed.). 'Children on Display: Children's History, Socialism, and Photography', special issue of *Jahrbücher für Geschichte Osteuropas* 67, no. 1 (2019).

Winkler, Martina. 'Kindheitsgeschichte. Version 1.0', *Docupedia-Zeitgeschichte*, 17 October 2016, http://docupedia.de/zg/Winkler_kindheitsgeschichte_v1_de_2016 (4 May 2024).

Yurchak, Alexei. *Everything Was Forever, until It Was No More: The Last Soviet Generation* (Princeton, NJ: Princeton University Press, 2006).

Zaprudnik, Jan. *Belarus: At a Crossroads in History* (Boulder: Westview Press, 1993).

Zaretsky, Natasha. *Radiation Nation: Three Mile Island and the Political Transformation of the 1970s* (New York: Columbia University Press, 2018).

Zatravkin, S. N., E. A. Vishlenkova. *'Kluby' i 'getto' sovetskogo zdravookhraneniia* (Moscow: ShIKO, 2022).

Zwierlein, Cornel, Rüdiger Graf, Magnus Ressel (eds.). 'The Production of Human Security in Premodern and Contemporary History', special issue of *Historical Social Research* 4 (2010).

Index

Studies in Environment and History (continued from page ii)

Other Books in the Series